Human Perspect
Industry 4.0 Organizations

Sustainability is a global issue and, with the advent of further legislation to make organizations "greener," companies are keen to promote sustainable performance to survive in the market. Using the facets of Industry 4.0, companies can become leaner and cleaner and measure their performance more effectively. This book reveals the sustainable innovations that organizations are undertaking because of Industry 4.0.

This book focuses on sophisticated aspects of how to make products tailor-made to suit specific requirements. It seeks to understand the status of sustainable performance that is impacted by different aspects related to human factors and concludes with detailing the future needs of businesses and potential trends. The book allows the reader to develop a deeper view of sustainability and organizational problems and to bridge the gap between theory and practice. Each chapter contains a self-contained study of a business and the decisions made to improve performance and is supported with tables, charts, and illustrations, and a wide list of bibliographic references.

Human Perspectives of Industry 4.0 Organizations will be of interest to students, graduates, researchers, and practitioners in the fields of logistics, supply chain management, management, leadership, organization, and sustainability, plus those interested in Industry 4.0 more generally. It will appeal to students in graduate programs covering sustainable aspects of business, management, supply chain management, and industrial engineering.

Sonia Umair is an accomplished academician, researcher, and assistant professor with over a decade of experience in the field of management. She is currently working as an assistant professor at Al-Zahra College for Women, Muscat, Oman. She holds a Ph.D. in Management from Universiti Putra Malaysia, Malaysia. Her academic and research interest lies in the fields of HRM, sustainability, business ethics, leadership, and organizational behavior. She has worked as a research collaborator and research assistant at the Universiti Putra Malaysia. She has published extensively in reputable journals and conferences, contributing valuable insights to the field.

Umair Waqas is Assistant Professor and Program Chair of Supply Chain Management and Logistics in the College of Business at the University of Buraimi, Oman. He is a member of several committees, including the member college of the business board, member college of business postgraduate committee, member teaching learning and curriculum committee, and member moderation and examination committee. He is also responsible for curriculum revision and progression of his department.

He received Ph.D. in Management (specialization: Supply Chain Risk Management) in 2020 from the School of Business and Economics, Universiti Putra Malaysia, Malaysia. He has experience of 10 years in the field of research and education.

Beata Mrugalska, Ph.D., D.Sc., Eur.Erg. is an Associate Professor and head of the Division of Applied Ergonomics, Institute of Safety and Quality Engineering, Faculty of Management Engineering, Poznan University of Technology in Poland. She holds an M.Sc. (2001) in Management and Marketing from the Faculty of Mechanical Engineering and Management at the Poznan University of Technology, and a PhD (2009) in Machine Construction and Operation from the Faculty of Computer Science and Management at the Poznan University of Technology. She was awarded a D.Sc. (dr habil.) degree in Mechanical Engineering by the Faculty of Mechanical Engineering of the Poznan University of Technology, Poland (2019). Since 2018, she has been a board member of the Center for Registration of European Ergonomists (CREE). She is responsible for promoting the professional title of EuroErgonomist in the international scientific and industrial environment. She is also a board member of the Polish Academy of Arts and Sciences, Commission of Ergonomics, and a member of the Ergonomics Committee at the Polish Academy of Sciences, Branch Poznan. She has over 120 publications focused on human factors in modern organizational management concepts. She serves as a member of 8 editorial boards of international journals and a Guest Editor of 5 special issues of journals. Beata Mrugalska was honored with the Top Peer Reviewer by Publons for her contribution to the preparation of reviews of papers.

Ibrahim Rashid Al Shamsi is Associate Professor and Dean in the College of Business, University of Buraimi, Oman. His areas of specialization include business, human resource management, and other management-related subjects. He obtained his Master's in Business Administration from the UK and Ph.D. in Business Administration from Malaysia. He also serves indexed journals as a Reviewer and is involved in various academic and research tasks. Before joining academia, he worked for 20 years with the Directorate General of Health Services, Al Buraimi Governorate, Ministry of Health, where he held important positions as Head of the Administration Department, Director of Administration and Finance, and Director of Planning and Studies.

Human Perspectives of Industry 4.0 Organizations

Reviewing Sustainable Performance

Edited by
Sonia Umair
Umair Waqas
Beata Mrugalska
Ibrahim Rashid Al Shamsi

Routledge
Taylor & Francis Group
NEW YORK AND LONDON

Designed cover image: Getty Images

First edition published 2025
by Routledge
605 Third Ave., 21st Floor, New York, NY 10158 USA

and by Routledge
4 Park Square, Milton Park, Abingdon, Oxon, OX14 4RN

Routledge is an imprint of Taylor & Francis Group, LLC

ISBN: 978-1-032-59862-8 (hbk)
ISBN: 978-1-032-61682-7 (pbk)
ISBN: 978-1-032-61681-0 (ebk)

DOI: 10.1201/9781032616810

Typeset in Times
by SPi Technologies India Pvt Ltd (Straive)

Contents

Contributors

Danyal Aghajani
LUT Kouvola Unit, Lappeenranta University of Technology
Kauppalankatu, Kouvola, Finland

Mohammad Sultan Ahmad Ansari
Modern College of Business and Science
Oman

Shahreyar Ansari
The Islamia University of Bahawalpur
Bahawalpur, Pakistan

Zeeshan Asim
College of Business
University of Buraimi
Al Buraimi, Oman
and
Institute of Business and Management
Karachi, Pakistan
and
INTI International University
Persiaran Perdana BBN, Putra Nilai
Nilai, Negeri Sembilan

Azan Al Azzani
College of Business
University of Buraimi
Al Buraimi, Oman

Ujjal Bhuyan
Department of commerce
Jagannath Barooah College
Assam, India

Magdalena Dalewska
Poznan University of Technology
Poznan, Poland

Mir Dost
University of Winchester
United Kingdom

Waseem Fatima
Tawam International School AL Buraimi
Sultanate of Oman

Muhammad Zulkifl Hasan
Faculty of Information Technology
Department of Computer Science
University of Central Punjab
Lahore, Pakistan

Asghar Hayyat
Department of Business Administration
Ghazi University
Dera Ghazi Khan, Pakistan

Muhammad Zunnurain Hussain
Department of Computer Science
Bahria University Lahore Campus
Pakistan

Shoaib Irshad
Department of Business Studies
Namal University, Mianwali

Hyder Kamran
University of Buraimi
Al Buraimi, Oman

Shad Ahmad Khan
College of Business
University of Buraimi
Oman

Zaheer Ahmed Khan
Mazoon College Muscat
Oman

Muhammad Latif Khan
Department of Mechanical Engineering
Global College of Engineering & Technology
Ruwi, Sultanate of Oman

Beata Mrugalska
Poznan University of Technology
Poznan, Poland

Devarajanayaka Muniyanayaka
College of Business
University of Buraimi
Al Buraimi, Oman

Ijaz Nawaz
National College of Business and Management Lahore, Pakistan
Harbanspura Campus, Lahore, Pakistan

Maria Nemilentseva
LUT Kouvola Unit
Lappeenranta University of Technology
Kauppalankatu, Kouvola, Finland

Pooyan Rahmanivahid
Department of Mechanical Engineering
Global College of Engineering & Technology
Ruwi, Sultanate of Oman

Muhammad Mustafa Raziq
College of Business Administration
University of Sharjah: Sharjah
Sharjah, United Arab Emirates

and

NUST Business School
National University of Science and Technology (NUST)
Islamabad, Pakistan

Muhammad Rehan Saleem
NUST Business School
National University of Science and Technology (NUST)
Islamabad, Pakistan

Rohani Salleh
Management and Humanities Department
Universiti Teknologi PETRONAS
Seri Iskandar, Malaysia

Ibrahim Rashid Al Shamsi
College of Business
University of Buraimi
Al Buraimi, Oman

Shahryar Sorooshian
Department of Business Administration
University of Gothenburg
Gothenburg, Sweden

Adeel Tariq
LUT Kouvola Unit
Lappeenranta University of Technology
Kauppalankatu, Kouvola, Finland

and

NUST Business School
National University of Science and Technology (NUST)
Islamabad, Pakistan

Waqas Tariq
School of Finance and Economics
Institute of Industrial Economics
Jiangsu University
Zhenjiang, P. R. China

and

Department of Commerce
University of Sialkot
Sialkot, Pakistan

Marko Torkkeli
LUT Kouvola Unit
Lappeenranta University of Technology
Kauppalankatu, Kouvola, Finland

Sonia Umair
Department of Managerial and Financial Sciences
Al-Zahra College for Women
Oman

Umair Waqas
College of Business
University of Buraimi
Al Buraimi, Oman

Zia-ur-Rehman
Department of Business Administration
Ghazi University
Dera Ghazi Khan, Pakistan

1 Revolutionizing Pharmaceutical Traceability

Exploring the Role of Innovative Leadership in a New Venture Digital Traceability Deployment

Shahreyar Ansari
The Islamia University of Bahawalpur, Bahawalpur, Pakistan

Muhammad Rehan Saleem
National University of Science and Technology (NUST), Islamabad, Pakistan

Adeel Tariq
Lappeenranta University of Technology Kauppalankatu, Kouvola, Finland
National University of Science and Technology (NUST), Islamabad, Pakistan

Muhammad Mustafa Raziq
University of Sharjah, Sharjah, United Arab Emirates
National University of Science and Technology (NUST), Islamabad, Pakistan

Mir Dost
University of Winchester, United Kingdom

DOI: 10.1201/9781032616810-1

1.1 INTRODUCTION

Operations and supply chains (SCs) have undergone numerous significant changes as a result of Industry 4.0 and digital technology (Zhou et al., 2023), and for the economy's survival and development, SC management is essential (Waqas et al., 2023). Traceability of the products has been recognized as a valuable and useful process in this regard (Noronen, 2023). Besides internal demand, external pressure from stakeholders, governments, and customers is pushing businesses to become experts in traceability solutions (Hastig & Sodhi, 2020; Noronen, 2023). Moreover, corporate performance is greatly enhanced by the product's traceability (Khan et al., 2021). Furthermore, a boost in digital technologies has further accentuated the need for digital traceability to gain several benefits (Zhou et al., 2023). Some industries have employed traceability procedures and systems for materials; however, their use has been mostly or solely associated with quality management, not directly with products or digital traceability systems (Garcia-Torres et al., 2019). Although the digital product traceability demand has been acknowledged, several businesses, including the pharmaceutical industry, have not fully adopted it.

It is pertinent to consider digital traceability solutions for corporates in developing countries as consumers are seeking greater openness in product traceability and becoming more aware of the significance and needs of critical products such as pharmaceuticals (Rejeb et al., 2019). Traceability has become a necessity rather (Jin & Zhou, 2014). Thus, the adoption of digital traceability based on digital technologies such as blockchain makes end-to-end product visibility possible and aids in the mapping of supply networks (Hastig & Sodhi 2020). This highlights the significance of digital traceability solutions in developing countries to procure genuine products and provide several benefits to consumers (Joshi & Sharma, 2022). Particularly, the construction of digital traceability processes and systems is lacking in the industrial sector, and the literature does not offer straightforward managerial implications. Companies must therefore create original solutions in the absence of industry-wide standards or norms (Hjaltadóttir & Hild, 2021).

Strong leadership is necessary for the successful execution of digital traceability efforts. When top management shows their commitment to digital transformation, businesses are more inclined to spend money on traceability (Asad et al., 2021; Hackius & Petersen, 2017; Majid et al., 2023). Recent studies and data show a large growth in global investment in innovation capital, underscoring the necessity for strong innovative leadership (IL) (Al-Hyari, 2023; Khalili, 2017). IL promotes new ideas to boost innovation, progress, and performance. It entails anticipating issues and taking proactive steps to resolve them, as well as establishing a friendly workplace that achieves team goals (Al-Hyari, 2023). Considering its significance, successful and growing firms need innovative leadership (Abbas & Asghar, 2010; Majid et al., 2023). IL may foster innovation, game-changing ideas, and ongoing adaptability by encouraging firms to explore new ground (Abbas & Asghar, 2010). To properly comprehend the role of IL in the healthcare sector, SC management, and digital traceability, further study is needed. This will likely clarify how IL creates new healthcare goods and solutions (Bag et al., 2021; Mehmood et al., 2021).

According to Carmeli et al. (2010), IL improves company performance by increasing strategic fit (internal/external). To be more precise, IL covers how to properly motivate people to take initiative, create an open performance evaluation system, and create a culture that values good relationships (Safdar et al., 2019). The organization is likely to become more creative as a result of having good IL (Carmeli et al., 2010). Healthcare and traceability-related innovations probably are crucial for developing cutting-edge products and services as well as cutting-edge solutions like digital traceability (Bag et al., 2021). The introduction of innovative goods and services and the development of creative solutions, such as digital traceability, are expected to be largely dependent on IL in the healthcare and traceability sectors. Thus, the significance of IL cannot be undermined considering their critical role in the development of traceability solutions for the enterprise.

In this regard, literature contains scant evidence pertinent to IL's role in the implementation of traceability solutions for products and services. A thorough investigation explaining the role of IL in effectively implementing and introducing traceability solutions through a new venture is lacking. Thus, this study aims to address this gap and specifically focus on a research objective to understand the role of IL in introducing digital traceability solutions in pharmaceutical organizations.

This study contributes to two streams of literature: leadership and SC. Firstly, from the SC perspective, we contributed to this stream by outlining the business needs for digital traceability solutions, where we highlighted an essential element vital to the success of any installation. Secondly, we contributed to a body of research on IL in SC (Bag et al., 2021; Hastig & Sodhi, 2020) and have added knowledge on IL roles for effectively establishing digital traceability solutions. We emphasize that strong leadership is needed to promote and implement digital traceability systems. Thus, building on theoretical support, this study shows that digital traceability systems require leadership.

In doing so, this research focuses on a case study of Pharma Trax, a Pakistani company to revolutionize pharmaceutical traceability. Pharma Trax was developed to address pharmaceutical SC authenticity and transparency issues in SC. It was led by the Chief Innovation Officer (CIO), an enthusiastic, innovative leader who uses technology to solve issues in new ways. Pakistan's pharmaceutical industry struggled to track items along the SC, like many others. Without a clear traceability system, patient safety, product quality, regulations, and corporate productivity suffer. Pharma Trax emerged as a pathfinder in the middle of these difficulties, providing digital traceability solutions that promised to change the business environment.

1.2 LITERATURE REVIEW

As highlighted by researchers, SC traceability through manufacturer-to-consumer visibility creates a variety of values and possibilities (Hastig and Sodhi 2020). Traceability refers to the capacity to research an entity's background, use, or location using documented identifications (ISO, 1994). Specifically, when addressing a physical item, traceability can take geographical data, the history of the manufacturing process, the origin of the parts, and distribution into consideration (Noronen 2023). There are two types of traceability: internal and external. Internal traceability, according to the SC, refers to the specified traceable activities within a single process step (Schuitemaker & Xu, 2020). Internal traceability aims to improve productivity and reduce costs by keeping records of the product/services within a single facility, operation, or company (Schuitemaker & Xu, 2020). The record includes information regarding the raw products and processes before the product/service is delivered. External traceability discusses occurrences or actions that can be traced apart from the process phases (Schuitemaker & Xu, 2020). External traceability is used to identify where products have gone, what they have gone into, and what objects they have come into touch with. External traceability occurs between firms and is dependent on cooperation among all SC participants. Traceability can be achieved in one of two ways: downstream refers to traveling toward distributors and customers, whereas upstream refers to going back to the manufacturer (Aljabhan & Obaidat, 2023). As a result, each SC partner should be able to trace the product inside as well as outside to ensure traceability across the SC (Noronen 2023).

Moreover, traceability efforts have also been acknowledged to have a positive impact on the sustainability initiative (Biswas et al., 2023; Noronen, 2023). As traceability ensures and evaluates sustainability claims made about commodities and goods along the SC, it guarantees ethical behavior and respect for people and the environment (Compact, 2014; Noronen, 2023). Moreover, traceability procedures boost information exchange about all facets of sustainability and give SC partners a way to become more visible and transparent. Therefore, traceability methods have wider beneficial implications for SC and economic viability in addition to several other benefits (Cousins et al., 2019).

1.2.1 Digital Traceability

The idea of traceability is not new, and it has been used by many industries to increase SC transparency and accountability. Nevertheless, in the age of Industry 4.0, digital traceability solutions have become effective instruments that make use of cutting-edge technologies to offer real-time monitoring, verification, and transparency throughout the SC (Bahramian Dehkordi et al., 2023; Feng et al., 2020). Moreover, traceability systems have changed from being paper-based to being IT-driven to allow more efficient and timely information handling (Noronen, 2023; Razak et al., 2021). According to Razak, Hendry, and Stevenson (2021), the most helpful and pertinent traceability data must meet the following criteria: they must be complete, accurate, timely, easy to access, and secure. These criteria can be met by using current and developing technological solutions (Kazancoglu et al., 2023). The technological solution is selected in accordance with the business' requirements for both the width and depth of knowledge and the goals for the traceability system (Noronen, 2023; Razak et al., 2021). Business operations can be made more sustainable and resilient to uncertainty and future SC disruptions by utilizing both new and existing technologies (Kazancoglu et al. 2023).

Unlike traditional traceability solutions, digital traceability requires organizations to create traceability solutions for a complete SC using radical technologies such as radiofrequency identification (RFID), blockchain, and big data to monitor a product's origin and destination (Hastig & Sodhi 2020). As digital transformation advances, by connecting data from the initial plant producer via a producer and distributor to the end users, digital traceability expands this internal traceability to span the entire product life cycle (Aung & Chang, 2014). With the use of smart labeling, distinctive digital identification, distributed digital storage, smart contracts, and electronic certifications, traceability systems are further improved (Hastig & Sodhi, 2020), thus, enhancing and solidifying the integrity and transparency of food producers. Therefore, understanding the role of leadership in the enactment and enablement of digital traceability systems is crucial and of higher importance to further the discussion and provide practical implications. Leaders have a responsibility to research the technology's potential to comprehend its relevance and how digital traceability raises the quality of products for consumers.

1.2.2 Leadership and Digital Traceability

For digital technology adoption to be effective, leadership is essential both within a company and throughout SCs. Leadership roles inspire team members in the SC and support any technology that comes from outside or within the SC (Chow, 2018; Majali et al., 2022). Leadership promotes trust, stimulates traceable investments, and assures a fair allocation of the costs associated with uncertainty. An efficient leader takes on the duty of proactively setting traceability standards (Majid et al. 2023; Mehmood et al. 2021), dealing with any investment reluctance and eventually improving traceability acceptance throughout the chain. Furthermore, to overcome underinvestment in traceability, unstable situations highlight the requirement of a leadership role (Charlier & Valceschini, 2008). The challenges from the standpoint of individual enterprises require firm-focused leadership when aiming for chain-wide traceability deployment (Charlier & Valceschini 2008). To successfully deploy traceability, the focus of firm's leadership role includes coordinating relational objectives, addressing data security problems, and managing potential consequences (Roy, 2021).

Moreover, the presence of pioneering leaders who have already adopted digital technologies encourages other businesses to devote resources (Hackius & Petersen, 2017; Hastig & Sodhi, 2020). This pattern can be seen across some sectors, including the agri-food industry, where pioneers of digital technology adoption like Carrefour and Wal-Mart have encouraged others to follow suit (Hastig & Sodhi, 2020; Kim & Laskowski, 2018). These industry innovative leaders are essential in bringing together SC participants, exchanging knowledge, and working together to successfully adopt digital technology solutions for digital traceability (Bateman, 2015). Moreover, this study responds to the call where academics have called further attention to understanding how leaders might create distinctive business propositions based on traceability among diverse configurations and further explore the phenomenon using qualitative techniques (Roy, 2021).

1.3 METHODS

The study is qualitative in nature and is based on a case study of a new venture, for anonymity and confidentiality purposes named Pharma Trax, which introduced a traceability solution in the market. The investigation is being done in an exploratory manner to better understand how Pharma Trax introduced digital traceability solutions to the market. The exploratory nature of the research method and study design allows for flexibility, when novel information is uncovered (Gohar et al., 2022; Rashid et al., 2022). The researcher for this study collected data using in-depth interviews with the CIO of the company and other key personnel where necessary and using company documents and other available sources. It included a series of interviews online and offline until researchers reached the data saturation point, thus, several interviews were conducted by the researchers to collect and validate the information collected during the process.

Semi-structured interviews helped explain how innovative leadership improved Pharma Trax's digital traceability system implementation (Rashid et al., 2022). Semi-structured interviews also provided for a deliberate balance between guided inquiries and unconstrained investigations, allowing for a methodical assessment of various important topics. Interview questions addressed questions related to digital traceability solutions, their introduction, how they overcame traceability implementation hurdles, and IL's role. Sample questions include such as the following:

> "What were the main issues or resistance that Pharma Trax faced while introducing these products to the market, and how did IL techniques allow the company to get beyond these challenges?"

This question focused on the executive team's tactics to overcome challenges and incorporate digital traceability systems efficiently. Several interviews were conducted by the researchers to collect and validate the information collected during the process. After discussing numerous parties, researching relevant publications, and extensively studying how IL affects Pharma Trax's implementation of digital traceability solutions, the researchers were able to strengthen their conclusions.

1.4 FINDINGS

Our case study shows the intricate interaction between IL and Pharma Trax's digital traceability system installation. Interviews and analysis led to these conclusions. Our research highlights the complicated interaction between leadership styles, implementation challenges, and organizational dynamics in the following areas.

1.4.1 Pharma Trax: From Inspiration to Inception

Numerous businesses have been inspired to venture into uncharted territory and adopt cutting-edge technologies in their pursuit of innovation and sustainable performance (Allioui & Mourdi, 2023). Among them is Pharma Trax, a Pakistani business that set out on a mission to transform the traceability system for the pharmaceutical industry. Pharma Trax was founded to address the crucial problems of openness and authenticity within the pharmaceutical SC, and it specializes in developing and implementing traceability solutions.

This idea of traceability came to the leadership of Zauq (alternate name for protecting confidentiality) group, the parent company of Pharma Trax, after an incident occurred in Bangladesh, where a fire at a plaza destroyed garment products. There was confusion over the total loss as the affected company had weighed the products differently from the insurance company. The opportunity to provide traceability in the textile industry was realized; a venture was started for this purpose, and it proved to be a success. At that time, there was a scant concept of transparency and traceability in Pakistan. People were more concerned about getting their products. They rarely deliberated on where products came from. Sensing this opportunity, the leadership of Zauq Group worked on providing traceability solutions in the textile industry. This venture proved to be a success. Based on their success in the textile industry, Zauq Group's management saw an opportunity to diversify into another industry where their expertise in developing technological solutions could bring great benefits. The Pharma Trax project was launched in 2013, marking the beginning of their journey to revolutionize traceability in the Pakistani pharmaceutical industry. It was led by its leadership, mainly the CIO, a visionary entrepreneur who is passionate about using technology to create novel solutions.

1.4.2 Inspiration behind Pharma Trax

Pharma Trax was created in response to the urgent issues facing Pakistan's healthcare and pharmaceutical industries, where a sizeable portion of the population struggles with inadequate access to necessary pharmaceuticals (Junaidi, 2015). Importing large quantities of medications to satisfy local demand led to ongoing drug shortages, which were aggravated by political upheaval and dishonest billing methods that increased the sector's turbulence (Haq, 2014).

Zauq Group saw an opportunity to address these pressing problems by utilizing its expertise in traceability solutions. Concerns about the quality of the products used in Pakistan's healthcare system were raised due to variations in the efficacy of pharmaceuticals obtained through different channels, as demonstrated by the disparities in outcomes between locally sourced and medications imported from the UAE.

The tragic episode at the Punjab Institute of Cardiology (PIC), which resulted in several deaths from tainted medication, highlighted the need for an immediate transparent traceability system to protect patient safety (Chaudary, 2013). Similar to the misuse of ephedrine, which resulted in its prohibition after allegations of widespread usage among youngsters, traceability systems are essential for keeping track of the production, distribution, and use of drugs (Dawn, 2012). These occurrences underlined the critical role that traceability plays in avoiding the entry of counterfeit medications into the market and the irrational price increases brought on by erroneous shortage claims.

The leadership of Zauq Group saw the potential effectiveness of a comparable strategy within the pharmaceutical business after learning from their successful use of traceability solutions in the textile industry. Their goal with Pharma Trax was to create a transparent traceability system that would ensure the safety, efficacy, and appropriate distribution of medicines. This would help to address industry-wide problems and promote an improved healthcare system for patients throughout Pakistan.

The management and CIO planned to establish a new business that would specialize in creating automated traceability systems to bring Pharma Trax's vision to life.

Both internal and external talent needed to be hired as part of the recruitment process. The traceability application's internal development and thorough testing were given top priority to ensure a smooth phased rollout and minimize any system disruptions.

1.4.3 Traceability Solutions in the Pharmaceutical Industry

Pakistan's pharmaceutical sector is significant in terms of investment as well as nature. The production of various pharmaceutical goods is carried out by more than 750 pharmaceutical enterprises (Jannat et al., 2023). Due to the recent spate of acquisitions in this sector, some of the companies are significantly larger than others. Due to their scale, several businesses dominate the market; currently, the top 50 corporations own roughly 90% of the market share in this sector (Jannat et al., 2023). Pakistan's 2020 pharmaceutical product exports were valued at US$235.75 million, according to the United Nations Commodity Trade Statistics Database (UN COMTRADE) on global commerce. The most recent update to Pakistan's pharmaceutical product export data, historical chart, and statistics was made in October 2022 (Jannat et al., 2023). Despite this expansion, Pakistan still heavily depends on imported bulk medicine active components, which is a result of the industry's explosive growth since 2010 and the rising use of Good Manufacturing Practices (GMP), which correspond to national and international standards (PRIME, 2017). With local players like Getz Pharma making significant strides and earning the World Health Organization's accreditation for original drug production, the pharmaceutical industry has seen a notable rise in the prominence of multinational subsidiaries like GSK and Abbott (GETZ Pharma, 2023).

1.4.4 Digital Tracing and Tracking at Pharma Trax

Pharma Trax is dedicated to ensuring the authenticity and exclusivity of pharmaceutical products, effective recall management, and region-based product usage tracking to address major industry concerns. Pharma Trax has established itself as a leader in the implementation of comprehensive tracking, tracing, and verification capabilities for pharmaceutical products, enhancing production standards and regulatory compliance. Pharma Trax was the first company in Pakistan to employ a 2D Data Matrix in this capacity.

Working on traceability solutions is explained by the CIO as:

> To implement the traceability system, each product package is given a special alphanumeric serial number (barcode), allowing for easy tracking throughout the SC. Through the installation of tracking devices in vehicles, real-time transportation monitoring is made possible, assuring the safe and open transportation of pharmaceutical products.

Pharma Trax intends to improve transparency and security in the pharmaceutical SC by integrating this cutting-edge traceability and tracking technology, discouraging unlawful activity, and empowering law enforcement agencies to stop the spread of illicit drug production. Additionally, the system gives customers insight into the progress of the product, which promotes trust and responsibility in the sector.

1.4.5 Leadership Role in Fostering Success

Fostering an innovative spirit at Pharma Trax was a priority for the CIO. To create an environment that focused on training and development, the CIO implemented several initiatives: leadership adopted an innovative culture throughout the organization. Their motto, "learning how to learn," reflected their commitment to continuous learning and improvement. This culture encouraged employees to think outside the box, share ideas, and embrace new approaches to problem-solving. Recognizing the complexity of developing traceability solutions for the pharmaceutical industry, leadership allocated a significant portion of their earnings to the Research and Development (R&D) department. They invested approximately US$300,000 over 18 months to develop the necessary solutions. The R&D department played a crucial role in driving innovation within the organization.

Unlike the hierarchical structure at Zauq Group, the CIO decided to implement a flat structure at Pharma Trax. The changed structure was explained by the CIO as

> There were no separate offices for managers, and the office space was designed as a big open hall with tables where direct communication could take place between employees and managers. Each department worked in the same hall, facilitating fast and easy information sharing.

Leadership also emphasized hiring the right personnel who had good communication and marketing skills. They wanted to assure their customers that high-quality products and timely deliveries would be provided. All the employees were given lectures and made to attend the seminars regarding customer dealings.

CIO also explained the method they used to enhance the creativity of their personnel:

> All employees gathered for a weekly meeting where they were divided into diverse groups with different leaders each week. Each group was given a topic or task to find solutions to a problem. During the meeting, each group presented what they had learned, allowing every employee to learn from each other. The other groups also provided opinions and insights on different scenarios.

Moreover, the CIO created an online open platform for employees to share ideas, and if their ideas were selected, they were rewarded with a special bonus. These initiatives aimed to create a collaborative and innovative environment at Pharma Trax, where employees were encouraged to learn, share ideas, and contribute to the company's growth and success. It resulted in the success of the start-up by establishing the right culture for introducing traceability solutions.

1.4.6 Challenges Addressed by the Leadership

IL is also required to address the challenges to establishing the new venture, for this reason, Pharma Trax leadership concurred with the obstacles and gained the trust of stakeholders in the pharmaceutical industry. As the CIO elaborated:

> *Pharma Trax* conducted workshops, seminars, and training sessions to raise awareness about the advantages of their traceability solution. They showcased case studies and success stories from other industries to demonstrate the positive impact of traceability on efficiency, safety, and regulatory compliance.

Pharma Trax approached influential and reputable companies within the pharmaceutical industry and offered them a pilot project or a trial period to demonstrate the effectiveness of their traceability solution. By partnering with respected companies and showcasing successful implementations, Pharma Trax built trust and credibility among other industry players.

Pharma Trax gained credibility and support by establishing strategic alliances and partnerships with industry groups, governing bodies, and other critical stakeholders. By working together with these entities, they can address any concerns, align their objectives, and collectively promote the adoption of traceability in the industry. It has been established in the literature that collaboration facilitates researchers to gain success in developing countries (Temel et al., 2013). Leadership focuses on understanding the specific pain points and challenges faced by different pharmaceutical companies and tailoring their traceability solution to address those needs. By highlighting the unique benefits and cost savings that their solution can bring to each company, they can appeal to their interests and incentivize them to adopt the technology. As explained by the CIO,

> *Pharma Trax* emphasizes the regulatory compliance benefits of their traceability solution. They can highlight how the system helps companies meet regulatory requirements, track, and trace products effectively, and maintain quality control.

Demonstrating that their solution aligns with industry standards and regulations can alleviate concerns and build trust. It is important for leadership to carefully analyze the industry landscape, consider the specific needs and concerns of stakeholders, and tailor their strategies accordingly. They proactively addressed objections, providing evidence of the system's effectiveness and showcasing the long-term benefits of adopting Pharma Trax's traceability solution.

1.5 IMPLICATIONS

1.5.1 Theoretical Implications

Digital traceability solutions in SC are gaining importance; however, little research has investigated the role of IL in implementing and enhancing digital traceability solutions in the literature. In this chapter, we explore the influence of IL and explain its impact on digital traceability solutions. In doing so, this chapter has important implications for leadership and traceability literature. The case study findings reveal the essential role of IL needed for implementing novel practices in SC such as digital traceability solutions. The role of IL is an important leverage for transformational changes in the organization as IL emphasizes the value of establishing a culture of innovation, agility, and response to market dynamics and technical improvements. According to Carmeli et al. (2010), digital traceability requires (IL), as it stimulates initiative and innovation, and makes it simpler to implement new ideas, making it vital for strategy alignment.

The case study also highlighted that IL is crucial to the organization's digital traceability strategy. This conclusion has important theoretical implications for leadership and traceability experts. Companies must adopt new technologies to compete in today's fast-paced economy; here, the role of IL is highly significant. IL understands the market dynamics, considers stakeholders in the process, and can implement new solutions accordingly. This is similar to earlier research showing how leadership improves digital skills and firm technical advancement (Chatterjee et al., 2023; Chen et al., 2023). The organization requires visionary IL who can balance technology's short-term rewards with its long-term aims and competitive position.

This study shows that IL is crucial to addressing internal and external difficulties related to new solutions and initiatives. According to the research findings, IL can start new businesses, grasp complicated problems, and find creative solutions. This is theoretically crucial because leadership practices may make new enterprises more marketable and help them handle challenges with in-depth knowledge, improving their sustainability and competitiveness. Leadership is crucial for problem-solving and firm profitability, as has been shown. This is in line with the existing literature that emphasizes the role of leadership in effectively navigating through challenges and ensuring the long-term success of the business (Probst et al., 2011).

1.5.1.1 Managerial Implications

The chapter places a strong emphasis on the essential role that IL commitment plays in the effective adoption of digital traceability solutions, and it provides important implications for managers and policymakers. Firstly, an organization shall understand the significant role of IL in the implementation of novel solutions in the organization. With the help of IL, digital traceability solutions can be implemented in a way that promotes uniformity among SC participants. So, leaders should commit to allocating funds and promoting the use of digital technologies for traceability, this dedication shall not be limited to only implementing traceability solutions; instead, it shall be prevalent throughout the entire SC. Secondly, to successfully adopt digital traceability solutions across the complete SC, IL can enable a creative culture where employees can be able to understand the significance of novel solutions, information exchange, and coordinated efforts. With the active participation of employees in implementation of novel solutions, organizations can actively achieve their goals from implementation and afterward, in terms of long-term success.

Thirdly, IL can also facilitate organizations in selecting technologies or digital solutions (such as IoT, blockchain, and artificial intelligence) that suit the organization's needs and traceability solutions. For traceability, businesses should seek out and develop visionary, tech-savvy ILs who can navigate the overall digital transferability process. However, organizations shall also utilize the role of IL to investigate technologies not only for digital traceability solutions but also for improving other processes in SC to gain enhanced benefits. Lastly, IL can play a significant role in implementing traceability systems not only for operational needs but also to promote sustainability initiatives by giving customers verified information about the items' ethical standards and place of origin. They can also share the best practices for wider implementation in the industry to promote sustainable practices and effectively utilize digital traceability solutions.

1.6 CONCLUSION

Traceability of products has emerged as a crucial element for assuring openness, accountability, and sustainability in the continually changing landscape of SC management. The widespread use of traceability systems across industries has been accelerated by external demands as well as the advantages they provide to organizations. To demonstrate how IL can bring about radical change within an industry, this chapter has examined the story of Pharma Trax, a Pakistani pharmaceutical venture. Pharma Trax was successful in changing Pakistan's pharmaceutical SC by leveraging its experience and using an adaptive strategy to implement digital traceability solutions. For companies looking to use technology and leadership to create good change in the quest for openness, authenticity, and effectiveness, Pharma Trax example offers insightful information. The results highlight the crucial part that leadership commitment plays in catalyzing investments and activities pertaining to digital traceability. The chapter deepens the previous theoretical frameworks by highlighting the value of IL in the context of digital traceability. It also promotes a more comprehensive awareness of the leadership's function in enhancing product traceability using digital technology. This research serves as a benchmark for organizations looking to utilize the promise of digital traceability solutions through IL.

ACKNOWLEDGMENT

We are very thankful to the CIO of Pharma Trax, for his help in collecting the data and support for the completion of this manuscript.

APPENDIX

Interview Questions

1. Can you explain the original rationale for integrating digital traceability systems into Pharma Trax and how it was set up for market entry?
2. Given the difficulties encountered in the early phases of integration and market adoption, how did innovative leadership assist the successful implementation of digital traceability solutions?

3 Can you elaborate in detail on the role of innovative leadership in influencing and managing a new solution such as the introduction of Pharma Trax digital traceability solutions to the market?
4 Can you explain in detail the process followed by Pharma Trax to set up digital traceability systems from both internal and external perspectives?
5 What were the main issues Pharma Trax encountered as they launched digital traceability solutions, and how did innovative leadership strategies help the company overcome these difficulties?
6 What were the primary problems or opposition encountered during the process of introducing these solutions to the market, and how did creative leadership tactics help to overcome these difficulties?
7 How have digital traceability solutions changed the organizational culture at Pharma Trax, and how has the company's visionary leadership helped to create a culture that is open to constant change and innovation?
8 What strategic actions does Innovative leadership intend to take to strengthen Pharma Trax's digital traceability solutions and maintain its position as a market leader in this area?
9 Would you like to elaborate on other areas necessary for innovative leadership to consider for the success of digital ventures in the market?

REFERENCES

Abbas, W., & Asghar, I. (2010). The role of leadership in organizatinal change: relating the successful organizational change with visionary and innovative leadership.

Al-Hyari, H. (2023). Job Security as a Mediating Variable between Innovative Leadership and Innovative Work Behavior among Employees. *Journal of System and Management Sciences*, *13*(1), 532–574.

Aljabhan, B., & Obaidat, M. A. (2023). Privacy-Preserving Blockchain Framework for Supply Chain Management: Perceptive Craving Game Search Optimization (PCGSO). *Sustainability*, *15*(8), 6905.

Allioui, H., & Mourdi, Y. (2023). Unleashing the Potential of AI: Investigating Cutting-Edge Technologies That Are Transforming Businesses. *International Journal of Computer Engineering and Data Science (IJCEDS)*, *3*(2), 1–12.

Asad, M., Asif, M. U., Bakar, L. J. A., & Sheikh, U. A. (2021, December). Transformational leadership, sustainable human resource practices, sustainable innovation and performance of SMEs. In *2021 International Conference on Decision Aid Sciences and Application (DASA)* (pp. 797–802). IEEE.

Aung, M. M., & Chang, Y. S. (2014). Traceability in a food supply chain: Safety and quality perspectives. *Food Control*, *39*, 172–184.

Bag, S., Gupta, S., Choi, T.-M., & Kumar, A. (2021). Roles of innovation leadership on using big data analytics to establish resilient healthcare supply chains to combat the COVID-19 pandemic: A multimethodological study. *IEEE Transactions on Engineering Management*.

Bahramian Dehkordi, B., Podmetina, D., & Torkkeli, M. (2023). Blockchain as a Sustainability Booster in Supply Chain Management. *Handbook of Sustainability Science in the Future*.

Bateman, A. H. (2015). Tracking the value of traceability. *Supply Chain Management Review*, *9*, 8–10.

Biswas, D., Jalali, H., Ansaripoor, A. H., & De Giovanni, P. (2023). Traceability vs. Sustainability in supply chains: The implications of blockchain. *European Journal of Operational Research*, *305*(1), 128–147.

Carmeli, A., Gelbard, R., & Gefen, D. (2010). The importance of innovation leadership in cultivating strategic fit and enhancing firm performance. *The Leadership Quarterly*, *21*(3), 339–349.

Charlier, C., & Valceschini, E. (2008). Coordination for traceability in the food chain. A critical appraisal of European regulation. *European Journal of Law and Economics*, *25*, 1–15.

Chatterjee, S., Chaudhuri, R., Vrontis, D., & Giovando, G. (2023). Digital workplace and organization performance: Moderating role of digital leadership capability. *Journal of Innovation & Knowledge*, *8*(1), 100334.

Chaudary, A. (2013, March 22). *WHO says drug caused PIC deaths.* https://www.dawn.com/news/797093/who-says-drug-caused-pic-deaths

Chen, S.-L., Su, Y.-S., Tufail, B., Lam, V. T., Phan, T. T. H., & Ngo, T. Q. (2023). The moderating role of leadership on the relationship between green supply chain management, technological advancement, and knowledge management in sustainable performance. *Environmental Science and Pollution Research*, *30*(19), 56654–56669.

Chow, C. (2018). Blockchain for Good? Improving supply chain transparency and human rights management. *Governance Directions*, *70*(1), 39–40.

Compact, U. G. (2014). A guide to traceability-a practical approach to advance sustainability in global supply chains. *UN Global Compact*, *21*, 1–45.

Cousins, P. D., Lawson, B., Petersen, K. J., & Fugate, B. (2019). Investigating green supply chain management practices and performance: The moderating roles of supply chain ecocentricity and traceability. *International Journal of Operations & Production Management*, *39*(5), 767–786.

Dawn. (2012, September 16). *Drug 'ephedrine': Uses and abuses.* https://www.dawn.com/news/749723

Feng, H., Wang, X., Duan, Y., Zhang, J., & Zhang, X. (2020). Applying blockchain technology to improve agri-food traceability: A review of development methods, benefits and challenges. *Journal of Cleaner Production*, *260*, 121031.

Garcia-Torres, S., Albareda, L., Rey-Garcia, M., & Seuring, S. (2019). Traceability for sustainability–literature review and conceptual framework. *Supply Chain Management: An International Journal*, *24*(1), 85–106.

GETZ Pharma. (2023, October). *About Us. GETZ Pharma.* https://getzpharma.com/about-us/

Gohar, M., Abrar, A., & Tariq, A. (2022). The role of family factors in shaping the entrepreneurial intentions of women: A case study of women entrepreneurs from Peshawar, Pakistan. *The Role of Ecosystems in Developing Startups: Frontiers in European Entrepreneurship Research*, *40*, 40–63.

Hackius, N., & Petersen, M. (2017). Blockchain in logistics and supply chain: trick or treat?. In Digitalization in Supply Chain Management and Logistics: Smart and Digital Solutions for an Industry 4.0 Environment. In *Proceedings of the Hamburg International Conference of Logistics (HICL)* (vol. 23, pp. 3–18). Berlin: epubli GmbH.

Hastig, G. M., & Sodhi, M. S. (2020). Blockchain for supply chain traceability: Business requirements and critical success factors. *Production and Operations Management*, *29*(4), 935–954.

Haq, S. (2014, September 30). *Industry woes: Drug shortage feared as consignments stuck at ports.* https://tribune.com.pk/story/769399/industry-woes-drug-shortage-feared-as-consignments-stuck-at-ports

Hjaltadóttir, R. E., & Hild, P. (2021). Circular Economy in the building industry European policy and local practices. *European Planning Studies*, *29*(12), 2226–2251.

ISO, B. (1994). *9000-1: Quality management and quality assurance standards-guidelines for selection and use*. British Standards Institution.

Jannat, A., Shafiq, N., Hanif, M., Riasat, M., & Rafique, S. (2023). A brief insight to Pakistan's Pharmaceutical industry-A critical review study. *Pakistan Journal of Medical & Health Sciences, 17*(2), 2–2.

Jin, S., & Zhou, L. (2014). Consumer interest in information provided by food traceability systems in Japan. *Food Quality and Preference, 36*, 144–152.

Joshi, S., & Sharma, M. (2022). Digital technologies (DT) adoption in agri-food supply chains amidst COVID-19: An approach towards food security concerns in developing countries. *Journal of Global Operations and Strategic Sourcing, 15*(2), 262–282.

Junaidi, I. (2015). *No law to stop pharma companies from importing raw materials.* Retrieved from DAWN: https://www.dawn.com/news/print/1195030.

Kazancoglu, I., Ozbiltekin-Pala, M., Mangla, S. K., Kumar, A., & Kazancoglu, Y. (2023). Using emerging technologies to improve the sustainability and resilience of supply chains in a fuzzy environment in the context of COVID-19. *Annals of Operations Research, 322*(1), 217–240.

Khalili, A. (2017). Creative and innovative leadership: Measurement development and validation. *Management Research Review, 40*(10), 1117–1138.

Khan, S. A. R., Zia-ul-haq, H. M., Umar, M., & Yu, Z. (2021). Digital technology and circular economy practices: An strategy to improve organizational performance. *Business Strategy & Development, 4*(4), 482–490.

Kim H. M., & Laskowski, M. (2018). *Agriculture on the blockchain: Sustainable solutions for food, farmers, and financing. Supply chain revolution.* New York, NY: Barrow Books, AMACOM.

Majali, T., Alkaraki, M., Asad, M., Aladwan, N., & Aledeinat, M. (2022). Green transformational leadership, green entrepreneurial orientation and performance of SMEs: The mediating role of green product innovation. *Journal of Open Innovation: Technology, Market, and Complexity, 8*(4), 191.

Majid, F., Raziq, M. M., Memon, M. A., Tariq, A., & Rice, J. L. (2023). Transformational leadership, job engagement, and championing behavior: Assessing the mediating role of role clarity. *European Business Review, 35*(6), 941–963.

Mehmood, M. S., Jian, Z., Akram, U., & Tariq, A. (2021). Entrepreneurial leadership: The key to develop creativity in organizations. *Leadership & Organization Development Journal, 42*(3), 434–452.

PRIME. (2017). *Pakistan's Pharmaceutical Industry.* https://primeinstitute.org/wp-content/uploads/2021/05/July-2017-Annual-Report-on-the-Pharmaceuticals-Industry-of-Pakistan.pdf

Noronen, E. (2023). Enhancing sustainability through traceability: a case study on design of Valmet's purchased item traceability system (Publication No. 2023053150983) [Master's thesis, Lappeenranta-Lahti University of Technology LUT]. LUTPub.

Probst, G., Raisch, S., & Tushman, M. L. (2011). Ambidextrous leadership: Emerging challenges for business and HR leaders. *Organizational Dynamics, 40*(4), 326–334.

Rashid, R., Badir, Y. F., Tariq, A., & Afsar, B. (2022). The role of product lines in determining the degree and speed of integration: Evidence from the pharmaceutical MNCs in cross-border acquisitions. *European Journal of International Management, 17*(1), 27–59.

Razak, G. M., Hendry, L. C., & Stevenson, M. (2021). Supply chain traceability: A review of the benefits and its relationship with supply chain resilience. *Production Planning & Control, 34*, 1114–1134.

Rejeb, A., Keogh, J. G., & Treiblmaier, H. (2019). Leveraging the internet of things and blockchain technology in supply chain management. *Future Internet, 11*(7), 161.

Roy, V. (2021). Contrasting supply chain traceability and supply chain visibility: Are they interchangeable? *The International Journal of Logistics Management, 32*(3), 942–972.

Safdar, U., Tariq, A., Saudagar, T., & Razzaq, R. (2019, July). Inclusive leadership and creative self-efficacy: Testing the moderating and mediating mechanisms. In *Academy of Management Proceedings* (Vol. 2019, No. 1, p. 17550). Briarcliff Manor, NY 10510: Academy of Management.

Schuitemaker, R., & Xu, X. (2020). Product traceability in manufacturing: A technical review. *Procedia CIRP*, *93*, 700–705.

Temel, S., Mention, A.-L., & Torkkeli, M. (2013). The impact of cooperation on firms' innovation propensity in emerging economies. *Journal of Technology Management & Innovation*, *8*(1), 54–64.

Waqas, U., Abd Rahman, A., Ismail, N. W., Kamal Basha, N., & Umair, S. (2023). Influence of supply chain risk management and its mediating role on supply chain performance: Perspectives from an agri-fresh produce. *Annals of Operations Research*, *324*(1–2), 1399–1427.

Zhou, X., Zhu, Q., & Xu, Z. (2023). The role of contractual and relational governance for the success of digital traceability: Evidence from Chinese food producers. *International Journal of Production Economics*, *255*, 108659.

2 Supply Chain 4.0 A Source of Sustainable Initiative across Food Supply Chain

Trends and Barriers

Zeeshan Asim
University of Buraimi, Al Buraimi, Oman
Institute of Business and Management, Karachi, Pakistan
INTI International University, Putra Nilai, Negeri Sembilan

Shahryar Sorooshian
University of Gothenburg, Gothenburg, Sweden

Ibrahim Rashid Al Shamsi,
Devarajanayaka Muniyanayaka, and Azan Al Azzani
University of Buraimi, Al Buraimi, Oman

2.1 INTRODUCTION

The advancement of technology is widely recognized as one of the primary drivers behind the current sustainability challenges that the world is facing. As technology becomes more sophisticated, there is a growing demand for resources, resulting in a significant increase in the volume of greenhouse gas emissions being released into the environment (Ojo et al., 2018).

It is commonly believed that this phenomenon is responsible for the increase in global warming, which is the primary cause of climate change. However, when used correctly, technological advancements and digitalization can play a crucial role in addressing the pressing sustainability issues that pose a significant threat to society (Nations, 2023). The integration of the latest Industry 4.0 technologies into supply chains and logistics, commonly referred to as Supply Chain 4.0 at the operational level, has the potential to make a significant contribution toward achieving the Sustainable Development Goals (SDGs) advocated and pursued on a daily basis (Chetna et al., 2022).

DOI: 10.1201/9781032616810-2

Industry 4.0 is a term used to describe the data-driven revolution in various sectors. Information technology is utilized to connect data, people, processes, services, and systems, leading to smarter industries and ecosystems with a strong focus on the environment (Ye et al., 2023). The integration of Industry 4.0 in supply chains can lead to a reduction in both lead time and expenses. It can also foster coherence among the decision-making processes of different members of the supply chain and improve the overall performance of the production process (Waqas et al., 2023). Ultimately, this contributes to the achievement of sustainable development objectives (Sajid Ullah et al., 2023).

Effective supply chain management is the starting point for achieving sustainability across various business functions. By ensuring a well-coordinated supply chain, from the acquisition of materials to the delivery of finished products to end-users, the objective of sustainability can be achieved (Esmaeilian et al., 2020). Sustainable production management through the supply chain is crucial in the food manufacturing industry due to the complex nature of food production. This involves taking into account important factors such as food safety, food waste, and food security (Esmaeilian et al., 2020). By integrating specific components of Industry 4.0, it becomes possible to effectively and sustainably manage the supply chain of the food industry (Lahane et al., 2023). Further improvement and innovation of Industry 4.0 elements can greatly enhance the sustainability of managing food processing supply chains (Ojo et al., 2018).

Nowadays, ensuring that the large global population has access to sustainable nourishment is a challenging task. This challenge arises because we have exceeded the Earth's natural capacity to replenish its resources by 30% due to global consumption (Anish et al., 2022). Over the past 55 years, the demand for food has tripled, and it is expected that the global population will exceed nine billion by 2050, leading to a further increase in demand (Anish et al., 2022). The food chain supply (FSC) is grappling with several environmental challenges, including greenhouse gas (GHG) emissions, climate change, soil contamination, groundwater depletion, and forest degradation (Kumar et al., 2020). Quality, time constraints, temperature limitations, perishability problems, infrastructure reliance, and other related challenges all pose limitations to the FSC's ability to ensure sustainability (Anish et al., 2022). The escalating impact of these challenges, coupled with a rising demand from consumers and society, is compelling companies, governments, and policymakers to establish sustainable food supply chains (SFSCs). Sustainability is of particular concern in developing nations, which currently house more than 80% of the world's population and account for 99% of population growth (Population Bureau, 2020).

The circular economy has become increasingly significant as a solution to overcome significant challenges, despite the obstacles faced. While strategies based on the circular economy, such as circular supply chains, sharing economy, recycling, reuse, and remanufacture, have gained momentum, they are still not widely implemented in FSCs (Sharma et al., 2019).

The idea of circular economy can be utilized in different aspects of FSC (chains) to mitigate excessive consumption, spoilage, and the depletion of natural resources (Tseng et al., 2019). The concept considers waste products as nutrient sources that can be reintegrated into the food systems, with technology playing a significant role in this process. Since the emergence of Industry 4.0, most sectors across the FSC have adopted technologies to streamline their operational activities. For example, real-time simulations allow companies to replicate their operational activities, AI-based autonomous systems enable intelligent decision-making, 3D printing technology facilitates rapid prototyping, cloud systems provide real-time information access and control, and blockchain ensures secure operational information storage in a decentralized manner. Additionally, the Internet of Things (IoT) has had a significant impact on the food subsector by enabling the integration of physical processes, calculations, and networking into cyber-physical systems (Abdo et al., 2023). Hence, advancements in Industry 4.0 are predicted to have a significant impact on the longevity and adoption of circular economy principles in FSCs. However, to fully leverage these technologies and enhance the sustainability of FSCs worldwide, strategic innovation is necessary to ensure the successful integration of these technologies.

2.2 SUSTAINABILITY PRACTICES AND FSC

The food processing journey begins at the farm, where essential components are cultivated or produced, and it concludes when the food is consumed by the end consumer (Ojo et al., 2017).

It is essential to oversee the entire FSC to ensure the safety of food items, especially since human lives are at risk in the food industry. Managing the complex and demanding FSC, which includes perishable goods and safety concerns related to agricultural products, requires the implementation of an efficient approach at all levels of food processing and production (Yadav et al., 2022). The presence of a thoughtfully crafted strategy spanning the entire supply chain is of paramount importance in safeguarding the integrity of food products. Food enterprises must operate in a sustainable manner to foster customer loyalty and trust. Furthermore, consumers are increasingly conscious of the health advantages and potential risks linked to their dietary choices (Mishra et al., 2018).

Ensuring both an ample supply and excellent product quality within an FSC necessitates a meticulously designed and organized supply chain strategy. Proficient coordination and oversight play a pivotal role in delivering top-notch food items to end consumers through a multifaceted FSC involving various participants. Moreover, the increasing focus on safeguarding the environment has elevated the significance of sustainability when choosing suppliers. Social, environmental, and economic considerations are now taken into account in Figure 2.1 (Zimmer et al., 2015).

Maintaining the quality and nutritional value of raw materials and finished goods in the FSC network is crucial. However, it can be a challenging task due to the storage and procurement processes involved (Suryaningrat, 2016). To ensure that fresh food products maintain their quality and natural features, it is essential to have suitable technology in place, including proper temperature control throughout the FSC. The use of cold supply chain management (CSCM) techniques has been effective in achieving this goal within the food industry (Mahla et al., 2020). Integrating Industry 4.0 technology completely into managing the cold supply chain for food products would be a significant advancement. This integration would greatly improve the quality and value of the products while also reducing costs. The term "cold supply

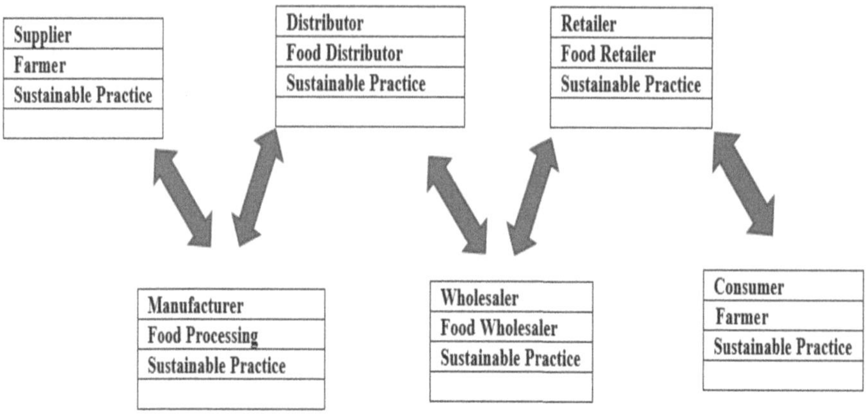

FIGURE 2.1 Food supply chain with sustainable practices.

(Adapted from Dani (2015)).

chain" refers to a series of logistical processes used to handle perishable agricultural products, starting from the moment they are harvested and continuing until they are delivered to the ultimate receiver (Kumar et al., 2023b).

To preserve the quality and safety of agricultural materials and processed foods during their procurement and transportation within the FSC network, it is essential to implement temperature control measures. This is crucial for maintaining the integrity of the products and ensuring that they remain undamaged (Ojo et al., 2017). Sustainable supply chain management (SSCM) relates to the creation of corporate value through the planning, execution, and control of business operations that prioritize the social, ecological, and economic aspects of the organization. The primary objective of SSCM is to attain long-term sustained performance and improve production and overall supply chain management practices (Stindt, 2017). The statement pertains to generating financial benefits within a business while simultaneously reducing adverse social and environmental impacts throughout the supply chain network. This is crucial across all industries as it enhances a company's corporate social responsibility (CSR) initiatives, resulting in improved logistics performance, greater efficiency, and better resource utilization (Beske et al., 2014). Sustainability in the FSC primarily involves the efficient management and recovery of raw materials to produce finished products in food processing and manufacturing settings (Sgarbossa & Russo, 2017). According to the latest statistics from the United Nations, the increasing global population will require a twofold increase in food production by the year 2050 to meet the demand (Jararweh et al., 2023).

2.3 INTERNET OF THINGS

As the number of connected devices continues to grow, concerns about privacy, security, and the ethical use of data have become increasingly pressing (Hameed & Alomary, 2019). Especially health, banking, and supply chain sectors, as well as governments and regulatory bodies, are taking steps to ensure that IoT devices are secure and that users have control over their data, which is essential for the continued growth and success of IoT (Hameed & Alomary, 2019).

The application of IoT technology in the food sector is substantial. In recent years, IoT, coupled with artificial intelligence and extensive data analytics, has emerged as a crucial cornerstone for the FSC (Sanjeev & Vinay, 2022a). IoT enhances the visualization and traceability of the FSC and allows companies to confront uncertainty across their value chain (Nugroho et al., 2023). Therefore, the smart degree of supply chain processes enables the alignment of supply with demand (Raouf et al., 2023). Food safety is an essential concern during the transition process across the value chain. Fresh food undergoes several chemical changes during this time, and any misconduct can directly lead to food safety issues (Sanjeev & Vinay, 2022a).

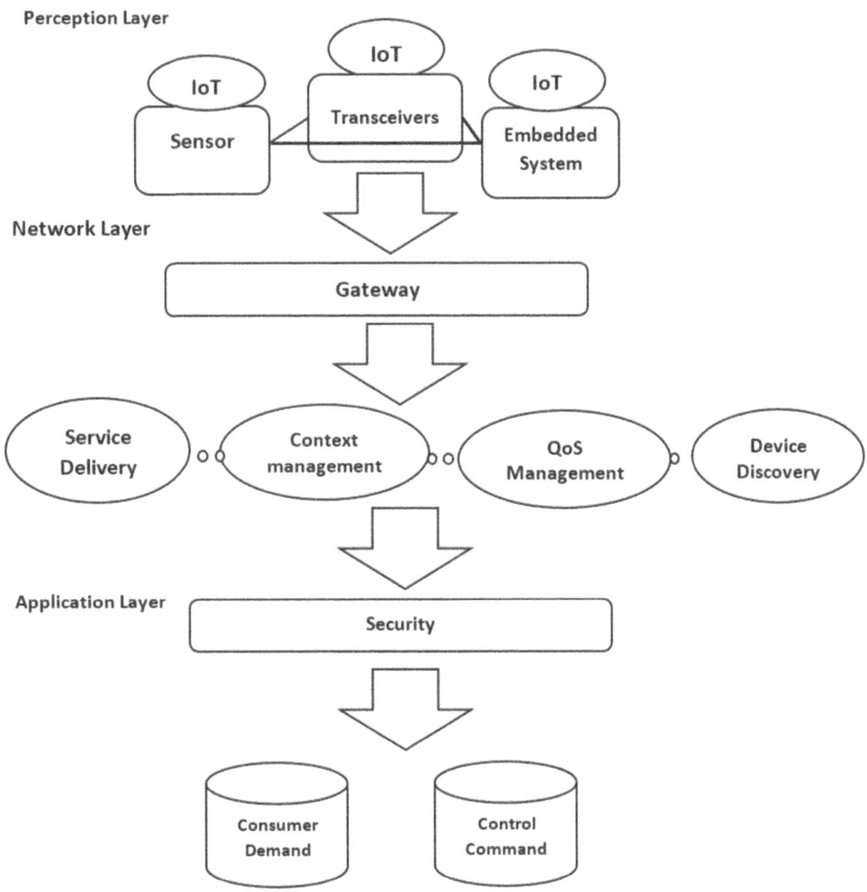

FIGURE 2.2 IoT technology structure.

(Adapted from Sharma and Singh (2022)).

The basic IoT technology structure is shown in Figure 2.2. The IoT technology infrastructure traditionally comprises three layers: the physical layer (also known as the perception layer), the network layer, and the application layer. The application layer directly connects with consumer commands (Sanjeev & Vinay, 2022a; Stankovic, 2014). The first layer is the physical layer, which is responsible for collecting data from multiple sensing devices, such as temperature sensors, motion sensors, and light sensors (Senthil et al., 2022). This layer may also include gateways or edge devices that preprocess the data before it is sent to the network layer. Similarly, the network layer is responsible for transmitting all the information collected by the physical layer to the cloud or data center for processing. This layer could include wired or wireless networks such as Wi-Fi, Bluetooth, Zigbee, or cellular networks. It also includes protocols that enable communication between devices and cloud services (Sanjeev & Vinay, 2022a; Stankovic, 2014). Likewise, the network layer bears the responsibility of conveying all data gathered by the physical layer to the cloud or data center for further processing. This tier encompasses both wired and wireless networks, such as Wi-Fi, Bluetooth, Zigbee, or cellular networks. It also encompasses the protocols that facilitate communication between devices and cloud services (Sanjeev & Vinay, 2022a; Stankovic, 2014). In the grand scheme of things, the structure of an IoT system is meticulously crafted to facilitate the gathering, transmission, and examination of data sourced from a diverse array of outlets. This orchestration aims to enhance workflows, boost productivity, and catalyze inventive advancements, as shown in Figure 2.2.

2.4 IOT ADOPTION IN FOOD SUPPLY CHAIN

Businesses across different sectors are considering IoT as a means of enhancing their operational productivity. Businesses across different sectors are considering IoT as a means of enhancing their operational productivity (Carcary et al., 2018). The impact of IoT can be increased by combining it with related technologies such as cloud computing, advanced internet infrastructure, extensive data processing, robotics, and semantic technologies. (Vermesan & Friess, 2013). The FSC is a growing field of study that involves the cooperation of various stakeholders across all intermediaries, starting with farmers, processors, manufacturers, distributors, and retailers (Bourlakis & Matopoulos, 2010). According to a recent study, 14% of global food production is wasted before reaching its intended destination. This food waste occurs at all stages of the FSC, from production to processing to distribution to consumption. There are many factors that contribute to food waste, including spoilage, pest infestation, and improper handling and storage (FAO, 2020). The main causes of these losses were inadequate processing and packaging methods, substandard transportation and distribution systems, and insufficient storage infrastructure and approaches used in the FSC (Aamer et al., 2021; Mogale et al., 2020). Lack of visualization, traceability, and transparency across intermediaries due to insufficient data availability creates more complexity among intermediaries in documenting food loss during each stage of the supply chain (Aamer et al., 2021; Cattaneo et al., 2020). The significant availability of data allows companies to measure and track food loss across the value chain. It also enables companies to improve all performance parameters in the food supply chain by utilizing and sharing data (Aamer et al., 2021).

Some of the previous studies explore the concern regarding food waste and safety issues in different frames of reference. For instance, Jensen et al. (2013) investigate food safety and waste issues from a supply chain integration perspective. Similarly, Dora (2019) explores similar food waste issues from a supply chain collaboration perspective. Also Irani and Sharif (2016) address similar issues in communication across the supply chain, while Aktas et al. (2018) discuss in frame of consumer behavior.

Therefore, companies are still looking for significant solutions to resolve data availability issues. Due to this concern, some studies encourage technology adoption in the food supply chain for measuring food safety performance (Patidar et al., 2018). Although smart technologies such as the Internet of Things (IoT), machine learning, and blockchain have become increasingly important for enhancing the traceability and visibility of the FSC, their practical implementation in this area is still not well understood. Recent studies have emphasized the importance of deploying these technologies in the FSC to improve its operational efficiency (Antonucci et al., 2019; Chains, 2021).

IoT allows companies to revamp and upgrade their existing food supply chains for data-driven decision-making. The FSC is a complex system involving many stakeholders and covering a wide range of locations. It is crucial for setting standards for food quality, safety, and efficient operation (Chains, 2021). Regarding data accessibility, the Internet of Things (IoT) has the potential to profoundly impact every aspect of the FSC, enhancing food safety, efficiency, traceability, and transparency across the entire system. The comprehensive IoT architecture across the sustainable supply chain in terms of data availability for food safety, efficiency, traceability, and transparency is represented in Figure 2.3 (Sanjeev & Vinay, 2022a).

2.4.1 Role of IoT at Supplier Level

In the realm of the food supply chain, unique characteristics of food products in their early stages, such as perishability, seasonality, limited shelf life, extended production cycles, stringent government regulations on food safety, supply unpredictability, variations in consumer preferences, and consistent requirements for storage, processing, transportation, and distribution, differentiate agro-supply chain management from other supply chains. Smart farming entails the monitoring of various agricultural factors to improve crop yields, reduce costs, and optimize process inputs (Nukala et al., 2016). These components encompass factors such as environmental conditions, growth conditions, soil characteristics, irrigation water, pest and fertilizer management, weed control, and the controlled environment within a greenhouse. IoT devices provide invaluable information about various physical parameters, enhancing cultivation techniques in agricultural environments. Wireless sensor networks (WSNs) play a critical role in IoT technology since most IoT applications across different markets depend on wireless data transmission (Zhang & Kovacs, 2012). Similarly, unmanned aerial systems (UAS) have substantial potential in precision

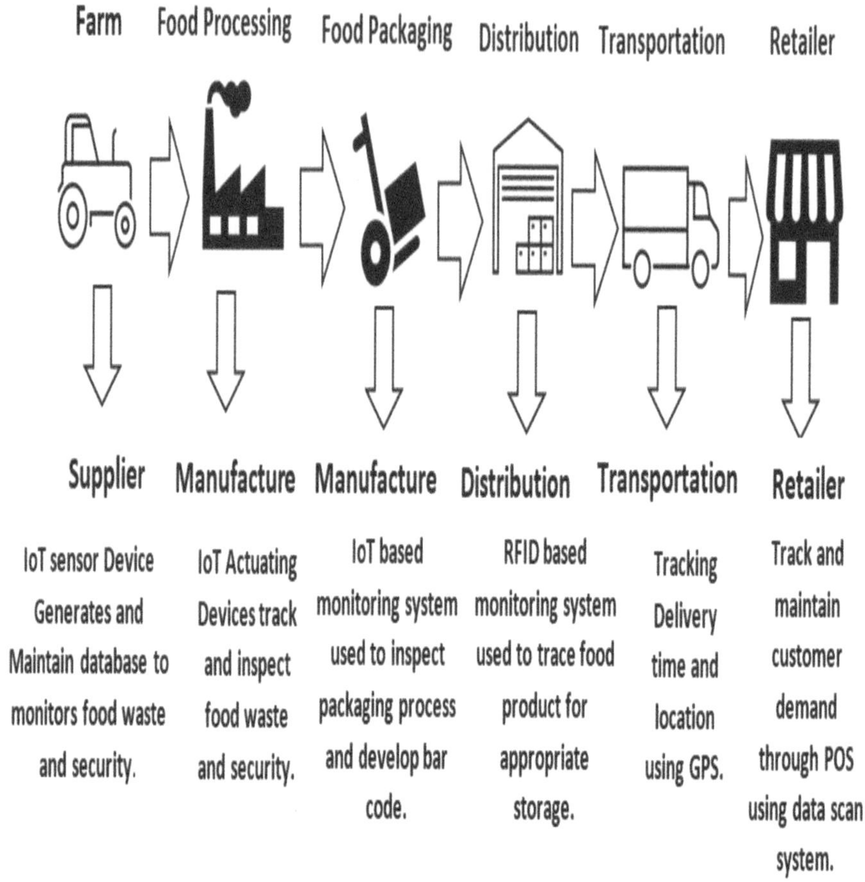

FIGURE 2.3 IoT architecture.

(Adapted from Sanjeev and Vinay (2022a)).

farming. They can function as either sensing or communication platforms, rendering them innovative technology.

2.4.2 Role of IoT during Food Processing

To effectively manage large-scale food processing with multiple continuously operating manufacturing lines, a substantial number of resources are required. Throughout the processing, sensors within each process generate various data regarding machinery, employees, inventory, and finished products (Sanjeev & Vinay, 2022a). By utilizing IoT technology, this data can be processed in real time to improve planning and decision-making (Sanjeev & Vinay, 2022a). By utilizing IoT technology, this data can be processed in real time to improve planning and decision-making (Sanjeev & Vinay, 2022a).

2.4.3 Role of IoT in Cold Chain

The cold chain is considered a distinct supply chain for food safety. It involves controlling temperatures at every stage, encompassing the acquisition of raw materials and resources, their storage, transportation, processing, sales, and consumption (Ricke & Atungulu, 2018). Globally applicable uniform tracing codes are necessary to trace food supply and demand across the entire supply chain network. This enables enterprises throughout the network to establish cold chain traceability mechanisms with information integration at each phase, ultimately optimizing the cost of screening and tracing activities to obtain better data regarding food security and waste (Ricke & Atungulu, 2018).

2.4.4 Role of IoT during Food Transportation

Ensuring food safety heavily depends on the controlled transportation of food. Certain food items require specific temperature, humidity, and air-controlled conditions during transportation (Sanjeev & Vinay, 2022a). Consistently tracking such conditions is necessary throughout the supply chain, and any deviation from the desired condition can result in food decomposition. IoT technology allows us to monitor all the activities involved in food transportation. For instance, all the critical information gathered during transportation can be handled through RFID technology (Sanjeev & Vinay, 2022a).

2.5 BARRIERS IN ADOPTING IOT

As smart devices generate an increasing amount of data, it becomes challenging to determine what information should be retained and for how long (Fernandez-Gago et al., 2017; Sanjeev & Vinay, 2022a). During the data integration across several intermediaries, different types of systems were involved that had several characteristics of data, including multiple electronic data interchange, which becomes a significant challenge among stakeholders across the FSC to create homogeneity among data (Fernandez-Gago et al., 2017). In terms of data confidentiality, trust, privacy, and data ownership, these are considered potential challenges across stakeholders in the complex food supply chain (Kim & Laskowski, 2017). Large investment is required to adopt the IoT for full integration (Sanjeev & Vinay, 2022a). There is unavailability of simulation capability to test the real-time data (Sanjeev & Vinay, 2022a). There is an urgency for technical availability of a flexible network to support and streamline operations.

2.6 BASIC THEORY ON BLOCKCHAIN

In recent times, a plethora of innovative technological advancements have made their way into our everyday routines, with blockchain emerging as a noteworthy illustration (Leng et al., 2018). Blockchain has been implemented for a wide variety of purposes, such as cryptocurrency, digital transaction administration, and countless others. One area that remains unexplored is the potential use of blockchain technology (BT) in agriculture and food industries (Patel et al., 2022).

The primary domains where BT could prove to be highly advantageous are the food supply chain and information and communication technology. For instance, blockchain can be utilized to remotely monitor farming conditions and optimize irrigation systems based on these conditions. Blockchain is an electronic distributed ledger that stores data in digital format, providing assurance of the accuracy and security of the recorded information. The ability to create trust without requiring a trusted third party makes BT particularly valuable for enhancing the security and reliability of the food supply chain (Leng et al., 2018). Unlike the conventional internet framework, the blockchain network functions in a decentralized fashion, void of a singular server to house all data (Creydt & Fischer, 2019). Instead, a multitude of nodes or computers interconnect within the blockchain network, allowing simultaneous data storage. Prior to inclusion within the network, each new piece of data undergoes comprehensive scrutiny by all participants and is only incorporated into the blockchain following verification grounded in consensus, as shown in Figure 2.4 (Creydt & Fischer, 2019).

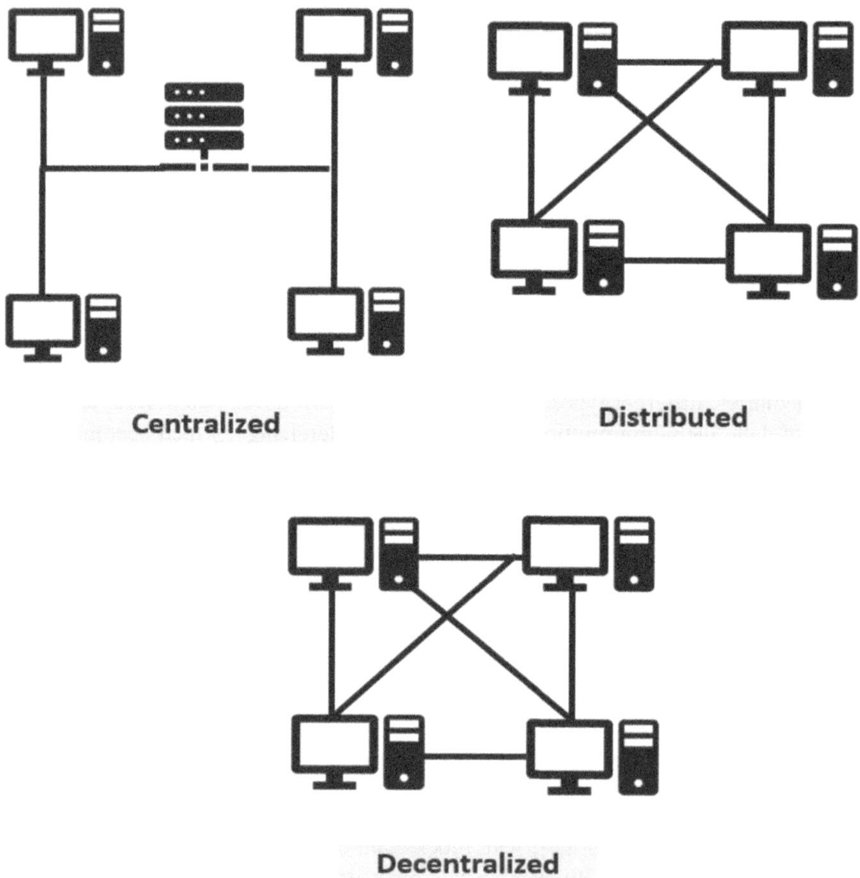

FIGURE 2.4 Blockchain is a distributed ledger.

(Adapted from Creydt and Fischer (2019)).

In simple terms, blockchain is referred to as a system that keeps records and maintains an unalterable shared database of transactions among various parties (Pandey et al., 2022). A decentralized network that links several nodes from various sources without mutual trust. Each node's ledger records and maintains transactions to guarantee constant synchronization (Patel et al., 2022). The network operates through peer-to-peer connections where nodes belong to different networks that are not trusted by each other. Each node maintains a ledger recording all transactions and guarantees their synchronization (Luzzani et al., 2021). The fundamental attributes of BT include its capacity to preserve the integrity of data, establish decentralization, and ensure high reliability. Moreover, blockchain has the benefit of dispensing with the need for a third party to facilitate communication and transactions among nodes (Pandey et al., 2022).

BT revolutionizes the maintenance of administrative control and digital regulation. Information is converted into digital codes and securely stored in shared databases within blockchains, leading to increased transparency and significantly reducing the likelihood of data being destroyed or altered—thus ensuring immutability (Esmaeilian et al., 2020). The potential of BT resides in its ability to create a transparent and decentralized system where every agreement, payment, and transactional activity is securely recorded in a digital format. These records can be authenticated and seamlessly shared among a network of individuals, machines, and organizations. As a result, the reliance on intermediaries, including brokers, bankers, and lawyers, is significantly reduced, leading to a more efficient and cost-effective ecosystem (Lansiti & Lakhani, 2017). Intermediaries are responsible for guaranteeing the correctness and confirmation of transactions in a variety of businesses. However, BT represents a paradigm change by replacing human and conventional actors with computer codes (Esmaeilian et al., 2020). The usual features of blockchain often vary depending on the specific platform employed (Barton, 2018).

1. *Shared ledger*: A data structure that is dispersed within a specific location and mutually accessed by various individuals.
2. *Permissioning*: Transactions that are both secure and verified, guaranteeing the confidentiality and openness of information.
3. *Smart Contracts*: The database incorporates business terms and executes them using transactions.
4. *Consensus*: Relevant users validate transactions to ensure the data remains unalterable and can be traced back.

BT guarantees security, upkeeps digital records, limits the involvement of third-party intermediaries in transactions, reduces transaction expenses, and enhances the efficiency of supply chain operations.

Numerous parties participate in the FSC, encompassing cultivators, providers, processors, distributors, sellers, and consumers. The FSC commences at the farm level and concludes at the point of consumption by the individual (Patidar et al., 2023). It entails three primary components: (i) the movement of goods, (ii) the exchange of data, and (iii) the circulation of financial resources within the entire supply chain (Patidar et al., 2023). These streams are interlinked. The entirety of these flows and their associated data are methodically recorded within the conventional supply chain utilizing ledger systems.

2.7 BLOCKCHAIN ADOPTION IN SUSTAINABLE FOOD SUPPLY CHAIN

Consumer confidence has been eroded over the past decade due to various incidents related to food safety. However, advanced technologies such as the Internet of Things (IoT) and big data have been employed to address these concerns. Nevertheless, the issue of data manipulation has left many people feeling helpless, resulting in significant trust issues between businesses operating both upstream and downstream (Kumar et al., 2023a).

The utilization of blockchain can enhance the reliability of products and strengthen relationships among network participants, as it functions as a shared and immutable database (Saurabh & Dey, 2021). Furthermore, customers will have the opportunity to acquire knowledge about and participate in the food distribution process during this period. Simultaneously, this can help reduce the incidence of "unmanageable risk" within the agricultural and food sectors. When combined with BT, stakeholders can achieve more significant outcomes (Kumar et al., 2023a). There has been significant interest in the potential of blockchain to enhance supply chain management, especially in how the Ethereum blockchain can be applied in the soybean supply chain. This example illustrates how it can eliminate the need for a central authority while still facilitating the tracing, tracking, and completion of commercial transactions (Salah et al., 2019).

There has been significant interest in the potential of blockchain to enhance supply chain management, especially in how the Ethereum blockchain can be applied in the soybean supply chain. This example illustrates how it can eliminate the need for a central authority while still facilitating the tracing, tracking, and completion of commercial transactions (Chandan et al., 2023). The implementation of BT in the food supply chain enhances trust, security, traceability, and decentralization. The use of blockchain as a data structure in food supply chains establishes information symmetry, thereby adding value to the overall system. Given the increasing prevalence of blockchain, numerous scholars are exploring the application of relevant technologies for traceability systems in SFSCs. Industry 4.0 technologies have the potential to offer customizable and flexible solutions that leverage information to improve the sustainability of FSCs. Accurate, accessible, accountable, and reliable information streams are anticipated to significantly shape the development of contemporary SFSCs through I4.0 technologies (Sharma et al., 2018).

The use of BT has the potential to improve accountability and transparency in SFSC transactions. Harnessing BT carries the potential to enhance accountability and transparency in SFSC transactions. BT, as a cutting-edge advancement, possesses the capability to institute reliable protocols, ensuring both transparency and security in FSC processes. It holds the promise of elevating transparency and accountability in SFSC transactions (Kamble et al., 2020). This emerging technology harbors substantial potential for creating trust-enforcing protocols that guarantee both the transparency and security of FSC activities (Kamble et al., 2020).

By employing BT, one can trace a product's source and access this data within a few days, a task that would typically consume a considerable amount of time. As a result, the process of ascertaining a product's origin is streamlined through the utilization of BT. Consequently, pinpointing the origin of contamination within an FSC becomes more feasible, enhancing the management of product recalls in the event of a food safety concern (Anish et al., 2022).

To address traceability concerns, food organizations are employing Internet of Things (IoT) and blockchain technologies (Balamurugan et al., 2022). Traceability is a critical concern within the FSC as it impacts food quality, enhances safety measures, and improves supply chain performance (Alabi & Ngwenyama, 2022). Recently, BT has gained significant attention in various industrial applications due to its ability to securely record events in a decentralized manner, eliminating the requirement for a centralized authority that is trusted (Saurabh & Dey, 2021). Based on the preceding discussion, BT has the potential to effectively address concerns related to food security, safety, and quality. Nevertheless, it is worth noting that widespread adoption of this technology has not yet been realized (Khan et al., 2020).

New studies such as Lei and Ngai (2023), Taranya et al. (2023), and Zou et al. (2023) have highlighted the existence of analytical frameworks for implementing blockchain in the food sector, while also emphasizing the lack of empirical and analytical research on the subject (Soon, 2022). The comprehensive blockchain architecture plays a pivotal role in ensuring a sustainable supply chain. It achieves this by enhancing data availability for food safety, improving efficiency, enabling traceability, and increasing transparency, as shown in Figure 2.5. The utilization of decentralized ledger technology brings together entities within the FSC network, eliminating the need for intermediaries. Decentralized ledger technology establishes connections among all participants within the FSC network, fostering trust and openness in a collaborative environment through a range of distinctive features. Due to its decentralized operation, the technology is not governed by a single entity. An agreement signifies a collective decision reached by all stakeholders regarding the execution of a transaction. This distinctive feature and capability enable the tracing of any transaction. Whenever a transaction occurs, the distributed ledger is promptly updated, allowing all nodes to access and view it. The exceptional characteristics of this technology contribute to its enhanced reliability and trustworthiness. Unlike traditional databases, which can be altered or edited, BT ensures data security through distributed ledger techniques. Once data is created within the blockchain, it becomes immutable and cannot be changed or modified. By utilizing this technology, smart contracts are implemented, necessitating the creation of contract logic programs customized to the unique agreements between multiple parties involved in the food supply chain. Specific terms and conditions are in place during the establishment of the smart contract.

2.7.1 The Role of Blockchain at Cultivator Level

The conveyance of encoded data holds immense significance across all facets of contemporary living. Individuals involved in agriculture seek robust protection for their private information, desiring its impervious storage and non-disclosure to external

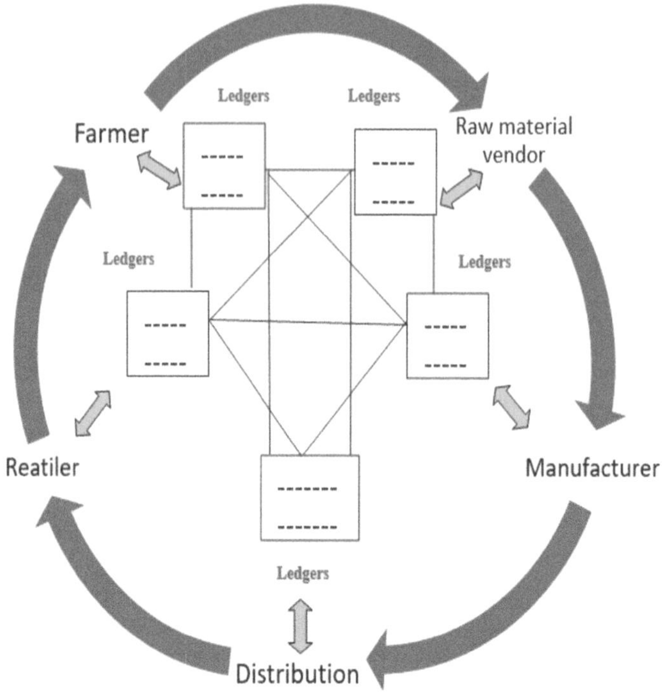

FIGURE 2.5 Blockchain architecture.

(Adopted across Sustainable Food Supply Chain (Patidar et al., 2023)).

entities, both corporate and individual levels (Mohammed, 2019). As previously noted, this network proves highly effective in expediting and safeguarding monetary transactions—similarly, it has the potential to address issues tied to funding agricultural initiatives and contribute to the broader advancement of the industry as a whole (Lin et al., 2020). Leveraging blockchain technologies, proprietors of small farms can locate investors and enhance their enterprises through the utilization of the Initial Coin Offering (ICO) initiative. This funding method resembles crowdfunding but employs cryptocurrencies instead of conventional payment (Alobid et al., 2022; Ante et al., 2018). BT is also present on various exchanges. Although the functioning remains consistent, stock markets can now take advantage of blockchain's capabilities. Agricultural practitioners can readily trade forthcoming contracts for commodities like crops, livestock, fruits, vegetables, and other farm products at predetermined prices (Kamilaris et al., 2019). This ensures that farmers are aware of their costs, and consumers are shielded from unforeseen price fluctuations (Kamilaris et al., 2019).

2.7.2 The Role of Blockchain at Food Traceability

While significant progress has been made in enhancing food quality and safety, industries remain deeply dedicated to ensuring the safety of food (Sanjeev & Vinay, 2022a).

Nevertheless, instances of foodborne diseases are not uncommon (Sanjeev & Vinay, 2022a). For this, speeding up tracking food to the end consumer is essential. However, within existing frameworks, tracking the product's path from farm to consumer frequently consumes days or even weeks. This tracking duration is further extended if the food products belong to a global market network (Antonucci et al., 2019). The potential of BT lies in significantly diminishing this period of traceability and expediting food recalls. Since the blockchain network houses comprehensive data regarding each participant within the FSC, this technology accelerates the entire traceability process remarkably (Antonucci et al., 2019).

2.7.3 The Role of Blockchain at Food Safety

The adoption of BT assists in pinpointing and eliminating origins of contamination, thereby expediting the response to foodborne disease outbreaks (Creydt & Fischer, 2019). IoTs are employed to capture crucial data like temperature and relative humidity during transportation and storage, securely storing the data within the protective confines of the blockchain network (Sanjeev & Vinay, 2022a). Consequently, robust cold chains can be established, enabling heightened oversight of food safety (Fuertes, 2016).

2.7.4 The Role of Blockchain in Controlling Food Scam

Although there have been advancements and the incorporation of sophisticated technology to detect and prevent issues within the FSC, cases of worldwide food fraud still endure (Sanjeev & Vinay, 2022a). In response, businesses are utilizing digital remedies to construct and execute comprehensive management systems that assure the authenticity of food. By recording all supply chain data within the blockchain system, any attempt to manipulate information becomes impracticable. Consequently, this facilitates successful regulation of food fraud (Jan Mei et al., 2023).

2.7.5 The Role of Blockchain toward Consumer Expectation

BT excels in enhancing transparency that is readily accessible to consumers. With heightened customer expectations, establishing transparency within the system becomes crucial, and this goal can be achieved by integrating pertinent data directly onto the packaging. This can be executed by employing either one-dimensional barcodes or two-dimensional QR codes, conveniently scannable by consumers using their smartphones (Hashri & Gusti Bagus, 2018).

2.8 BARRIERS IN ADOPTING BLOCKCHAIN

BT presents itself as a potential technological remedy for establishing a food traceability structure. Nevertheless, the limitations imposed by boundaries indicate that executing such a traceability framework mandates a meticulously structured and standardized supply chain encompassing all stakeholders, whether internal or external (Behnke & Janssen, 2020).

It's not only the technology that matters; what's crucial is the standardization of both internal and external traceability processes (requiring organizational changes), along with ensuring consistency in master data across all members of the supply chain (Lone & Naaz, 2012). Resolving these significant challenges is essential to achieve a level of traceability that brings transparency to the entire supply chain, benefiting all stakeholders, including the end consumer (Behnke & Janssen, 2020). Data governance needs to ensure consistent data definitions and authorization for data creation, access, and alteration (Behnke & Janssen, 2020; Lone & Naaz, 2012). Creating a framework for data governance isn't solely the obligation of contributors involved in a particular blockchain project; it demands agreement, ideally within a sector and perhaps spanning a whole industry. In the absence of such an accord, suppliers would be compelled to adhere to separate interface standards, which would ultimately diminish the economic efficiency of BT (Wang et al., 2020).

2.9 CONCLUSION

This chapter presents an overview of the IoT and blockchain, exploring their potential applications within the framework of an SFC. It also highlights existing uses of IoT and blockchain in FSC, outlining how these technologies are integrated. Recognizing that both technologies offer more benefits than drawbacks, it is advised that a larger number of supply chain stakeholders embrace IoT technologies. Additionally, the chapter emphasizes the significance of Industry 4.0 and elucidates how these components will facilitate the emergence of business opportunities and the advancement of innovative products. Conversely, this section also illustrates that deficiencies in real-time data awareness or accessibility have adverse effects on supply chain operations. A crucial stride toward effectiveness and transparency is the meticulous observation and assessment of each participant's engagement. The presence of accurate information plays a pivotal role in instigating changes across the diverse spectrum of functions within FSCs.

REFERENCES

Aamer, A. M., Al-Awlaqi, M. A., Affia, I., Arumsari, S., & Mandahawi, N. (2021). The internet of things in the food supply chain: adoption challenges. *Benchmarking: An International Journal*, *28*(8), 2521–2541. https://doi.org/10.1108/BIJ-07-2020-0371

Abdo, H., Sandeep, J., Hana, T., Guillermo, G.-G., Nour, A. A., Gulden, G., … José, M. L. (2023). Food processing 4.0: Current and future developments spurred by the fourth industrial revolution. *Food Control*, *145*, 1–15. https://doi.org/10.1016/j.foodcont.2022.109507

Aktas, E., Sahin, H., Topaloglu, Z., Oledinma, A., Huda, A. K. S., Irani, Z., … Kamrava. (2018). A consumer behavioural approach to food waste. *Journal of Enterprise Information Management*, *31*(5), 658–673. https://doi.org/10.1108/JEIM-03-2018-0051

Alabi, M. O., & Ngwenyama, O. (2022). Food security and disruptions of the global food supply chains during COVID-19: Building smarter food supply chains for post. *British Food Journal*, *125*(1), 167–185. https://doi.org/10.1108/BFJ-03-2021-0333

Alobid, M., Abujudeh, S., & Szucs, D. I. (2022). The role of Blockchain in revolutionizing the agricultural sector. *Sustainability*, *14*(7), 4313. https://doi.org/10.3390/su14074313

Anish, K., Sachin, K. M., & Pradeep, K. (2022). Barriers for adoption of Industry 4.0 in sustainable food supply chain: A circular economy perspective. *International Journal of Productivity and Performance Management Decision*. https://doi.org/10.1108/IJPPM-12-2020-0695

Ante, L., Sandner, P., & Fiedler, I. (2018). Blockchain-based ICOs: Pure hype or the dawn of a new era of startup financing. *Journal of Risk and Financial Management*, *11*(80). https://doi.org/10.3390/jrfm11040080

Antonucci, F., Figorilli, S., Costa, C., Pallottino, F., Raso, L., & Menesatti, P. (2019). "A review on blockchain applications in the agri-food sector", *Journal of the Science of Food and Agriculture*, *99*(14), 6129–6138. https://doi.org/10.1002/jsfa.9912

Balamurugan, S., Ayyasamy, A., & Joseph, K. S. (2022). IoT-Blockchain driven traceability techniques for improved safety measures in food supply chain. *International Journal of Information Technology*. https://doi.org/10.1007/s41870-020-00581-y

Barton, D. (2018). The Future of Finance: How FinTech, AI & Blockchain Will Shape Our Future. *IBM Watson*.

Behnke, K., & Janssen. (2020). Boundary conditions for traceability in food supply chains using blockchain technology. *International Journal of Information Management*, *52*, 1–10. https://doi.org/10.1016/j.ijinfomgt.2019.05.025

Beske, P., Land, A., & Seuring, S. (2014). Sustainable supply chain management practices and dynamic capabilities in the food industry: A critical analysis of the literature. *International Journal of Production Economics*, *152*, 131–143. https://doi.org/10.1016/j.ijpe.2013.12.026

Bourlakis, M., & Matopoulos, A. (2010). Trends in food supply chain management. *Delivering Performance in Food Supply Chains*, *185*, 511–527. https://doi.org/10.1533/9781845697778.6.511

Carcary, M., Maccani, G., Doherty, E., & Conway, G. (2018, September 24–25). Exploring the determinants of IoT adoption: Findings from a systematic literature review. In *Perspectives in Business Informatics Research. 17th International Conference*, BIR 2018, Stockholm, Sweden.

Cattaneo, A., Sanchez, M. V., Torero, M., & Vos, R. (2020). Reducing food loss and waste: Five challenges for policy and research. *Food Policy*, *98*. https://doi.org/10.1016/j.foodpol.2020.101974

Chains, C. F. S. (2021). Impact on value addition and safety. *Trends Food Science Technology*, *114*, 323–332.

Chandan, A., John, M., & Potdar, V. (2023). Achieving UN SDGs in food supply chain using blockchain technology. *Sustainability*, *15*, 2109. https://doi.org/10.3390/su15032109

Chetna, C., Puneet, K., Rakesh, A., Peter, R., & Amandeep, D. (2022). Supply chain collaboration and sustainable development goals (SDGs). Teamwork makes achieving SDGs dream work. *Journal of Business Research*, *147*, 290–307. https://doi.org/10.1016/j.jbusres.2022.03.044

Creydt, M., & Fischer, M. (2019). Blockchain and more - algorithm driven food traceability. *Food Control*, *105*, 45–51. https://doi.org/10.1016/j.foodcont.2019.05.019

Dani, S. (2015). *Food Supply Chain Management and Logistics: From Farm to Fork*. Kogan Page Limited.

Dora, M. (2019). Collaboration in a circular economy: learning from the farmers to reduce food waste. *Journal of Enterprise Information Management*, *33*(4), 769–789. https://doi.org/10.1108/JEIM-02-2019-0062

Esmaeilian, B., Sarkis, J., Lewis, K., & Behdad, S. (2020). Blockchain for the future of sustainable supply chain management in Industry 4.0 *Resources, Conservation & Recycling*, *16*. https://doi.org/10.1016/j.resconrec.2020.105064

FAO. (2020). *The state of food security and nutrition in the world 2020* (Transforming food systems for affordable healthy diets, Issue.

Fernandez-Gago, G., Moyano, F., & Lopez, J. (2017). Modelling trust dynamics in the Internet of Things. *Information Science*, *396*, 72–82. https://doi.org/10.1016/j.ins.2017.02.039

Fuertes, G. (2016). Intelligent packaging systems: sensors and nanosensors to monitor food quality and safety. *Journal of Sensors*, *2*, 1–8. https://doi.org/10.1155/2016/4046061

Hameed, A., & Alomary, A. (2019). Security issues in IoT: a survey. *2019 International conference on innovation and intelligence for informatics, computing, and technologies (3ICT)*, Sakhier, Bahrain.

Hashri, H., & Gusti Bagus, B. N. (2018). *Blockchain Based Traceability System in Food Supply Chain International Seminar on Research of Information Technology and Intelligent Systems (ISRITI)*, Yogyakarta, Indonesia. https://doi.org/10.1109/ISRITI.2018.8864477

Irani, Z., & Sharif, A. M. (2016). Sustainable food security futures: Perspectives on food waste and information across the food supply chain. *Journal of Enterprise Information Management*, *29*(2). https://doi.org/10.1108/JEIM-12-2015-0117

Jan Mei, S.-S., Shingai, N., & Lisa, J. (2023). Food fraud and mitigating strategies of UK food supply chain during COVID-19. *Food Control*, *148*. https://doi.org/10.1016/j.foodcont.2023.109670

Jararweh, Y., Fatima, S., Jarrah, M., & AlZu'bi, S. (2023). Smart and sustainable agriculture: Fundamentals, enabling technologies, and future directions. *Computers and Electrical Engineering*, *110*. https://doi.org/10.1016/j.compeleceng.2023.108799

Jensen, Munksgaard, & Arlbjørn. (2013). Chasing value offerings through green supply chain innovation. *European Business Review*, *25*(2), 124–146. https://doi.org/10.1108/09555341311302657

Kamble, S. S., Gunasekaran, A., & Sharma, R. (2020). Modeling the blockchain enabled traceability in agriculture supply chain. *International Journal of Information Management*, *52*, 1–16. https://doi.org/j.ijinfomgt.2019.05.023

Kamilaris, A., Fonts, A., & Prenafeta-Boldu, F. X. (2019). The rise of blockchain technology in agriculture and food supply chains. *Trends in Food Science & Technology*, *91*, 640–652. https://doi.org/10.1016/j.tifs.2019.07.034

Khan, P. W., Byun, Y.-C., & Park, N. (2020). IoT-blockchain enabled optimized provenance system for food industry 4.0 using advanced deep learning. *Sensors*, *20*(10). https://doi.org/10.3390/s20102990

Kim, H., & Laskowski, M. (2017). Agriculture on the blockchain: sustainable solutions for food, farmers, and financing. *SSRN Electron Journal*. https://doi.org/10.2139/ssrn.3028164

Kumar, A., Mangla, S. K., & Kayikci, Y. (2020). Investigating enablers to improve transparency in sustainable food supply chain using F-BWM", *Intelligent and Fuzzy Techniques: Smart and Innovative Solutions. INFUS 2020 Conference*, Istanbul, July 21–23. http://shura.shu.ac.uk/information.html

Kumar, M., Choubey, V. K., Raut, R. D., & Jagtap, S. (2023a). Enablers to achieve zero hunger through IoT and blockchain technology and transform the green food supply chain systems. *Journal of Cleaner Production*, *405*. https://doi.org/10.1016/j.jclepro.2023.136894

Kumar, N., Tyagi, M., & Sachdeva, A. (2023b). A sustainable framework development and assessment for enhancing the environmental performance of cold supply chain. *Management of Environmental Quality: An International Journal*, *34*(4), 1077–1110. https://doi.org/10.1108/MEQ-03-2022-0046

Lahane, S., Paliwal, V., & Kant, R. (2023). Evaluation and ranking of solutions to overcome the barriers of Industry 4.0 enabled sustainable food supply chain adoption. *Cleaner Logistics and Supply Chain*, *8*. https://doi.org/10.1016/j.clscn.2023.100116

Lansiti, M., & Lakhani, K. (2017). The truth about Blockchain. *Harvard Business Review*, 1–3. https://hbr.org/2017/01/the-truth-about-blockchain

Lei, C. F., & Ngai, E. W. T. (2023). Blockchain from the information systems perspective: Literature review, synthesis, and directions for future research. *Information and Management*, *60*(7). https://doi.org/10.1016/j.im.2023.103856

Leng, K., Bi, Y., Jing, L., Fu, H. C., & Van Nieuwenhuyse, I. (2018). Research on agricultural supply chain system with double chain architecture based on blockchain technology. *Future Generation Computer Systems*, *86*, 641–649. https://doi.org/10.1016/j.future.2023.10.001

Lin, W., Huang, X., Fang, H., Wang, V., Hua, Y., Wang, J., … Yau, L. (2020). Blockchain technology in current agricultural systems: From techniques to applications. *IEEE Access*, *8*, 143920–143937. https://doi.org/10.1109/ACCESS.2020.3014522

Lone, A., & Naaz, R. (2012). Applicability of Blockchain smart contracts in securing Internet and IoT: A systematic literature review. *Computer Science Review*, *39*, 100360. https://doi.org/10.1016/j.cosrev.2020.100360

Luzzani, G., Grandis, E., Frey, M., & Capri, E. (2021). Blockchain technology in wine chain for collecting and addressing sustainable performance: An exploratory study. *Sustainability*, *13*(22), 12898. https://doi.org/10.3390/su132212898

Mahla, B., Anup, S., Babak, A., Yahua, Z., Alice, W., & Anming, Z. (2020). Sustainable cold supply chain management under demand uncertainty and carbon tax regulation. *Transportation Research Part D*, *80*. https://doi.org/10.1016/j.trd.2020.102245

Mishra, S., Singh, S. P., Johansen, J., Cheng, Y., & Farooq, S. (2018). Evaluating indicators for international manufacturing network under circular economy. *Management Decision*. https://doi.org/10.1108/MD-05-2018-0565

Mogale, D. G., Kumar, S. K., & Tiwari, M. K. (2020). Green food supply chain design considering risk and post-harvest losses: A case study. *Annals of Operations Research*, *295*(1), 257–284. https://doi.org/10.1007/s10479-020-03664-y

Mohammed, I. A. (2019). A systematic literature mapping on secure identity management using blockchain technology. *International Journal of Innovative Technology and Exploring Engineering.*, *6*, 86–91. https://doi.org/10.1016/j.jksuci.2021.03.005

Nations, U. (2023). *The Sustainable Development Goals Report 2023: Special edition* (Towards a Rescue Plan for People and Planet, Issue. https://wedocs.unep.org/20.500.11822/9814

Nugroho, G., Tedjakusuma, F., Lo, D., Romulo, A., Pamungkas, D. H., & Kinardi, S. A. (2023). Review of the application of digital transformation in food industry. *Journal of Current Science and Technology*, *13*(3), 774–790. https://doi.org/10.59796/jcst.V13N3.2023.1285

Nukala, R., Panduru, K., Shields, A., Riordan, D., Doody, D, & Walsh, J. (2016). *Internet of things: A review from farm to fork, 27th Irish Signals and Systems Conference (ISSC)*. https://doi.org/10.1109/ISSC.2016.7528456

Ojo, O., Shah, S., & Coutroubis, A. (2017). *An Overview of Sustainable Practices in Food Processing Supply Chain Environments, International Conference on Industrial Engineering and Engineering Management*, Singapore. https://doi.org/10.1109/IEEM.2017.8290200

Ojo, O., Shah, S., Coutroubis, A., & Jiménez, M. T. M. O. (2018). *Potential Impact of Industry 4.0 in Sustainable Food Supply Chain Environment, 2018 IEEE International Conference on Technology Management, Operations and Decisions (ICTMOD)*, Marrakech, Morocco. https://doi.org/10.1109/ITMC.2018.8691223

Pandey, V., Pant, M., & Snasel, V. (2022). Blockchain technology in food supply chains: Review and bibliometric analysis *Technology in Society*, *69*, 1–26. https://doi.org/10.1016/j.techsoc.2022.101954

Patel, D., Sinha, A., Bhansali, T., Usha, G., & Velliangiri, S. (2022). Blockchain in food supply chain. *Procedia Computer Science*, *215*, 321–330. https://doi.org/10.1016/j.procs.2022.12.034

Patidar, R., Agrawal, S., & Pratap, S. (2018). Development of novel strategies for designing sustainable Indian agri-fresh food supply chain. *Springer India, Sadhana - Academy Proceedings in Engineering Sciences*, *43*(10). https://doi.org/10.1007/s12046-018-0927-6

Patidar, S., Sukhwani, V. K., & Shukla, A. C. (2023). Modeling of critical food supply chain drivers using DEMATEL method and blockchain technology. *Journal of The Institution of Engineers (India): Series C, 104*(3), 541–552. https://doi.org/10.1007/s40032-023-00941-0

Population Reference Bureau. (2020). *9 billion world population by 2020 countries (70 million annually)*. https://www.prb.org/9billionworldpopulationby2050/#:~:text5Thepopulationoflessdeveloped

Raouf, M., Kinza, N. M., & Karima, A. (2023). IoT-based food traceability system: Architecture, technologies, applications, and future trends. *Food Control, 145*. https://doi.org/10.1016/j.foodcont.2022.109409

Ricke, S. C., & Atungulu, G. G. (2018). *Food and Feed Safety Systems and Analysis*. Academic Press.

Sajid Ullah, S., Oleshchuk, V., & Pussewalage, H. S. G. (2023). A survey on blockchain envisioned attribute based access control for internet of things: Overview, comparative analysis, and open research challenges. *Computer Networks, 235*. https://doi.org/10.1016/j.comnet.2023.109994

Salah, K., Nizamuddin, N., Jayaraman, R., & Omar, M. (2019). Blockchain-based soybean traceability in agricultural supply chain. *IEEE Access*, 7, 73295–73305. https://doi.org/10.1109/ACCESS.2019.2918000

Sanjeev, K. S., & Vinay, S. (2022a). *Digitization of the food industry enabled by Internet of Things, blockchain, and artificial intelligence*. Elsevier. https://doi.org/10.1016/B978-0-323-91158-0.00013-2

Saurabh, S., & Dey, K. (2021). Blockchain technology adoption, architecture, and sustainable agri-food supply chains. *Journal of Cleaner Production*. https://doi.org/10.1016/j.jclepro.2020.124731

Senthil, M. N., Ganesh Gopal, D., Puspita, C., Waleed, A., & Muthukumaran. (2022). Integration of IoT based routing process for food supply chain management in sustainable smart cities. *Sustainable Cities and Society, 76*. https://doi.org/10.1016/j.scs.2021.103448

Sgarbossa, F., & Russo, I. (2017). A proactive model in sustainable food supply chain: Insight from a case study. *International Journal of Production Economics* (183), 596–606. https://doi.org/10.1016/j.ijpe.2016.07.022

Sharma, R., Kamble, S. S., & Gunasekaran. (2018). Big GIS analytics framework for agriculture supply chains: A literature review identifying the current trends and future perspectives. *Computers and Electronics in Agriculture, 155*, 103–120. https://doi.org/10.1016/j.compag.2018.10.001

Sharma, S. K., & Singh, V. (2022). Digitization of the food industry enabled by Internet of Things, blockchain, and artificial intelligence. In *Current Developments in Biotechnology and Bioengineering* (pp. 421–445). Elsevier.

Sharma, Y. K., Mangla, S. K., Patil, P. P., & Liu, S. (2019). When challenges impede the process For circular economy-driven sustainability practices in food supply chain. *Management Decision, 57*(4). https://doi.org/10.1108/MD-09-2018-1056

Soon, J. M. (2022). Food fraud countermeasures and consumers: A future agenda. *Future Foods*. https://doi.org/10.1016/B978-0-323-91001-9.00027-X

Stankovic. (2014). Research directions for the Internet of Things. *IEEE Internet of Things Journal, 1*(1), 3–9. https://doi.org/10.1109/JIOT.2014.2312291

Stindt, D. (2017). A generic planning approach for sustainable supply chain management – How to integrate concepts and methods to address the issues of sustainability. *Journal of Cleaner Production, 153*. https://doi.org/10.1016/j.jclepro.2017.03.126

Suryaningrat, I. B. (2016). Raw material procurement on agroindustrial supply chain management: A case survey of fruit processing industries in Indonesia. *Agriculture and Agricultural Science Procedia, 9*, 253–257. https://doi.org/10.1016/j.aaspro.2016.02.143

Taranya, R. T., Rama, D. Y., & Kavita, G. (2023). *Logistics, traceability in food supply chain management. 4th International Conference on Design and Manufacturing Aspects for Sustainable Energy*, ICMED-ICMPC, Hyderabad. https://doi.org/10.1051/e3sconf/202339101075

Tseng, M. L., Anthony, S., Chiu, F., Chien, C. F., & Tan, R. R. (2019). Pathways and barriers to circularity in food systems. *Resources, Conservation and Recycling*, *143*, 236–237. https://doi.org/10.1016/j.resconrec.2019.01.015

Vermesan, O., & Friess, P. (2013). *Internet of Things: Converging Technologies for Smart Environments and Integrated Ecosystems*. River Publishers.

Wang, B., Luo, W., Zhang, A., Tian, Z., & Li, Z. (2020). Blockchain-enabled circular supply chain management: A system architecture for fast fashion. *Computers in Industry*, *123*. https://doi.org/10.1016/j.compind.2020.103324

Waqas, U., Abd Rahman, A., Ismail, N.W. et al. (2023). Influence of supply chain risk management and its mediating role on supply chain performance: perspectives from an agri-fresh produce. *Annals of Operations Research 324*, 1399–1427. https://doi.org/10.1007/s10479-022-04702-7

Yadav, V. S., Singh, A. R., Raut, R. D., Luthra, S., & Kumar, A. (2022). Exploring the application of Industry 4.0 technologies in the agricultural food supply chain: A systematic literature review. *Computers and Industrial Engineering*, *169*, 1–19.

Ye, G., Liu, X., Fan, S., Tan, Y., Zhou, Q., Zhou, R., & Zhou, X. (2023). Novel supply chain vulnerability detection based on heterogeneous-graph-driven hash similarity in IoT. *Future Generation Computer Systems*, *148*, 201–210. https://doi.org/10.1016/j.future.2023.06.006

Zhang, & Kovacs. (2012). The application of small unmanned aerial systems for precision agriculture: A review. *Precision Agriculture*, *13*, 6. https://doi.org/10.1007/s11119-012-9274-5

Zimmer, K., Fröhling, M., & Schultmann, F. (2015). Sustainable supplier management – A review of models supporting sustainable supplier selection, monitoring and development. *International Journal of Production Research*, *54*(5), 1412–1442. https://doi.org/10.1080/00207543.2015.1079340

Zou, Z., Jin, Z., Zheng, Y., Yu, D., & Lan, T. (2023). Optimized consensus for blockchain in internet of things networks via reinforcement learning. *Tsinghua Science and Technology*, *28*(6). https://doi.org/10.26599/TST.2022.9010045

3 A Review of Digital Platform and Circular Economy

Opportunities and Challenges for Developing Countries

Maria Nemilentseva
Lappeenranta University of Technology, Kouvola, Finland

Adeel Tariq
Lappeenranta University of Technology, Kouvola, Finland
National University of Science and Technology (NUST), Islamabad, Pakistan

Waqas Tariq
Jiangsu University, Zhenjiang, P. R. China
University of Sialkot, Sialkot Pakistan

Danyal Aghajani
Lappeenranta University of Technology, Kouvola, Finland

Marko Torkkeli
Lappeenranta University of Technology, Kouvola, Finland

3.1 INTRODUCTION

Concerns about resource overexploitation, environmental harm, and social inequity are driving more sustainable communities and economies (Adams et al., 2016; Tariq et al., 2017). Circular economy is a viable solution to sustainability issues due to the need for greener growth (Wang et al., 2023). According to Geissdoerfer et al. (2017), the circular economy principles outline a sustainable system that extends, expands, and closes material and energy cycles to decrease resource inputs and waste

DOI: 10.1201/9781032616810-3

outputs. According to Boukhatmi et al. (2023), the digital approach makes it possible to switch from linear to circular economies simpler.

Digital strategy enabled organizations to build digital platforms that can give businesses access to priceless outside resources (Liu et al., 2023) and improve their practices linked with the circular economy. Digitalization is crucial for promoting the circular economy transition through information sharing, as documented in the literature (Jäger-Roschko & Petersen, 2022). Since digital platforms mediate network transactions targeted at reducing resource inputs and waste outputs, they have recently been recognized as an important circular economy enabler (Ciulli et al., 2020). Current research examines the significance of digital platforms for the circular economy at various stages and industry sectors (Boukhatmi et al., 2023; Honic et al., 2021). However, research providing a holistic overview of the digital platforms' influence on the circular economy is still lacking. Moreover, rules on how to prepare products for reuse, consumer knowledge of second-life products, and supportive legislation are still lacking in developing countries where the adoption of circular practices is still in its nascent stages and requires further attention to promote circular economy practices at the advanced level.

Digital platforms are essential for many organizations to adopt circular economy principles, according to extant research. Digital platforms can improve resource efficiency, sustainable product management, end-user inclusion, and circular manufacturing (Boukhatmi et al., 2023; Ciulli et al., 2020; Jäger-Roschko & Petersen, 2022). Researchers have examined the relationship between new technology and the circular economy from economic, social, environmental, and ethical perspectives (Denu et al., 2023). Digital transformation may also give firms cutting-edge data-sharing and execution tools, which might help them create new business models and boost their competitiveness (Ghoreishi, 2023). Cruz and da Cruz (2023) examined digital alternatives for customer participation in the textile and apparel value chain's circular economy. Optimal garment care, usage, and disposal methods were established and implemented to extend clothing lifespan. These studies suggest that digital solutions can foster a circular economy, especially in developing nations, although little is known about digital platforms.

Our focus on developing countries is timely and pivotal as digital platforms are constantly evolving globally and their significance has been well recognized. By developing digital and circular economies, these developing countries are likely to address the digital and sustainable development challenges (Li et al., 2023). Improved digital infrastructure gives developing nations a competitive edge and helps them embrace circular economy principles (Singh et al., 2020). The sluggish adoption of digital technology and execution may hinder poor nations' circular economy efforts (Singh et al., 2020). Digital platforms are crucial yet frequently underestimated in underdeveloped nations and can promote circular economy initiatives (Goyal et al., 2018). For instance, in some emerging economies, digital platforms are used to meet societal goals, helping people and small enterprises to find peers, customers, or suppliers. Some platform developers can even be considered social entrepreneurs and enterprises aiming to promote sustainable development and lead to a circular economy (Letta, 2023). Thus, the main purpose of this research is to present a comprehensive assessment of the literature on digital platforms and the circular economy to advance theory and research and improve conceptual coherence and clarity. Moreover, this research aims to provide opportunities and challenges that developing countries can gain from the adoption of digital platforms for the circular economy.

Grounded on an extensive systematic literature review (SLR), this study offers some notable contributions to the dialogue on digital platforms and circular economy. First, the study based on bibliometric analysis identifies the main authors, leading academic journals, clusters in the field, and co-occurrence of the keywords. Second, we identified the main categories of the digital platforms described in the literature such as digital shared economy, digital supply chain platforms, and digital ecosystem platforms, among others. Third, we proposed a framework where we identified the outcomes of the digital platform in terms of a circular economy. Finally, the chapter provided the opportunities and challenges for developing countries and avenues for future exploration.

3.2 METHODS

We carried out an SLR to create an updated summary of the published work on the topic of digital platforms and circular economy and to quantitatively assess the literature. Extant literature has relied on SLR methods to provide a holistic overview of the published work in specific domains, and SLR has been recognized as one of the best methods for evaluating a body of literature (Bhutta et al., 2022; Tariq et al., 2017; Tranfield et al., 2003). Moreover, SLR has been preferred in the literature for several reasons: (1) first, SLRs are more impartial than narrative literature reviews, (2) second, SLRs allow for the creation of comprehensive conclusions that result from an organized, well-detailed procedure that enables replication, and (3) third, adopting a quantitative technique is necessary for SLRs as it enables the identification of research gaps as well as areas where there is research (Mariani et al., 2022; Tariq et al., 2017).

Papers from two major databases, Scopus and Web of Science (WOS), which are widely used in the social sciences and management studies, were gathered to provide the data (Tranfield et al., 2003). Moreover, these databases consist of a selection of the most significant academic journals and research papers in the field of social sciences (Bhutta et al., 2022). Moreover, these databases make it possible to combine and organize data from several sources (such as articles, book chapters, and conference papers) in any bibliographic format (Mariani et al., 2022). As shown in Figure 3.1, we conducted an initial advanced search using the same set of terms/keywords in the Scopus and WOS databases to find relevant articles related to digital platforms and sustainability.

We followed extant research to extract the relevant data from the database and for that purpose, followed an SLR process (Syed et al., 2021). First, we compiled a list of various keywords based on current (systematic) literature reviews and bibliometric analyses that were specifically concerned with digital platforms and circular economy (Tariq et al., 2017). To find relevant articles, we looked for phrase combinations in the Scopus database that are connected to digital platforms and the circular economy in the title, abstract, or keywords (Williams et al., 2021). As a result, 409 documents came, and based on the researcher's recommendation, we focused our

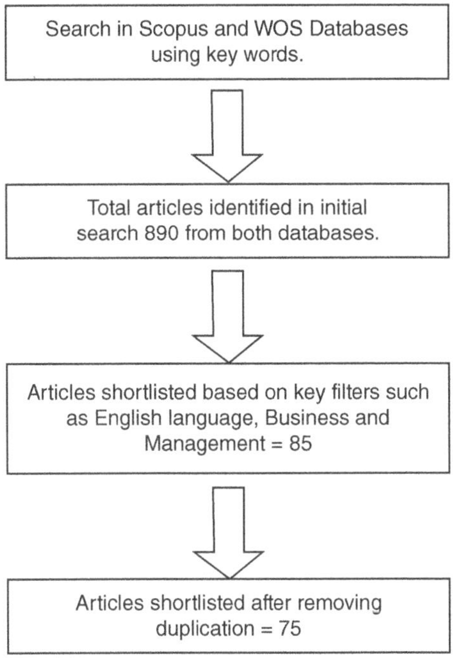

FIGURE 3.1 Systematic literature review steps (elaborated by the authors).

search to include only articles and review papers (Gaur & Kumar, 2018) written in the English language and related to the subject areas of business, management, and accounting, as well as economics, econometrics, and finance. Moreover, researchers also reviewed the title and abstract of the article to see its relevance to the research, in case, they couldn't decide from the title and abstract, they reviewed the introduction of the article to decide its selection for the review paper, which resulted in shortlisting of 40 articles.

Second, we utilized the same set of terms as Scopus to search papers from the WOS database core collection. It yielded 481 manuscripts; we also limited the articles to business management, finance, and economics fields. We also removed duplications from both databases and finally based on the relevance of the articles, we shortlisted from WOS a total of 35 articles. Both databases shortlisted articles to a total of 75 articles for this research. Figure 3.1 demonstrates this SLR process.

3.3 FINDINGS

We mapped the development of research publications about digital platforms and the circular economy till August 2023. The evolution of the prevalence of digital platforms and circular economy studies is depicted in Figure 3.2. As it is evident from Figure 3.2, in recent years, there has been a considerable increase in the volume of published research on digital platforms and circular economy, which suggests that academic interest in the field of digital platforms and circular sustainability is growing.

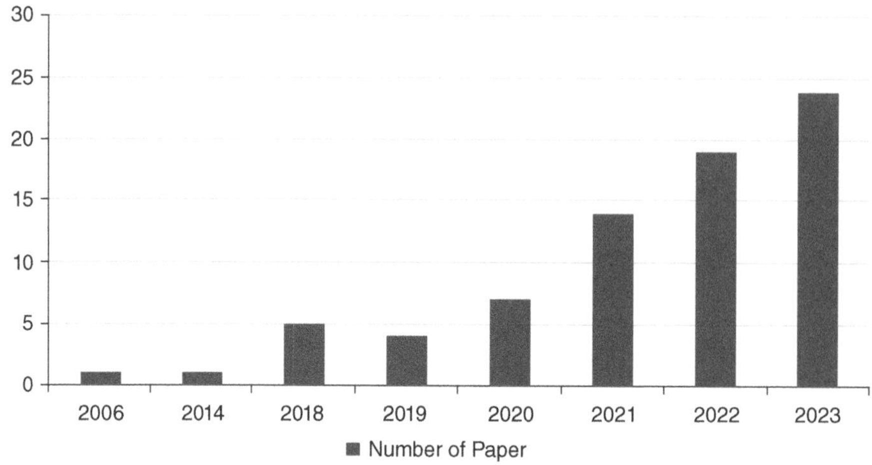

FIGURE 3.2 Articles published per year on digital platforms and circular economy (elaborated by the authors).

3.3.1 Bibliometric Analysis

To quantitatively evaluate and map research in the social sciences and the field of management, bibliometric analysis has become increasingly common in literature review papers in recent years. This strategy identifies knowledge gaps by exhibiting current scientific research trends and using several analytical methodologies (Nyantakyi et al., 2023). We depicted the digital platform and circular economy knowledge environment using the open-source VOSviewer (Version 1.6.17). Researchers liked the option's easy-to-use interface for accessing and evaluating bibliometric data. Use of readily available software improves study reproducibility, transparency, and dependability.

3.3.1.1 Bibliographic Coupling and Journal Co-citation Analysis

The acknowledgment of the work of other scientists, frequently in the form of citations, is integral to the advancement of scientific study. Therefore, techniques like bibliographic coupling prove very useful in recording this kind of acknowledgment. Bibliographic coupling is specifically based on the idea that works with similar publications are connected in terms of their substance (Donthu et al., 2021). Mariani et al. (2022) use it as a reliable method for tracking the evolution of a new research stream across time. In Table 3.1, we included the top research articles based on citations in the field of digital platforms and the circular economy.

Academic journals can be identified as having a significant percentage of publications in a given research domain by using bibliographic coupling and co-citation analyses. The network of scholarly publications with articles on digital platforms and circular sustainability is graphically depicted in Figure 3.3. The prevalence of several internationally renowned journals, such as *MIS Quarterly: Management Information Systems*, *International Journal of Production Economics*, and others are highlighted by this network.

TABLE 3.1
Leading Research Articles in the Field of Digital Platforms and Circular Economy

Authors	Title	Journal	Citations Count
Rai, Patnayakuni, and Seth (2006)	Firm Performance Impacts of Digitally Enabled Supply Chain Integration Capabilities	*MIS Quarterly: Management Information Systems*	1,390
Li, Dai, and Cui (2020)	The Impact of Digital Technologies on Economic and Environmental Performance in the Context of Industry 4.0: A Moderated Mediation Model	*International Journal of Production Economics*	304
Pencarelli (2020)	The Digital Revolution in the Travel and Tourism Industry	*Information Technology and Tourism*	176
Park and Li (2021)	The Effect of Blockchain Technology on Supply Chain Sustainability Performances	*Sustainability* (Switzerland)	118
Bechtsis, Tsolakis, Vlachos, and Srai (2018)	Intelligent Autonomous Vehicles in Digital Supply Chains: A Framework for Integrating Innovations towards Sustainable Value Networks	*Journal of Cleaner Production*	94

Source: Authors.

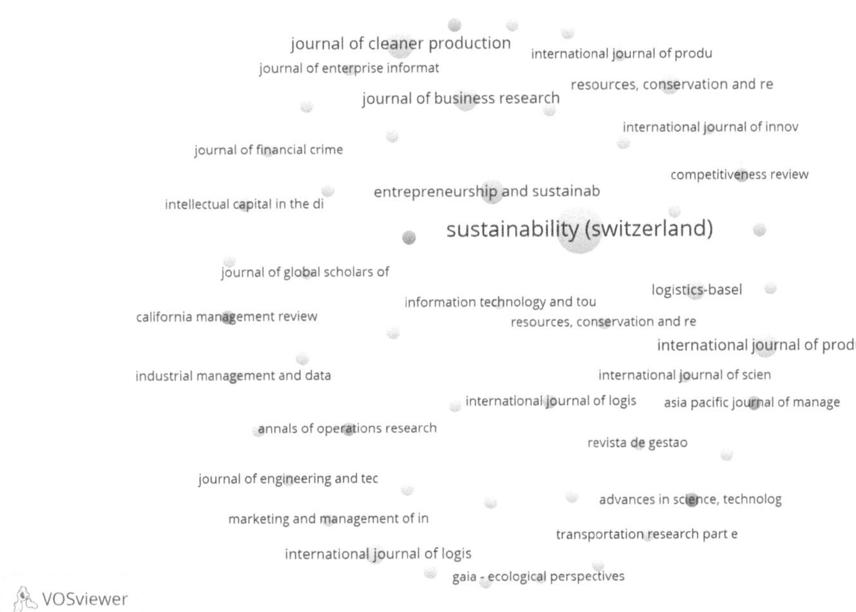

FIGURE 3.3 Leading academic journals on digital platforms and circular economy (elaborated by the authors).

3.3.1.2 Co-citation Analysis

Co-citation analysis is a reliable method for understanding the relationships between the references listed in various publications in a body of literature (Donthu et al., 2021). When a co-citation network is built at the reference level that is cited, it can be used to discover the connections in the semantic similarities network that are important for locating seminal work (Khanra et al., 2022. Figure 3.4 illustrates the network of authors within the field of digital platforms and circular economy literature.

3.3.1.3 Keyword Co-occurrence Analysis

To approximate the relationships between concepts and themes within the area, we performed a keyword co-occurrence analysis. Keyword co-occurrence analysis operates on the premise that words appearing together are interconnected by a thematic association. To understand how keywords and concepts have changed and advanced over time, we also charted their development. Figure 3.5 shows the networks of keyword co-occurrences found in the body of writing on digital platforms and circular economy studies. Co-word analysis was utilized to graphically represent literature-related phrase formation. Since it appears most often alongside other nodes, "sustainability" is the biggest node in the graph. It's also linked to "digital platform," "digitalization," "circular economy," and "business performance."

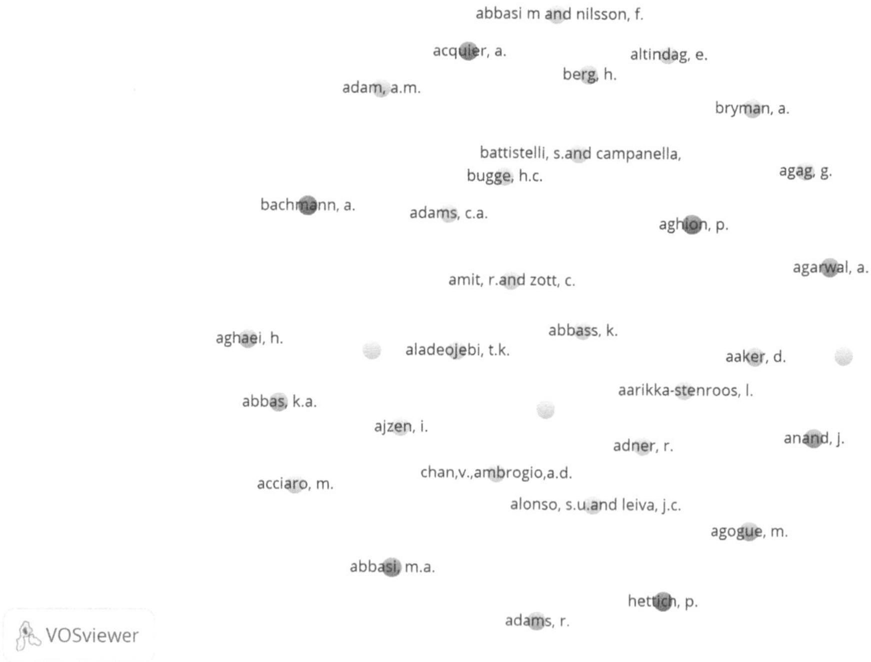

FIGURE 3.4 Clusters identified through bibliographic coupling (elaborated by the authors).

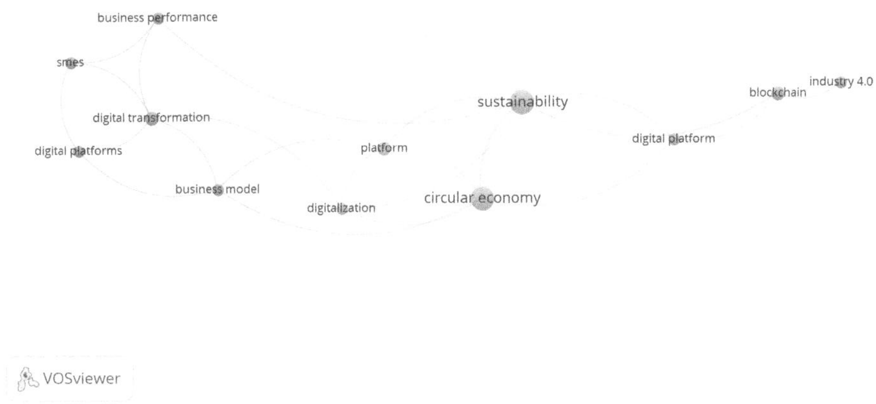

FIGURE 3.5 Co-occurrence of author keywords (elaborated by the authors).

3.3.2 Digital Platforms

Digitalization is crucial to the circular economy, according to study (Pagoropoulos, Pigosso, & McAloone, 2017). This paper defines a "circular economy" as an autonomous system that addresses resource depletion and rubbish disposal. The circular economy extends the lifespan of shared, repurposed, recycled, remanufactured, and reconditioned items. Circular economy principles have been changing traditional patterns of business. The circular economy is associated with supply chain operations, manufacturing processes, product and service lifecycles, and operational management reform (Homrich et al., 2018; Unterfrauner et al., 2019). Digital technology provides the instruments and processes necessary to facilitate and optimize various aspects of circular sustainability. A thorough analysis of the chosen articles revealed that digital platforms, including digital sharing platforms, digital supply chain platforms, digital technology, digital transformation, and digital ecosystem platforms, contribute to the circular economy in developing countries (Tables 3.2–3.6). Despite their various digitalization-related topics, the articles selected emphasize the importance of creating sustainability through a circular economy. As a result, we combined them into five themes, each containing a set of framework-defined subcategories. The categories are as follows:

1. Digital ecosystem platforms
2. Digital supply chain platforms
3. Digital sharing/technology platforms
4. Digital transformation
5. Digital bioeconomy platforms

It appears that the area with the most investigations was the digital ecosystem platforms. Figure 3.6 shows the digitalization in the circular economy for developing countries.

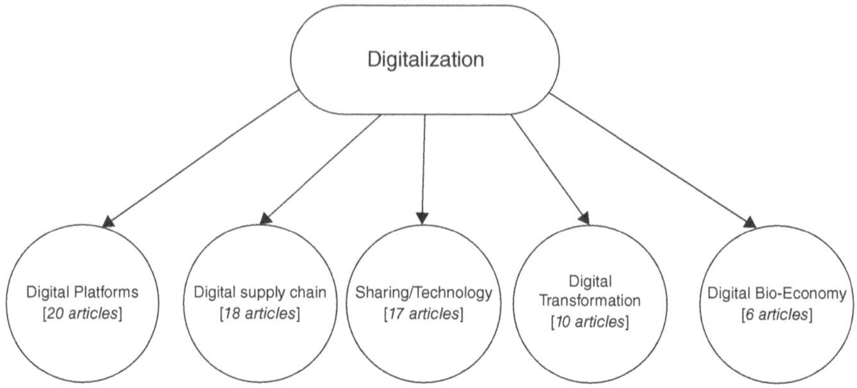

FIGURE 3.6 Digitalization in the circular economy for developing countries (elaborated by the authors).

3.3.2.1 Digital Ecosystems Platforms

As a network of businesses, people, and technology in a shared digital environment, a digital ecosystem platform creates and exchanges value (Gawer, 2022). Business strategies and industries rely heavily on these ecosystems in the digital age. Table 3.2 shows the categories, findings, and references connected to this concept. Business strategies and industries rely heavily on these ecosystems in the digital age. Table 3.2 shows the categories, findings, and references connected to this concept.

3.3.2.2 Digital Supply Chain Platforms

Supply chain management can be optimized by some concepts related to digital technologies and data-driven approaches (Al-Sartawi 2021). Integration of multiple digital tools, platforms, and data sources can enhance processes, improve visibility, and empower informed decision-making throughout the supply chain (Russo, Manzari et al., 2023). Table 3.3 shows the categories, findings, and references connected to this concept.

3.3.2.3 Digital Sharing/Technology Platforms

The process of making resources, information, and services available for sharing via digital tools, platforms, and technologies is termed "digital sharing" or "technology" (Sutherland & Jarrahi 2018). A sharing economy is considered a classical economic system based on the exchange of marketable private assets with a constructive – better utilization of goods and products. Thus, the sharing economy supports sustainable development (Li et al., 2019; Schwanholz & Leipold, 2020). The use of digital platforms is a key element of the sharing economy promoting internal and external sharing practices. In turn, internal sharing supports sustainable performance in all three dimensions, while external sharing – being more risky – affects only the environmental side (Li et al., 2019). By considering this idea, the effectiveness of resource usage, connections, and collaboration may be enhanced in the digital era. Table 3.4 shows the categories, findings, and references connected to this concept.

TABLE 3.2
Digital Ecosystem Platform Categories and Findings

Categories	Findings	References
Decentralized database	Transparency Reliability Traceability Effectiveness in managing the supply chain Contribute environmental, social, and governance (ESG) performance	Gruchmann (2022); Marino and Pariso (2021); Ngai et al. (2014); Park and Li (2021)
Sustainable procurement practices	Reduced pollutants Utilize energy and resources economically and effectively Boost interest in environmentally friendly products and services Adopting a dynamic perspective in performance evaluation Competitive advantage and business performance Enterprise's digital capability Reducing transaction expenses Minimizing market tension Boosting commercial exchange to intensify resource use	Aoki et al. (2019); Fu, Zha, and Zhou (2023); Grover, Kar, and Vigneswara Ilavarasan (2018); Puspita, Christiananta, and Ellitan (2020); Singh et al. (2020)
Blockchain	Blockchain engines Blockchain connectors Cryptocurrency-directed interoperability Data-driven dynamic capabilities	Chen, Li, and Zhang (2022); Punathumkandi, Sundaram, and Panneer (2021)
Optimization	Organization of production Optimize consumption to balance the load peaks Positive impacts on innovation culture and innovation performance Match the supply and demand in the supply chain Foster the circular economy transition Increase innovation performance in the presence of e-Receipt cloud solutions	Aversa et al. (2021); Blackburn, Ritala, and Keränen (2023); Boukhatmi, Nyffenegger, and Grösser (2023); Gavrila and de Lucas Ancillo (2021); Gitelman and Kozhevnikov (2023); Jain et al. (2022); Khattak (2022); Lehner and Elbert (2023); Wang et al. (2023)
Digital symbiosis	Facilitate the identification and exploitation of synergy opportunities Solves the challenges to matchmaking platforms Promotes behavioral intention Positive correlation between entrepreneurial awareness and the perceived quality of digital apps	Bivona (2022); Chan et al. (2023); Fellnhofer (2022); Krom, Piscicelli, and Frenken (2022)

Source: Authors.

3.3.2.4 Digital Transformation Platforms

The digital transformation has impacted many activities, products, and services of a company. The use of technology can help businesses in several sections such as enhancing productivity and creativity, increasing customer satisfaction, and rising competitiveness (Kraus, Jones et al. 2021). As technology advances quickly and

TABLE 3.3
Digital Supply Chain Platforms Categories and Findings

Categories	Findings	References
Digital industrialization	• Reducing pollution emissions • Green innovation upgrading • Strategic governance of product-related data • Digitization: software as a service, traceability	Ivanov, Dolgui, and Sokolov (2022); Piétron, Staab, and Hofmann (2023); Silva and Sehnem (2022); Tian (2023)
Digital supply chain platforms	• Think about how to innovate and rethink, repair, minimize, prevent, and recycle • The creative, resilient, and stable industries • Boost the sustainability • Competitive in closed-loop systems • Mediate the effects on both economic and environmental performance • Internationalization process • Data-driven demand supply • Strategies for completing orders and supplier cooperation • Minimize the average product storage and shipment fluctuations • Increasing the total net profits • Estimate market demand	(Abideen et al. (2021); Ding et al. (2023); Hong et al. (2021); Li, Dai, and Cui (2020); Li et al. (2023); Lyu and Jiang (2023); Rajala et al. (2018); Sharma et al. (2022); Unterfrauner et al. (2019)
Smart factories	• Optimizing energy and resource utilization • Reducing waste • Strong recycling • Circular business models before implementation • Vehicles (intelligent autonomous vehicles) in sustainable supply networks • Precise assessment of cloud service providers' reliability • Improving supply chain organizational performance • Influence on the performance of small and medium enterprises (SMEs)	Bechtsis et al. (2018); Khoruzhy et al. (2022); Lange et al. (2021); Pandey et al. (2023); Ramakrishna, Alzoubi, and Indiran (2023)

Source: Authors.

consumer expectations shift, the industry is undergoing a digital transition. Table 3.5 shows the categories, findings, and references connected to this concept.

3.3.2.5 Digital Bioeconomy Platforms

To develop novelty and sustainability in business, the "digital bioeconomy" is presented as a combination of bioeconomy and digital technologies (Lekkas, Panagiotakis et al., 2021). It focuses on the use of digital tools and technologies to optimize biological

TABLE 3.4
Digital Sharing/Technology Platform Categories and Findings

Categories	Findings	References
Digital sharing platforms	• Stewardship instead of ownership • Stakeholder-based perspective • Identify strongly with the circular economy • Positive impacts of internal sharing on economic, environmental, and social performances • External sharing has positive impacts on environmental performances • Increasing the idle capacity's utility • Support worthwhile and reliable communities • Minimizing damage to the ecosystem	Belhadi et al. (2023); Carlsson, Nevzorova, and Vikingsson (2022); Füller, Hutter, and Kröger (2021); Li, Li, and Wang (2022); Pérez-Pérez et al. (2023); Schwanholz and Leipold (2020); Tseng et al. (2021)
Digital technologies	• Reduces loyalty to the B2B sharing platform by users • Decrease in the electronic word-of-mouth (eWoM) value • Realize the economic and/or circular potentials • Maximize the lifespan of products • Using automated control, develop cyclical enterprise models • Procedure and resource optimization • Improve firm performance, task performance, and efficiency improvements • Avoid competitors' pressures • Maintaining legitimacy in high uncertainty	De Bruyne and Verleye (2023); Kumar and Chopra (2022); Ying Li et al. (2019); Rai, Patnayakuni, and Seth (2006); Vehmas et al. (2018); Xie and Wang (2023)
Information system platform	• Credibility to customers • Support purchasing decision • Finding proper service • Developing a circular economy blockchain architecture • Excellent quantity and caliber of growth in the finance-technology connection • Expand the supply chain process integration's higher-order capabilities	Mamedova et al. (2022); Sheikh et al. (2018); Tatli, Yavuz, and Ongel (2023); Yadav, Kar, and Kashiramka (2022)

Source: Authors.

resource production, processing, and utilization. Table 3.6 shows the category, findings, and references connected to this concept.

Circular economy principles are being advanced through digital platforms, fostering the adoption of sustainable practices across industries (Agrawal et al., 2022). By exchanging real-time data, these platforms improve resource management, reduce waste, and maximize material and product efficiency (Demestichas & Daskalakis, 2020). Through shared and repurposed assets, they promote a circular approach to the consumption and production of goods (Morseletto, 2020). By using digital technologies, supply chains become transparent, traceable, and collaborative, which

TABLE 3.5
Digital Transformation Platforms Categories and Findings

Categories	Findings	References
Circular economy	• A sustainable business model that collects value and creates a social benefit • Increases consumers' confidence • Value creation and gaining long-term competitive advantages with the use of the Industry 4.0 framework • Favorable effect on the ESG performance of the business • Business-related green innovation encourages the connection between corporate ESG performance and digital transformation strategy • Positive effect of digital business model maturity on sustainable business success • Chances of surviving during emergencies • Sustainable development in the years after the epidemic	Akarsu (2023); Akpan, Effiom, and Akpanobong (2023); Andrade and Gonçalo (2021); Gong (2023); Tunn et al. (2020); Zhao, Li, and Li (2023)
Data analytics	• Digital data • Automation • Digital communication network • Successful implementation • Improve efficiency, reduce costs, increase transparency • Marketing functions from the point of competitiveness	Kasperovica and Lace (2021); Lányi, Hornyák, and Kruzslicz (2021); Pencarelli (2020); Savastano et al. (2022)

Source: Authors.

TABLE 3.6
Digital Bioeconomy Platforms Categories and Findings

Category	Findings	References
Digital finance/ digital economy	Lowering the volatility of company finances to improve sustainability Fostering sustained improvement in performance The creation and spreading of knowledge Circularly reinvesting time, money, and effort into the community Increase the social value Increasing the long-term effectiveness of micro-, small-, and medium-sized enterprises (MSMEs) Moderating online purchase intention	Hu et al. (2023); Presch et al. (2020); Hairudinor and Rusidah (2023); Sun and Ertz (2021); Teichmann, Boticiu, and Sergi (2023); Watanabe, Naveed, and Neittaanmäki (2019)

leads to better decision-making and resource optimization. Moreover, they support the repair, refurbishment, and resale of products, aligning with circular economy ideals. To shift toward a more sustainable and circular future, digital platforms connect the gap area between producers and consumers (Agrawal et al., 2022).

3.3.3 Digital Platform Influence on the Circular Economy

It can be beneficial for organizations to adopt circular economy principles in multiple ways. For example, incorporating circular economy principles into supply chain management can sustainably improve business operations. By reducing their ecological footprint, organizations can also differentiate themselves from their competition and generate new revenue (Bechtsis et al., 2018; Boukhatmi et al., 2023; Gitelman & Kozhevnikov, 2023; Ivanov et al., 2022; Jain et al., 2022; Krom et al., 2022; Lange et al., 2021; Lehner & Elbert, 2023; Ying Li et al., 2020; Ying Li et al., 2019; Park & Li, 2021; Punathumkandi et al., 2021). These are the highlighted points in this subject:

- Matching supply and demand in the supply chain is crucial for efficiency.
- The transition to a circular economy purpose to reduce waste and promote sustainability.
- Strategies such as decentralization, transparency, and traceability contribute to supply chain efficiency and ESG performance.

3.3.3.1 Sustainability and Environmental Practices

Sustainable development has several challenges and opportunities in developing countries. Although reaching sustainability in these countries can be complicated, it is essential for the well-being of their environment and people in long-term periods (Tariq et al., 2019). To overcome barriers and build a sustainable future, these countries often need help from international partners (Belhadi et al., 2023; Fellnhofer, 2022; Füller et al., 2021; Li et al., 2023; Lyu & Jiang, 2023; Mamedova et al., 2022; Singh et al., 2020; Tian, 2023; Vehmas et al., 2018; Xie & Wang, 2023). These are the highlighted points in this subject:

- ISO 14001 plays a significant role in sustainable procurement practices.
- Sustainability initiatives aim to reduce pollution, make efficient use of resources, and promote sustainable products.
- Strategies focus on environmental benefits, such as reducing pollution emissions, as part of global initiatives like the Belt and Road Initiative.

3.3.3.2 Circular Business Models and Sustainability-Oriented Innovation

Creating sustainable businesses and reducing waste are closely related concepts of circular business models and innovation (Barros, Salvador et al., 2021). The circular business models rely on the collaboration of actors focused on the creation and capture of sustainable value, leading to both economic and noneconomic benefits (Blackburn et al., 2023). Innovation focused on sustainability involves changes in the organization's products and working processes aimed at creating not only economic but also environmental and social value. The flexibility to integrate and reconfigure digital platforms has a favorable impact on innovation focused on sustainability. In other words, increasing the capacities of digital platforms fosters innovation centered on sustainability. Companies using digital platforms get access to external knowledge and resources and are more ready for market changes capable of performing sustainably with the participation of other actors, thus fastening innovation processes (Wang et al., 2023). Using circular business models emphasizes resource efficiency, product longevity, and reducing environmental impacts in place of the "take-make-dispose" linear model (Abideen et al., 2021; Akarsu, 2023; Aoki et al., 2019; Blackburn et al., 2023; Gruchmann, 2022; Pencarelli, 2020; Piétron et al., 2023; Sun & Ertz, 2021; Tunn et al., 2020; Watanabe et al., 2019). These are the highlighted points in this subject:

- Circular business models aim to minimize transaction costs, reduce friction, and intensify resource use.
- Open innovation and digital applications can enhance sustainability-oriented innovation.
- Circular economy practices include disintermediation and decentralization.

3.3.3.3 Sustainable Supply Chain and Social Impact

In supply chain management, environmental and social issues are highlighted in developing sustainability in supply chains. Keeping a supply chain positive and minimizing its negative impacts is the key to reaching these purposes (Aversa et al., 2021; Carlsson et al., 2022; Chan et al., 2023; Chen et al., 2022; De Bruyne & Verleye, 2023; Ding et al., 2023; Gong, 2023; Hong et al., 2021; Hu et al., 2023; Khoruzhy et al., 2022; Kumar & Chopra, 2022; Lányi et al., 2021; Li et al., 2022; Ngai et al., 2014; Pandey et al., 2023; Presch et al., 2020; Rai et al., 2006; Rajala et al., 2018; Hairudinor & Rusidah, 2023; Savastano et al., 2022; Schwanholz & Leipold, 2020; Sharma et al., 2022; Sheikh et al., 2018; Teichmann et al., 2023; Unterfrauner et al., 2019; Zhao et al., 2023). These are the highlighted points in this subject:

- Optimal operational strategies incorporate economic, social, and environmental responsibilities.
- Companies focus on customer behaviors and preferences while promoting sustainable practices.
- Sustainable supply chains contribute to organizational performance and affect SMEs.

3.3.3.4 Sustainability and Business Performance

Circular economy and sustainable business performance link together. In addition to fostering social and economic exchange, value co-creation, and resource integration, digital platforms improve business excellence and sustainable performance (Agrawal et al., 2022). As these platforms are always dynamic, they create fruitful soil for the development of various firms, including SMEs (Suchek, Ferreira et al., 2022). Bringing value to customers and offering value-added services via digital platforms can help firms increase their competitiveness (Savastano et al., 2022). As an example, research conducted by Hong et al. (2021) illustrates the positive impact of digital platforms (supply chain service platforms) on firms' performance from both quality and economy perspectives. The same research highlights the significant role of digital platforms for small firms in emerging economies (in this case Chinese companies). Cheng, Ling, and Zhang (2022) also highlight that the data-driven dynamic capabilities of SMEs can be considered as a factor directly affecting their sustainable development performance and the quality of their responsible innovations. Sustainable performance also directly depends on digital regulatory and digital customer pressure. While talking about performance, the important and advantageous impact that supply chain abilities have on competitive advantage and corporate success must be mentioned (Puspita, Christiananta, & Ellitan, 2020). Developing sustainability in companies leads to several benefits such as economy efficiency, innovations, risk reduction, and increasing the value and reputation of them. By recognizing these opportunities correctly and utilizing them properly, the results can secure the well-being of the environment and society (Akpan et al., 2023; Andrade & Gonçalo, 2021; Kasperovica & Lace, 2021; Pérez-Pérez et al., 2023; Puspita et al., 2020; Ramakrishna et al., 2023; Silva & Sehnem, 2022; Tseng et al., 2021; Wang et al., 2023; Yadav et al., 2022). These are the highlighted points in this subject:

- Strategic orientation, supply chain capability, and innovation culture are linked to competitive advantage.
- Digital business models, when mature, positively impact business success and enterprise ESG performance.
- Effective digital transformation can enhance an enterprise's ESG performance and sustainability.

3.3.3.5 Economic Growth through Efficient Use of Resources

Innovation-driven strategies and innovations have positive effects on economic growth and sustainability (Mamedova et al., 2022). In turn, digital economies support innovations and circular economy initiatives that lead to creating several job opportunities and helping the economy of developing countries. Considering the development and widespread of digital technologies, supporting a circular and sharing economy, we can state that the product life cycle in these conditions is extended as it can receive a "second" life several times. Additionally, digital platforms can increase efficiency by considering the idea of reusing, recycling, and remanufacturing.

3.3.3.6 Collaboration, Consumer Engagement, and Transparency

As research in a variety of industries (like fashion and food) shows, digital platforms promote sustainable products and services (mostly in consumers' behavior) (Pirola, Boucher et al. 2020). As a result, people are more likely to buy products and services that are sustainable, that change habits, and choices, and, in the long term, help companies to become more sustainable in the case of production (Jain et al., 2022; Silva & Sehnem, 2022). Moreover, digital platforms' inner orchestration mechanisms pretend to provide and support trust and transparency between platform actors. Customers and other stakeholders of companies can get and use the opportunity to start

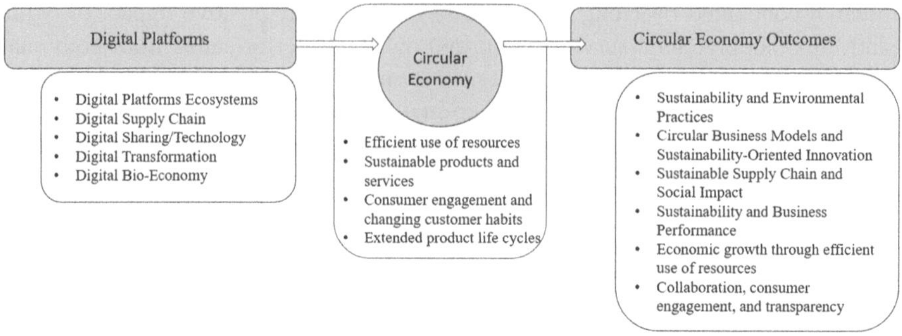

FIGURE 3.7 Digital platforms for circular economy and circular economy outcomes from 75 papers (elaborated by the authors).

cooperation. This process improves the product's quality and enhances novelty in productions (Kasperovica & Lace, 2021; Blackburn et al., 2023). Figure 3.7 illustrates the relation between digital platforms, which are classified into five categories; circular economy; and circular economy outcomes.

3.4 DISCUSSION

Based on the objectives of this research to understand digital platforms' potential for circular economy and sustainable platforms, this research has documented results from an SLR review which is in line with existing research where scholars have proposed that the application of various technologies within industries stimulates intentions toward circular economy (Jain et al., 2022). Digital platforms foster information flows between business areas and facilitate sustainable interactions between multiple actors (resource sharing, collaboration, and networking) that are likely to enhance companies' transition and improve performance toward a circular economy (Boukhatmi et al., 2023). Similarly, developing countries can utilize different types of digital platforms as highlighted in this research such as digital platforms for sustainable supply chains to enhance collaboration through shared information for improving performance in terms of circular economy.

Using digital platforms to stimulate the adoption of digital technologies improves resource management and operational efficiency in sectors like food technology as these platforms allow faster procedures, improve resource allocation, and lowers waste, which in turn encourages the incorporation of circular economy mechanisms (Silva & Sehnem, 2022). Thus, the adoption of digital technology should be encouraged in developing countries to facilitate the integration of circular economy mechanisms. This can result in greater resource efficiency, decreased waste, and sustainable practices. Moreover, digital platforms that help SMEs grow in tandem with government initiatives to support them and with a circular economy (such as in China) facilitate the creation of innovation-driven strategies and inventions. This positively affects economic growth and sustainability (Mamedova et al., 2022), Thus, digital platforms can support innovations and circular economy initiatives, simultaneously leading to the creation of new jobs and new business opportunities in developing economies. Digital platform, together with government support programs for SMEs offered by developing countries and potential support for circular economy at the state level, can stimulate and even foster economic growth in developing regions.

Digital marketing promotion via digital platforms at this point can support sustainable development and circular economy not only on a company's level but also to change habits and develop new consumption culture on an individual level (Hairudinor & Rusidah, 2023; Khattak, 2022; Blackburn et al., 2023; Wang et al., 2023; Jain et al., 2022; Vehmas et al., 2018; De Bruyne & Verleye, 2023). These platforms provide a fruitful basis for the development of more personalized services of better quality (Ding et al., 2023, Gitelman & Kozhevnikov, 2023; Blackburn et al., 2023; Silva & Sehnem, 2022). That also can shape the existing standards within the developing countries' environment and result even in changes in consumption patterns and the development of sustainable consumption culture in developing countries.

On the business level, the reduction of costs and improvement of business performance, together with fostering the development of sustainable products and services, are the benefits of digital platforms (Watanabe et al., 2019; Tunn et al., 2020; Bivona, 2022; Blackburn et al., 2023). The creation of new production channels together with the ability of these platforms to create new markets, match demand and supply, and stimulate network effects results in the constant development of supplier–customer relationships and even in accelerating transformation inside industries (e.g., finance) and the creation of new digital platform industries (Krom, Piscicelli, & Frenken, 2022). These advantages of digital platforms' use can be employed by developing countries with the aim not only to support local producers but also to answer the existing calls for sustainable development. Within emerging economies, an additional incentive for relying on digital platforms as they enhance technological application resulting in boosting innovations and having a positive effect on innovative culture within a variety of industries. Innovation culture improvements, in turn, can assist in overcoming challenges of the existing linear economy within a shorter period in developing countries.

3.4.1 Opportunities and Challenges

Circular economy promotion in different countries and regions is caused by economic benefits, demands from consumers, government regulations, and the ability to provide economic and environmental growth while assisting in the achievement of countries' sustainable development goals. Encouragement of circular economies can be enabled and facilitated by digitalization (Jäger-Roschko & Petersen, 2022). However, in developing countries, China was the first country to introduce circular economy policies on a state level in 2002 (Nham & Ha, 2022). Several other countries such as India, Malaysia, and others are also incorporating circular economy practices to gain several advantages and put less harm to the environment (Goyal et al., 2018; Singh et al., 2020). This shift for developing countries is critical as they aim to become sustainably developed, and there is a need to shift from linear economy to circular economy models within almost all sectors. This process is assisted and fostered by the integration of technologies (including Industry 4.0 technologies) by industries. This results in cost-effective production of environmentally friendly, competitive products in developing countries (Ahmed et al., 2022).

As it was discussed earlier, digital platforms can support circular economies. Their development and evolution require the participation of various actors. While collaborating and building platforms and developing them, various opportunities and challenges can appear within the developing countries setting (see Table 3.7). Additionally, we included the main potential outcomes or impacts of digital platforms toward circular economy and sustainability to demonstrate the digital platforms' long-term output. These results are based on an analysis of scientific papers covering experiences from both developed and developing countries and included in the literature review. We highlighted the key points opening the potential of digital platforms for sustainability that can be used as lessons for emerging countries.

3.5 PRACTICAL IMPLICATIONS

This SLR accumulating knowledge extracted from a range of studies can be useful for practitioners and policymakers. Among all, this research identifies challenges and opportunities for developing countries. There are a variety of possible outcomes of digital platforms toward circular economy and sustainability in general. The successful development of digital platforms and their further evolution requires orchestrated actions by different actors involved. Business entities, digital platform developers, and industry representatives need to choose an appropriate approach, be prepared for potential barriers, and develop plans of action for various scenarios. Policymakers need to find ways to respond to global challenges, meet societal needs and support digital platforms' evolution to their advantage, create funding programs, and stimulate companies, and industry clusters to become platform actors and move toward sustainable processes. It helps to create and support ecosystems and stimulate the development of regions in general. Businesses, including SMEs, play an active role in building and reshaping the platforms, strategically identifying and applying the most appropriate platforms, resources, and partners. Digital platform developers therefore need to analyze challenges and opportunities, identify platform characteristics, promote them, and create a basis for successful collaboration of its actors, and control platform development and growth. Finally, policymakers together with platforms' creators can support businesses, especially small enterprises by educating them about the importance of collaboration with different actors (including customers) and the opportunities and benefits they can get. Additionally, policymakers can develop and introduce schemes supporting the use of digital platforms, for example, through tax reductions for companies and entrepreneurs.

3.6 LIMITATIONS AND FUTURE RESEARCH

We aimed to present a holistic overview of the digital platforms' impact on the circular economy; however, the process of writing SLR always involves limitations. First, it reflects a static picture of a situation existing at a particular moment. Second, the growing interest in digital platforms and the circular economy, confirmed by the literature analysis, means that some relevant papers could stay outside our review

TABLE 3.7
Main Potential Opportunities, Challenges, and Outcomes of Digital Platforms for Circular Economies

Opportunities	Challenges	Outcomes
1. Digital platforms can facilitate varied actors to collaborate and integrate (Li et al., 2022; Rajala et al., 2018; Silva & Sehnem, 2022), which advances attempts to implement circular economies in developing countries.	1. Lack of commitment to sustainability (environmental consciousness) results in the need to educate companies, their staff, and in some cases their customers. Businesses may be reluctant to embrace digital platforms in this case to gain circular economy benefits (Kumar & Chopra, 2022; Lange et al., 2021; Fu et al., 2023; Krom et al., 2022).	1. Digital platforms are capable of enabling and facilitating circular economies in developing countries (Li et al., 2022; Hu et al., 2023; Tunn et al., 2020: Li et al., 2019; Silva & Sehnem, 2022).
2. Moreover, digital platforms can promote the growth of shared values, aiding in sustainable development and producing advantages for the circular economy in developing nations that are both economically and non-economically beneficial (Pencarelli, 2020; Li et al., 2020; Li et al., 2019; Andrade & Gonçalo, 2021; Blackburn et al., 2023).	2. In some regions and industries, there could be technical barriers, including a lack of equipment and a lack of knowledge on how to implement and use a technology within the exact setting (Gavrila & de Lucas Ancillo, 2021; Kumar & Chopra, 2022; Krom et al., 2022; De Bruyne & Verleye, 2023).	2. Digital platforms foster the creation of new production, marketing, and communication channels in a dynamic setting with constantly improving standards (Ding et al., 2023, Gitelman & Kozhevnikov, 2023; Blackburn et al., 2023; Silva & Sehnem, 2022).
3. Furthermore, based on existing experiences, developing countries may overcome the challenges of a linear economy and promote the adoption of circular economy principles by using digital platforms (Gavrila & de Lucas Ancillo, 2021; Pandey et al., 2023; Tunn et al., 2020; Li et al., 2019; Chan et al., 2023).	3. In addition to existing technical barriers, the complex and uncertain environment of digital innovation and digital platforms can be considered dangerous and not trustworthy. Information quality and data confidentiality concerns can act as data-sharing obstacles. The trust gaps between digital platform's actors result in difficulties with local partners' choices caused by a fear of economic loss and other stakeholder collaboration difficulties (Kumar & Chopra, 2022; Lange et al., 2021; Khoruzhy et al., 2022; Krom et al., 2022; Blackburn et al., 2023; Belhadi et al., 2023; Hong et al., 2021).	3. Digital platforms are capable of creating new markets matching supply and demand and even can cause the formation of digital platforms industries while enabling transactions supporting sustainable development (Watanabe et al., 2019; Tunn et al., 2020; Bivona, 2022; Blackburn et al., 2023).

(*Continued*)

TABLE 3.7 (CONTINUED)

Opportunities	Challenges	Outcomes
4. The development of circular value, which is essential for sustainable development in developing countries, is likely to be facilitated by the strategic coordination of digital platform capabilities for circular business models (Blackburn et al., 2023; Zhao et al., 2023; Mamedova et al., 2022).	4. There is a need for and even a lack of support for digital platforms and their actors from the government. Among all, it also refers to the lack of funding and supportive programs (Krom et al., 2022; De Bruyne & Verleye, 2023; Chan et al., 2023; Chen et al., 2022; Fellnhofer, 2022; Sheikh et al., 2018; Perez-Perez et al., 2023; Yadav et al., 2022; Tian. 2023).	4. On the industrial level, digital platforms can support the development of regional companies, and exact territories, strengthen their position in the market, and enhance industry symbiosis and development (Krom, Piscicelli, & Frenken, 2022).
5. Digital platforms influence shifts in consumer behavior, which is essential for promoting the concepts of the circular economy and likely to benefit developing countries (Kumar & Chopra, 2022; Hairudinor & Rusidah, 2023; Jain et al., 2022).	5. In some areas and industries supply and demand are imbalanced, and this lowers the efficiency of a digital platform (Blackburn et al., 2023; Gitelman & Kozhevnikov, 2023; Gong, 2023).	5. Support of digital and sustainable innovations can result in improved firms' performance and the development of products and services of better quality. Additionally, digital platforms support overall economic growth (Li et al., 2022; Hairudinor & Rusidah, 2023; Gitelman & Kozhevnikov, 2023; Wang et al., 2023).
6. In the context of the circular economy in developing nations, leveraging the potential of digital technologies and supply chain platforms is essential to creating a dynamic and open environment that eventually improves both economic and environmental performance (Nham & Ha, 2022; Park & Li, 2021; Li et al., 2020; Ding et al., 2023; Ramakrishna et al., 2023; Silva & Sehnem, 2022).	6. Existing heterogeneity of regions and countries results in irregular distribution of need in and development of digital platforms (Marino & Pariso, 2021). It is caused by the uneven application of digital technologies and innovation culture within the developing countries context (Andrade & Gonçalo, 2021; Hu et al., 2023; De Bruyne & Verleye, 2023; Gong, 2023; Perez-Perez et al., 2023)	6. Organized distribution of information and resources together with efficient management of ecosystems leads toward circular value creation and positive network effects, including increased economic and resource efficiency and improvement of quality of life (Hu et al., 2023; Singh et al., 2020; Blackburn et al., 2023).

7. Increasing the production of digital information in businesses and encouraging quick responses to this data to increase the influence of the circular economy in developing nations (Li et al., 2022; Silva & Sehnem, 2022; Ivanov et al., 2022; Gong, 2023).	7. Risks of fraud and money laundering exist (Tseng et al., 2021; Teichmann et al., 2023.	7. Digital platform's transformation trajectories take into account socio-cultural, ethical, technology, and economic perspectives, resulting in sustainable production and consumption together with the formation of new innovative cultures in developing countries (Hairudinor & Rusidah, 2023; Khattak, 2022; Blackburn et al., 2023; Jain et al., 2022; Vehmas et al., 2018; De Bruyne & Verleye, 2023; Mamedova et al., 2022).
8. To increase the influence of a circular economy, accelerate sustainable development in developing nations, digital platforms can assist local invention and cultivate an innovative culture inside the nation or region (Hairudinor & Rusidah, 2023; Khattak, 2022; Blackburn et al., 2023; Wang et al., 2023; Jain et al., 2022; Vehmas et al., 2018; De Bruyne & Verleye, 2023).		8. Dynamic development of digital platforms, including effective knowledge and information sharing and management, constant development of customer–supplier relationships fosters digital transformation strategies development and application by companies and improves their performance, technology innovation performance, and ESG performance. Additionally, it boosts SMEs innovation performance, in general (Hairudinor & Rusidah, 2023; Li et al., 2022: Li et al., 2019; Rai et al., 2006; Zhao et al., 2023; Khattak, 2022; Fuller et al., 2021; Mamedova et al., 2022; Wang et al., 2023).

Source: Authors.

as we were trying to narrow them down. We included research articles from two databases, but some non-scientific research articles also could be valuable for the research. However, the authors have thoroughly checked the initial dataset to identify the most relevant articles.

To investigate further the impact of digital platforms on circular economy and sustainable performance, researchers can conduct research within the exact sectors of economics or with the focus on the exact regions or groups of developing countries aiming to reveal the existing trends and forecast upcoming ones. Moreover, scholars can develop a system of recommendations on how to make the process of digital transformation and evolvement and application of digital platforms more efficient within the exact environment. Additionally, cross-country and cross-industry studies can become a good supplement to existing literature. Additional interest can arise across platform competition, such as sharing of economy practices within the developing markets and governments and policymakers' initiatives, participation and role within the digital platform's environment and support of the circular economy, sustainable performance, and prevention of risks of failures on various levels.

3.7 CONCLUSION

In this research, we conducted a comprehensive literature review focusing on digital platforms' influence on the circular economy and highlighted how developing countries can benefit in terms of opportunities, challenges they face, and outcomes. Based on 75 articles obtained from SCOPUS and WOS databases, this book chapter provided an extensive bibliometric analysis and also developed main categories derived from literature on digital platforms. Moreover, we developed a framework identifying outcomes of the digital platform toward circular economy in developing countries. Through the investigation of the scientific literature, it can be concluded that digital platforms' evolution positively influences the circular economy, creating a fruitful basis for collaboration within various countries, regions, and industries. Among all, digital platforms incentivize movements from linear toward circular economy, sharing economy, and strengthening of industrial symbiosis (especially within exact regions with strong industry components). This research defined the main digitalization categories, including digital ecosystem platforms, digital supply chain, digital sharing/technology, digital transformation, and digital bioeconomy, and identified the main results of digital platforms in the circular economy within the supply chain and circular economy, sustainability and environmental practices, circular business models and innovation, sustainable supply chain and social impact, and sustainability and business performance. This study also highlights more detailed outcomes toward circular economy, including extending product life cycle, efficient use of resources, economic growth, engaging customers and forming new sustainable culture, fostering innovation and developing innovative culture. Digital platforms with application of digital technologies enhance data-driven decision-making and provide trust and transparency along with saving costs for better and sustainable business performance.

REFERENCES

Abideen, A. Z., Sundram, V. P. K., Pyeman, J., Othman, A. K., & Sorooshian, S. (2021). Digital twin integrated reinforced learning in supply chain and logistics. *Logistics*, *5*(4), 84.

Adams, R., Jeanrenaud, S., Bessant, J., Denyer, D., & Overy, P. (2016). Sustainability-oriented innovation: A systematic review. *International Journal of Management Reviews*, *18*(2), 180–205.

Agrawal, R., Wankhede, V. A., Kumar, A., Upadhyay, A., & Garza-Reyes, J. A. (2022). Nexus of circular economy and sustainable business performance in the era of digitalization. *International Journal of Productivity and Performance Management*, *71*(3), 748–774.

Ahmed, Z., Mahmud, S., & Acet, H. (2022). Circular economy model for developing countries: Evidence from Bangladesh. *Heliyon*, *8*(5), e09530.

Akarsu, T. N. (2023). Digital transformation towards a sustainable circular economy: Can it be the way forward? *Journal of Information Technology Teaching Cases*. https://doi.org/10.1177/20438869231178036

Akpan, I. J., Effiom, L., & Akpanobong, A. C. (2023). Towards developing a knowledge base for small business survival techniques during COVID-19 and sustainable growth strategies for the post-pandemic era. *Journal of Small Business & Entrepreneurship*, 1–23. https://doi.org/10.1080/08276331.2023.2232649

Al-Sartawi, M. (2021). *Big Data-Driven Digital Economy: Artificial and Computational Intelligence*. Springer.

Andrade, C. R. D. O., & Gonçalo, C. R. (2021). Digital transformation by enabling strategic capabilities in the context of "BRICS". *Revista de Gestão*, *28*(4), 297–315.

Aoki, K., Obeng, E., Borders, A. L., & Lester, D. H. (2019). Can brand experience increase customer contribution: How to create effective sustainable touchpoints with customers?. *Journal of Global Scholars of Marketing Science*, *29*(1), 51–62.

Aversa, P., Haefliger, S., Hueller, F., & Reza, D. G. (2021). Customer complementarity in the digital space: Exploring Amazon's business model diversification. *Long Range Planning*, *54*(5), 101985.

Barros, M. V., R. Salvador, G. F. do Prado, A. C. de Francisco and C. M. J. Piekarski (2021). Circular economy as a driver to sustainable businesses. *Cleaner Environmental Systems*, *2*, 100006.

Bechtsis, D., Tsolakis, N., Vlachos, D., & Srai, J. S. (2018). Intelligent Autonomous Vehicles in digital supply chains: A framework for integrating innovations towards sustainable value networks. *Journal of Cleaner Production*, *181*, 60–71.

Belhadi, A., Kamble, S., Benkhati, I., Gupta, S., & Mangla, S. K. (2023). Does strategic management of digital technologies influence electronic word-of-mouth (eWOM) and customer loyalty? Empirical insights from B2B platform economy. *Journal of Business Research*, *156*, 113548.

Bhutta, U. S., Tariq, A., Farrukh, M., Raza, A., & Iqbal, M. K. (2022). Green bonds for sustainable development: Review of literature on development and impact of green bonds. *Technological Forecasting and Social Change*, *175*, 121378.

Bivona, E. (2022). Determinants of performance drivers in online food delivery platforms: A dynamic performance management perspective. *International Journal of Productivity and Performance Management*, 72, 2497–2517.

Blackburn, O., Ritala, P., & Keränen, J. (2023). Digital platforms for the circular economy: Exploring meta-organizational orchestration mechanisms. *Organization & Environment*, *36*(2), 253–281.

Boukhatmi, Ä., Nyffenegger, R., & Grösser, S. N. (2023). Designing a digital platform to foster data-enhanced circular practices in the European solar industry. *Journal of Cleaner Production*, *418*, 137992.

Carlsson, R., Nevzorova, T., & Vikingsson, K. (2022). Long-Lived Sustainable Products through Digital Innovation. *Sustainability*, *14*(21), 14364.

Chan, H. L., Cheung, T. T., Choi, T. M., & Sheu, J. B. (2023). Sustainable successes in third-party food delivery operations in the digital platform era. *Annals of Operations Research*, 1–37. https://doi.org/10.1007/s10479-023-05266-w

Chen, Y., Li, J., & Zhang, J. (2022). Digitalisation, data-driven dynamic capabilities and responsible innovation: An empirical study of SMEs in China. *Asia Pacific Journal of Management*, 1–41. https://doi.org/10.1007/s10490-022-09845-6

Ciulli, F., Kolk, A., & Boe-Lillegraven, S. (2020). Circularity brokers: Digital platform organizations and waste recovery in food supply chains. *Journal of Business Ethics*, *167*, 299–331.

Cruz, E. F., & da Cruz, A. R. (2023). Digital solutions for engaging end-consumers in the circular economy of the textile and clothing value chain-A systematic review. *Cleaner and Responsible Consumption*, *11*(1), 100138.

De Bruyne, M. J., & Verleye, K. (2023). Realizing the economic and circular potential of sharing business models by engaging consumers. *Journal of Service Management*, *34*(3), 493–519.

Demestichas, K., & Daskalakis, E. (2020). Information and communication technology solutions for the circular economy. *Sustainability*, *12*(18), 7272.

Denu, M., David, P., Landry, A., & Mangione, F. (2023). *Emerging digital technologies to support circular manufacturing systems implementation: A literature review*. Premier congrès de la SAGIP.

Ding, S., Ward, H., Cucurachi, S., & Tukker, A. (2023). Revealing the hidden potentials of Internet of Things (IoT)-An integrated approach using agent-based modelling and system dynamics to assess sustainable supply chain performance. *Journal of Cleaner Production*, *421*, 138558.

Donthu, N., Kumar, S., Mukherjee, D., Pandey, N., & Lim, W. M. (2021). How to conduct a bibliometric analysis: An overview and guidelines. *Journal of Business Research*, *133*, 285–296.

Fellnhofer, K. (2022). Entrepreneurial alertness toward responsible research and innovation: Digital technology makes the psychological heart of entrepreneurship pound. *Technovation*, *118*, 102384.

Fu, F., Zha, W., & Zhou, Q. (2023). The impact of enterprise digital capability on employee sustainable performance: From the perspective of employee learning. *Sustainability*, *15*(17), 12897.

Füller, J., Hutter, K., & Kröger, N. (2021). Crowdsourcing as a service–from pilot projects to sustainable innovation routines. *International Journal of Project Management*, *39*(2), 183–195.

Gaur, A., & Kumar, M. (2018). A systematic approach to conducting review studies: An assessment of content analysis in 25 years of IB research. *Journal of World Business*, *53*(2), 280–289.

Gavrila, S. G., & de Lucas Ancillo, A. (2021). Spanish SMEs' digitalization enablers: E-Receipt applications to the offline retail market. *Technological Forecasting and Social Change*, *162*, 120381.

Gawer, A. J. I. (2022). Digital platforms and ecosystems: Remarks on the dominant organizational forms of the digital age. *Innovation*, *24*(1), 110–124.

Geissdoerfer, M., Savaget, P., Bocken, N. M., & Hultink, E. J. (2017). The Circular Economy–A new sustainability paradigm? *Journal of Cleaner Production*, *143*, 757–768.

Ghoreishi, M. (2023). The role of digital technologies in a data-driven circular business model: A systematic literature review. *Journal of Business Models*, *11*(1), 78–81.

Gitelman, L., & Kozhevnikov, M. (2023). New business models in the energy sector in the context of revolutionary transformations. *Sustainability*, *15*(4), 3604.

Gong, S. (2023). Digital transformation of supply chain management in retail and e-commerce. *International Journal of Retail & Distribution Management*. https://doi.org/10.1108/IJRDM-02-2023-0076

Goyal, S., Esposito, M., & Kapoor, A. (2018). Circular economy business models in developing economies: Lessons from India on reduce, recycle, and reuse paradigms. *Thunderbird International Business Review*, *60*(5), 729–740.

Grover, P., Kar, A. K., & Vigneswara Ilavarasan, P. (2018). Analyzing whether CEOs can act as influencers for sustainable development goals. In Arpan Kumar Kar, Shuchi Sinha & M. P. Gupta (Eds.), *Digital India: Reflections and Practice* (pp. 117–131). Cham: Springer.

Gruchmann, T. (2022). Theorizing the impact of network characteristics on multitier sustainable supply chain governance: A power perspective. *The International Journal of Logistics Management*, *33*(5), 170–192.

Hairudinor, & Rusidah, S. (2023). The role of digital marketing in the Sustainable performance of Indonesian MSMEs: Do the online purchase intention and actual purchase decision matter? *Transnational Marketing Journal*, *11*(1), 17–30.

Homrich, A. S., Galvão, G., Abadia, L. G., & Carvalho, M. M. (2018). The circular economy umbrella: Trends and gaps on integrating pathways. *Journal of Cleaner Production*, *175*, 525–543.

Hong, J., Guo, P., Deng, H., & Quan, Y. (2021). The adoption of supply chain service platforms for organizational performance: Evidences from Chinese catering organizations. *International Journal of Production Economics*, *237*, 108147.

Honic, M., Kovacic, I., Aschenbrenner, P., & Ragossnig, A. (2021). Material passports for the end-of-life stage of buildings: Challenges and potentials. *Journal of Cleaner Production*, *319*, 128702.

Hu, S., Zhu, Q., Zhao, X., & Xu, Z. (2023). Digital finance and corporate sustainability performance: Promoting or restricting? Evidence from China's listed companies. *Sustainability*, *15*(13), 9855.

Ivanov, D., Dolgui, A., & Sokolov, B. (2022). Cloud supply chain: Integrating industry 4.0 and digital platforms in the "Supply Chain-as-a-Service". *Transportation Research Part E: Logistics and Transportation Review*, *160*, 102676.

Jäger-Roschko, M., & Petersen, M. (2022). Advancing the circular economy through information sharing: A systematic literature review. *Journal of Cleaner Production*, *369*, 133210.

Jain, G., Kamble, S. S., Ndubisi, N. O., Shrivastava, A., Belhadi, A., & Venkatesh, M. (2022). Antecedents of Blockchain-enabled E-commerce Platforms (BEEP) adoption by customers–A study of second-hand small and medium apparel retailers. *Journal of Business Research*, *149*, 576–588.

Kasperovica, L., & Lace, N. (2021). Factors influencing companies' positive financial performance in digital age: A meta-analysis. *Entrepreneurship and Sustainability Issues*, *8*(4), 291.

Khanra, S., Kaur, P., Joseph, R. P., Malik, A., & Dhir, A. (2022). A resource-based view of green innovation as a strategic firm resource: Present status and future directions. *Business Strategy and the Environment*, *31*(4), 1395–1413.

Khattak, A. (2022). Hegemony of digital platforms, innovation culture, and e-commerce marketing capabilities: The innovation performance perspective. *Sustainability*, *14*(1), 463.

Khoruzhy, L. I., Petrovich Bulyga, R., Yuryevna Voronkova, O., Vasyutkina, L. V., Ryafikovna Saenko, N., Leonidovich Poltarykhin, A., & Aravindhan, S. (2022). A new trust management framework based on the experience of users in industrial cloud computing using multi-criteria decision making. *Kybernetes*, *51*(6), 1949–1966.

Kraus, S., P. Jones, N. Kailer, A. Weinmann, N. Chaparro-Banegas and N. J. Roig-Tierno (2021). Digital transformation: An overview of the current state of the art of research. *SAGE Open*, *11*(3), 21582440211.

Krom, P., Piscicelli, L., & Frenken, K. (2022). Digital Platforms for Industrial Symbiosis. *Journal of Innovation Economics & Management*, *3*, I124–XXVI.

Kumar, N. M., & Chopra, S. S. (2022). Leveraging blockchain and smart contract technologies to overcome circular economy implementation challenges. *Sustainability*, *14*(15), 9492.

Lange, K. P., Korevaar, G., Oskam, I. F., Nikolic, I., & Herder, P. M. (2021). Agent-based modelling and simulation for circular business model experimentation. *Resources, Conservation & Recycling Advances, 12*, 200055.

Lányi, B., Hornyák, M., & Kruzslicz, F. (2021). The effect of online activity on SMEs' competitiveness. *Competitiveness Review: An International Business Journal, 31*(3), 477–496.

Lehner, R., & Elbert, R. (2023). Cross-actor pallet exchange platform for collaboration in circular supply chains. *The International Journal of Logistics Management, 34*(3), 772–799.

Lekkas, D. F., Panagiotakis, I., & Dermatas, D. (2021). A digital circular bioeconomy–Opportunities and challenges for waste management in this new era. *Waste Management & Research, 39*(3), 407–408.

Letta, A. (2023). Covid-19 pandemic accelerated the adoption of digital solutions in developing nations. https://www.infine.lu/covid-19-pandemic-accelerated-the-adoption-of-digital-solutions-in-developing-nations/

Li, Y., Dai, J., & Cui, L. (2020). The impact of digital technologies on economic and environmental performance in the context of industry 4.0: A moderated mediation model. *International Journal of Production Economics, 229*, 107777.

Li, Y., Ding, R., Cui, L., Lei, Z., & Mou, J. (2019). The impact of sharing economy practices on sustainability performance in the Chinese construction industry. *Resources, Conservation and Recycling, 150*, 104409.

Li, Y., Hu, Y., Li, L., Zheng, J., Yin, Y., & Fu, S. (2023). Drivers and outcomes of circular economy implementation: Evidence from China. *Industrial Management & Data Systems, 123*(4), 1178–1197.

Li, Z., Li, H., & Wang, S. (2022). How multidimensional digital empowerment affects technology innovation performance: The moderating effect of adaptability to technology embedding. *Sustainability, 14*(23), 15916.

Liu, L., Fan, Q., Liu, R., Zhang, G., Wan, W., & Long, J. (2023). How to benefit from digital platform capabilities? Examining the role of knowledge bases and organisational routines updating. *European Journal of Innovation Management, 26*(5), 1394–1420.

Lyu, J., & Jiang, W. (2023). Internationalization pace, social network effect, and performance among China's platform-based companies. *Sustainability, 15*(10), 8252.

Mamedova, N. M., Bezveselnaya, Z. V., Ivleva, M. I., & Komarova, V. (2022). Environmental management for sustainable business development. *Entrepreneurship and Sustainability Issues, 9*(3), 134.

Mariani, M. M., Machado, I., Magrelli, V., & Dwivedi, Y. K. (2022). Artificial intelligence in innovation research: A systematic review, conceptual framework, and future research directions. *Technovation, 122*, 102623.

Marino, A., & Pariso, P. (2021). Digital government platforms: Issues and actions in Europe during pandemic time. *Entrepreneurship and Sustainability Issues, 9*(1), 462.

Morseletto, P. (2020). Targets for a circular economy. *Resources, Conservation and Recycling, 153*, 104553.

Ngai, E. W. T., Chau, D. C., Lo, C. W. H., & Lei, C. F. (2014). Design and development of a corporate sustainability index platform for corporate sustainability performance analysis. *Journal of Engineering and Technology Management, 34*, 63–77.

Nham, N. T. H. & Ha, L.T. (2022). Making the circular economy digital or the digital economy circular? Empirical evidence from the European region. *Technology in Society, 70*, 102023.

Nyantakyi, G., Atta Sarpong, F., Adu Sarfo, P., Uchenwoke Ogochukwu, N., & Coleman, W. (2023). A boost for performance or a sense of corporate social responsibility? A bibliometric analysis on sustainability reporting and firm performance research (2000–2022). *Cogent Business & Management, 10*(2), 2220513.

Pagoropoulos, A., Pigosso, D. C., & McAloone, T. C. (2017). The emergent role of digital technologies in the Circular Economy: A review. *Procedia CIRP*, *64*, 19–24.

Pandey, V., Sircar, A., Bist, N., Solanki, K., & Yadav, K. (2023). Accelerating the renewable energy sector through Industry 4.0: Optimization opportunities in the digital revolution. *International Journal of Innovation Studies*, *7*(2), 171–188.

Park, A., & Li, H. (2021). The effect of blockchain technology on supply chain sustainability performances. *Sustainability*, *13*(4), 1726.

Pencarelli, T. (2020). The digital revolution in the travel and tourism industry. *Information Technology & Tourism*, *22*(3), 455–476.

Pérez-Pérez, C., Benito-Osorio, D., Jimenez, A., & Bayraktar, S. (2023). The impact of country-level sustainability and digitalization on the performance: Sharing economy dashboard. *Journal of Organizational Change Management*, *36*(4), 621–634.

Piétron, D., Staab, P., & Hofmann, F. (2023). Digital circular ecosystems: A data governance approach. *GAIA*, *32*(S1), 40–46.

Pirola, F., X. Boucher, S. Wiesner and G. J. Pezzotta (2020). Digital technologies in product-service systems: A literature review and a research agenda. *Computers in Industry*, *123*, 103301.

Presch, G., Dal Mas, F., Piccolo, D., Sinik, M., & Cobianchi, L. (2020). The World Health Innovation Summit (WHIS) platform for sustainable development: From the digital economy to knowledge in the healthcare sector. In *Intellectual capital in the digital economy* (pp. 19–28). Routledge.

Punathumkandi, S., Sundaram, V. M., & Panneer, P. (2021). Interoperable permissioned-blockchain with sustainable performance, Sustainability (Switzerland). *Sustainability*, *13*(20), 11132.

Puspita, L. E., Christiananta, B., & Ellitan, L. (2020). The effect of strategic orientation, supply chain capability, innovation capability on competitive advantage and performance of furniture retails. *International Journal of Scientific & Technology Research*, *9*(3), 4521–4529.

Rai, A., Patnayakuni, R., & Seth, N. (2006). Firm performance impacts of digitally enabled supply chain integration capabilities. *MIS Quarterly*, *30*, 225–246.

Rajala, R., Hakanen, E., Mattila, J., Seppälä, T., & Westerlund, M. (2018). How do intelligent goods shape closed-loop systems? *California Management Review*, *60*(3), 20–44.

Ramakrishna, Y., Alzoubi, H., & Indiran, L. (2023). An empirical investigation of effect of sustainable and smart supply practices on improving the supply chain organizational performance in SMEs in India. *Uncertain Supply Chain Management*, *11*(3), 991–1000.

Russo, G., A. Manzari, B. Cuozzo, A. Lardo and F. J. Vicentini (2023). Learning and knowledge transfer by humans and digital platforms: Which tools best support the decision-making process? *Journal of Knowledge Management*, *27*(11), 310–329.

Savastano, M., Zentner, H., Spremić, M., & Cucari, N. (2022). Assessing the relationship between digital transformation and sustainable business excellence in a turbulent scenario. *Total Quality Management & Business Excellence*, *2022*, 1–22.

Schwanholz, J., & Leipold, S. (2020). Sharing for a circular economy? An analysis of digital sharing platforms' principles and business models. *Journal of Cleaner Production*, *269*, 122327.

Sharma, M., Luthra, S., Joshi, S., & Kumar, A. (2022). Developing a framework for enhancing survivability of sustainable supply chains during and post-COVID-19 pandemic. *International Journal of Logistics Research and Applications*, *25*(4–5), 433–453.

Sheikh, A. A., Rana, N. A., Inam, A., Shahzad, A., & Awan, H. M. (2018). Is e-marketing a source of sustainable business performance? Predicting the role of top management support with various interaction factors. *Cogent Business & Management*, *5*(1), 1516487.

Silva, T. H., & Sehnem, S. (2022). Industry 4.0 and the circular economy: Integration opportunities generated by startups. *Logistics*, *6*(1), 14.

Singh, P. K., Ismail, F. B., Wei, C. S., Imran, M., & Ahmed, S. A. (2020). A framework of e-procurement technology for sustainable procurement in iso 14001 certified firms in Malaysia. *Advances in Science, Technology and Engineering Systems Journal*, *5*(4), 424–431.

Suchek, N., Ferreira, J. J., & Fernandes, P. O. J. (2022). A review of entrepreneurship and circular economy research: State of the art and future directions. *Business Strategy and the Environment*, *31*(5), 2256–2283.

Sun, S., & Ertz, M. (2021). Dynamic evolution of ride-hailing platforms from a systemic perspective: Forecasting financial sustainability. *Transportation Research Part C: Emerging Technologies*, *125*, 103003.

Sutherland, W. and M. H. Jarrahi (2018). The sharing economy and digital platforms: A review and research agenda. *International Journal of Information Management*, *43*, 328–341.

Syed, F., Shah, S. H., Waseem, Z., & Tariq, A. (2021). Design thinking for social innovation: A systematic literature review & future research direction. *Proceedings of 1st International Conference on Business, Management & Social Sciences (ICBMASS).*

Tariq, A., Badir, Y., & Chonglerttham, S. (2019). Green innovation and performance: Moderation analyses from Thailand. *European Journal of Innovation Management*, *22*(3), 446–467.

Tariq, A., Badir, Y. F., Tariq, W., & Bhutta, U. S. (2017). Drivers and consequences of green product and process innovation: A systematic review, conceptual framework, and future outlook. *Technology in Society*, 51, 8–23.

Tatli, H. S., Yavuz, M. S., & Ongel, G. (2023). The Mediator Role of Task Performance in the Effect of Digital Literacy on Firm Performance. *Marketing and Management of Innovations*, *14*(2), 75–86.

Teichmann, F. M. J., Boticiu, S. R., & Sergi, B. S. (2023). Compliance concerns in sustainable finance: An analysis of peer-to-peer (P2P) lending platforms and sustainability. *Journal of Financial Crime*. https://doi.org/10.1108/JFC-11-2022-0281

Tian, J. (2023). The role of entrepreneurship, cooperative innovation, environmental investment in relationship between the Belt and Road Initiative and green innovation upgrading. *Management Decision*. https://doi.org/10.1108/MD-04-2023-0524

Tranfield, D., Denyer, D., & Smart, P. (2003). Towards a methodology for developing evidence-informed management knowledge by means of systematic review. *British Journal of Management*, *14*(3), 207–222.

Tseng, M. L., Bui, T. D., Lan, S., Lim, M. K., & Mashud, A. H. M. (2021). Smart product service system hierarchical model in banking industry under uncertainties. *International Journal of Production Economics*, *240*, 108244.

Tunn, V. S. C., Van den Hende, E. A., Bocken, N. M. P., & Schoormans, J. P. L. (2020). Digitalised product-service systems: Effects on consumers' attitudes and experiences. *Resources, Conservation and Recycling*, *162*, 105045.

Unterfrauner, E., Shao, J., Hofer, M., & Fabian, C. M. (2019). The environmental value and impact of the Maker movement—Insights from a cross-case analysis of European maker initiatives. *Business Strategy and the Environment*, *28*(8), 1518–1533.

Vehmas, K., Raudaskoski, A., Heikkilä, P., Harlin, A., & Mensonen, A. (2018). Consumer attitudes and communication in circular fashion. *Journal of Fashion Marketing and Management: An International Journal*, *22*(3), 286–300.

Wang, N., Wan, J., Ma, Z., Zhou, Y., & Chen, J. (2023). How digital platform capabilities improve sustainable innovation performance of firms: The mediating role of open innovation. *Journal of Business Research*, *167*, 114080.

Watanabe, C., Naveed, N., & Neittaanmäki, P. (2019). Digitalized bioeconomy: Planned obsolescence-driven circular economy enabled by Co-Evolutionary coupling. *Technology in Society*, *56*, 8–30.

Williams Jr, R. I., Clark, L. A., Clark, W. R., & Raffo, D. M. (2021). Re-examining systematic literature review in management research: Additional benefits and execution protocols. *European Management Journal*, *39*(4), 521–533.

Xie, W., & Wang, R. (2023). Application of data elements in the coupling of finance and technology on the digital electronic platform. *Electronic Commerce Research*, 1–26. https://doi.org/10.1007/s10660-023-09686-5

Yadav, H., Kar, A. K., & Kashiramka, S. (2022). How does entrepreneurial orientation and SDG orientation of CEOs evolve before and during a pandemic. *Journal of Enterprise Information Management*, *35*(1), 160–178.

Zhao, Q., Li, X., & Li, S. (2023). Analyzing the relationship between digital transformation strategy and ESG performance in large manufacturing enterprises: The mediating role of green innovation. *Sustainability*, *15*(13), 9998.

4 Exploring the Strategic Orientation Factors Influencing the Organizational Performance through Bibliometric Analysis

Mohammad Sultan Ahmad Ansari
Modern College of Business and Science

Shad Ahmad Khan
University of Buraimi, Oman

Ujjal Bhuyan
Jagannath Barooah University, Assam, India

4.1 INTRODUCTION

The strategic orientation (SO) of an organization refers to the overall approach and mindset that the organization adopts toward its goals and objectives. It can include the strategies and tactics used to achieve these goals, with the right attitudes/behaviors of the organization's leaders and employees. According to Ansari et al. (2018), emotional intelligence is a compassionate and cohesive force that supports in achieving organizational excellence. Research also has shown that a strong SO can have a significant impact on organizational performance (OP). Organizations can effectively align their strategies and goals with their resources and capabilities and can possibly succeed in achieving their goal. There are several models of SO on OP, but most focus on the key components that are essential for success (Kumar & Sharma, 2017; Evers et al., 2019), this includes the following:

a. Organizational culture: A strong organizational culture that values innovation, collaboration, and continuous improvement can help drive performance and achieve strategic goals (Kumar & Sharma, 2017; Evers et al., 2019).

 DOI: 10.1201/9781032616810-4

b. Strategic planning and execution: Effective strategic planning and execution involves setting clear goals and priorities, developing actionable plans, and monitoring progress toward achieving those goals (Kumar & Sharma, 2017; Evers et al., 2019).
c. Strategic vision and mission: Organizations with a clear and compelling vision and mission are better able to align their goals (Kumar & Sharma, 2017; Evers et al., 2019).
d. Customer focus: Organizations must understand, meet, and exceed the customers' expectations to possibly achieve their long-term objectives (Kumar & Sharma, 2017; Evers et al., 2019).
e. Resource allocation: The effective allocation of resources, including financial, human, and technological resources, is critical to achieving organizational goals (Kumar & Sharma, 2017; Evers et al., 2019).

In modern times, with the advent of digital technology, SO plays a crucial role in determining OP and success. The rapidly changing business landscape, technological advancements, globalization, and shifting customer preferences have made SO even more critical for businesses to thrive (Magd et al., 2023; Magd et al., 2021).

Contemporary organizations need to be customer-centric, meaning they must focus on understanding and fulfilling the customers' expectations (Al-Shammari, 2021). By gathering and analyzing customer feedback, behavior, and preferences, companies can tailor their offered services and products, and accordingly work out the marketing potential that can lead to improved customer satisfaction and loyalty (Saleem et al., 2023; Khan et al., 2020, 2021, 2023). Flexibility and malleability are vital in contemporary business settings (Evans & Bahrami, 2020). Organizations should be able to respond instantly to market dynamics, emerging trends, and aggressive threats. Agile companies can pivot their strategies and operations swiftly, gaining a competitive advantage over their slower-moving competitors (Day, 2011). In the interconnected world, forming strategic partnerships and collaborations with other organizations can open new opportunities and markets (Babu et al., 2020; He et al., 2020). Such partnerships can allow businesses to leverage each other's strengths and resources for mutual benefit (Ennis & Ennis, 2020).

In the digital era, data is abundant and can provide valuable insights into various aspects of business. Organizations that embrace data-driven decision-making processes can make informed choices, optimize operations, identify opportunities, and minimize risks (Khan, 2023; Kamal et al., 2022; Saleem et al., 2022). Embracing digital technologies and integrating them into various aspects of the business can significantly impact OP. Companies that invest in digital transformation often experience improved efficiency, better customer experiences, and streamlined operations (Naim & Khan, 2023; Khan & Magd, 2021; Magd & Khan, 2022). Harnessing social media to the advantage of the organization is another important consideration (Tirwa & Khan, 2022; Sharma, Khatri, & Khan, 2023a; Sharma et al., 2023a, 2023b).

Further, to remain competitive, organizations should adapt and adopt the art of innovation and creativeness (Sadegh Sharifirad, & Ataei, 2012; Hanifah et al., 2019).

Promoting employees and developing a culture of creativity with an aim of developing new processes for developing amazing products and services is essential for the modern organizations (Zareen & Khan, 2023; Bocar et al., 2022; Shafi et al., 2020).

One of the unignorable dimensions for the success of any organization in modern times is its orientation toward planetary sustainability and corporate social responsibility (CSR) (Wickert, 2021). With growing awareness of environmental and social issues, customers and stakeholders increasingly expect organizations to demonstrate a responsibility to sustainability and CSR. A SO can incorporate sustainable practices to support planet and improve the corporate standing in attracting environmentally friendly customers' (Khan & Rena, 2023; Khan & Gurung, 2019).

As markets become more uncertain, effective risk management is essential. A SO that considers potential risks and develops contingency plans can help organizations mitigate the impact of unexpected events and maintain continuity (Tirwa et al., 2022; Taylor, 2016). Further, to gauge the effectiveness of their strategies, organizations must establish performance metrics and monitor progress regularly (Epstein & Roy, 2003; Newman & Ford, 2021). This data can help them identify areas that require improvement and make necessary adjustments to their SO (Agyepong et al., 2020).

Overall, SO of an organization is a critical success factor (CSF) in determining its success. Organizations need to utilize their resources strategically considering the market situation in order to enhance their competitive position (Al-Ansaari et al., 2015). By focusing on key components such as vision, resource allocation, organizational culture, strategic planning and execution, and customer focus, organizations can achieve their goals and outperform their competitors. This study is also important as there is only one study, that is, Ramírez-Solis et al. (2022) that has explored the relationship between SO and OP. There was no other study published in Scopus database till this chapter was written to deal with these two concepts. Thus, this study utilizes bibliometric analysis to explore this concept in a better way.

4.2 RESEARCH METHODOLOGY

This study utilizes bibliometric analysis that is a quantitative research method used to analyze and measure patterns of publication, citation, and collaboration within a specific field of research or academic discipline (Khatri et al., 2022). It involves the systematic examination of bibliographic data from scholarly publications to gain insights into the structure, growth, and impact of academic literature. As the area of SO and OP is less explored, the co-occurrence analysis and cluster analysis will help the researcher to gain knowledge on the keywords that are predominantly used in this field (Khan et al., 2022a; Ali & Gölgeci, 2019). Thus, the bibliometric data was extracted from the Scopus database using the keywords, that is, "Strategy"; "Strategic Orientation"; "Organizational Performance" OR "Organizational Performance"; "Strategic Orientation" AND "Organizational Performance" OR "Organizational Performance". As organizational performance was found to be a less research area, the focus was made to understand strategy and SO through co-occurrence analysis and cluster analysis.

4.3 BIBLIOMETRIC ANALYSIS

4.3.1 Understanding Strategy

Strategy involves analyzing the current situation, setting clear targets, and outlining the steps needed to reach those targets effectively, and thus it is referred as a deliberate plan of action conceived to achieve specific goals or objectives (Bryson & George, 2020). Strategy guides decision-making, resource allocation, and prioritization to optimize outcomes in a dynamic and competitive environment (Ikram, Zhang, & Sroufe, 2020). Successful strategies adapt to changing circumstances and leverage strengths while mitigating weaknesses. Understanding strategy involves recognizing the importance of foresight, alignment, and execution, as well as continuous learning and improvement. In business, military, and various fields, strategy serves as a roadmap for sustainable success and gaining a competitive advantage. Thus, understanding "strategy", as a term needs exploration in terms of the keywords associated with it. For this purpose, a total of 2,346 documents were extracted from the Scopus database published in past 10 years; the total keywords approach was adopted by keeping the minimum occurrence as 61. A total of 30 keywords were extracted as presented in Table 4.1. For these extracted keywords, network visualization map was developed as presented in Figure 4.1. Also, to understand the grouping of the terms cluster analysis was performed (as presented in Table 4.2).

Cluster 1 consists of eight items and speaks about volume about the central theme of the strategy that is, decision-making, policy development, and governance approaches. It can be inferred that "conceptual framework" serves as the theoretical basis for understanding and analyzing these concepts. The other words like sustainability strategic approaches, stakeholders, and Europe speak about the contexts in which strategy has been used in the body of literature. Through Cluster 1, it is confirmed that strategy serves as a roadmap for sustainable success and gaining a competitive advantage for a business.

If we analyze Cluster 2, it mainly focuses on various aspects of business, entrepreneurship, and innovation, particularly in the context of the challenges set by the COVID-19 pandemic. "business" forms the core of the cluster, representing the overarching context. "Business development" indicates a drive toward growth and expansion that we can relate to as OP in our context. Inclusion of COVID-19 highlights the level of impact the pandemic caused to businesses. Further, this cluster also includes entrepreneurial-related terms, that is, "entrepreneur", "innovation", "open innovation", and. "small and medium-sized enterprises", the first three terms focus on creative solutions, collaboration, and individuals driving such innovation change. The term "small and medium-sized enterprises" indicates the focus (perhaps the most affected businesses during the pandemic). Strategy as a term also finds its place in this cluster. Together the keywords in this cluster depict a comprehensive perspective on business and entrepreneurial endeavors amid the backdrop of COVID-19, highlighting the importance of innovation and strategic thinking for small and medium-sized enterprises.

Cluster 3 specifically mentions the name of one country: China. Looking into the various keywords that have found their names in this cluster, that is, "development

TABLE 4.1
Co-occurrence Analysis for Strategy

S. No	Keyword	Occurrences	Total Link Strength
1	Innovation	2,341	3,285
2	Sustainability	359	1,065
3	Strategic approach	284	813
4	Sustainable development	251	764
5	China	226	627
6	Knowledge	123	340
7	Technological development	123	356
8	Corporate strategy	121	378
9	Manufacturing	115	328
10	Strategy	112	211
11	Research and development	107	285
12	Business	92	282
13	Competitiveness	90	269
14	Decision-making	90	253
15	Entrepreneur	89	268
16	Business development	86	273
17	Europe	85	236
18	Empirical analysis	83	238
19	Governance approach	82	263
20	Stakeholder	81	257
21	Economic development	78	238
22	Performance assessment	76	237
23	Conceptual framework	69	214
24	Open innovation	69	185
25	Small and medium-sized enterprises	69	217
26	Policy making	68	191
27	COVID-19	67	136
28	Development strategy	64	162
29	Learning	64	160
30	Economic growth	62	191

strategy", "economic development", "economic growth", and "sustainable development", it can be seen that how a country that has implemented a comprehensive development strategy to foster economic growth while ensuring environmental and social sustainability can be studied as a case for the other developing nations. This cluster gives a direction for the national policies and SO to gain economic growth, poverty reduction, environmental protection, and social equity. As China recognizes the importance of sustainable development, it strives to balance economic gains with environmental protection and social equity. By harmonizing these elements, China seeks to forge a path toward a more balanced, resilient, and inclusive development trajectory that aligns with global sustainability goals. This can be an inspiration for other developed and developing countries aspiring to improve on their national economic goals.

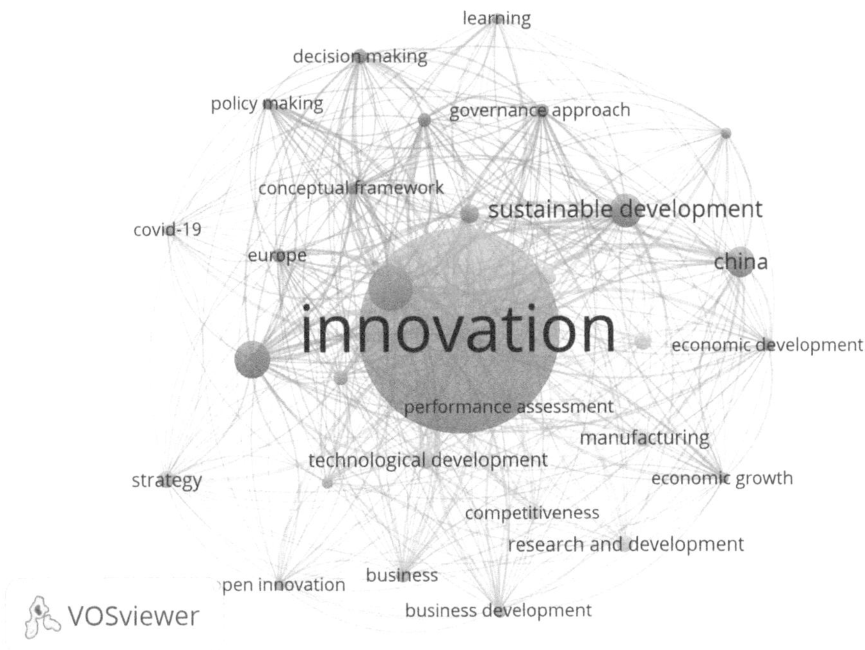

FIGURE 4.1 Network visualization map for Strategy.

(*Source*: extracted from Scopus database using VOSviewer).

TABLE 4.2
Cluster Analysis for Strategy

Cluster 1 (8 Items)	Cluster 2 (8 Items)	Cluster 3 (5 Items)	Cluster 4 (5 Items)	Cluster 5 (2 Items)	Cluster 6 (2 Items)
Conceptual framework	Business	China	Competi-tiveness	Knowledge	Research and development
Decision-making	Business development	Development strategy	Corporate strategy	Learning	Technological development
Europe	COVID-19	Economic development	Empirical analysis		
Governance approach	Entrepreneur	Economic growth	Manufacturing		
Policy making	Innovation	Sustainable development	Performance assessment		
Stakeholder	Open innovation				
Strategic approach	Small and medium-sized enterprises				
Sustainability	Strategy				

Cluster 4 appears to be more centric on individual organizations as it touches the crucial aspects of organizational success and performance evaluation. "Competitiveness" suggests the ability to excel in a competitive market, which necessitates a well-defined "corporate strategy" to achieve sustainable advantages. The inclusion of "empirical analysis" highlights the focus on data-driven research and evidence-based decision-making to comprehend and optimize OP. "Manufacturing" signifies a sector-specific dimension, indicating the relevance of industry-specific strategies for enhanced competitiveness. Further, "performance assessment" underscores the significance of measuring and evaluating organizational outcomes to ensure strategic objectives are met effectively. Together, these keywords point toward the importance of strategic planning, data-driven insights, and continuous evaluation to foster competitiveness and success in the manufacturing domain.

Cluster 5 has two keywords, that is, "knowledge" and "learning"; these two keywords represent a fundamental connection between acquiring, assimilating, and applying information. Emphasizing this cluster fosters continuous improvement, empowers decision-making, and facilitates personal and professional growth, making it a crucial aspect of individual and collective success.

Cluster 6 also has two keywords, that is, "research and development" and "technological development". These two keywords suggest a focus on innovation and progress in the technological realm. This combination reflects an emphasis on systematic inquiry and experimentation to drive advancements and improvements in various fields. These keywords imply a commitment to staying at the forefront of innovation, leveraging knowledge-driven strategies, and harnessing technology's transformative potential to achieve breakthroughs and drive sustainable growth.

4.4 STRATEGIC ORIENTATION

SO refers to an organization's overarching approach and mindset toward planning, decision-making, and goal-setting (Chevrollier & Kuijf, 2023). It involves aligning the organization's resources, capabilities, and actions with its long-term vision and mission (Barbosa, Castañeda-Ayarza, & Ferreira, 2020). A strategically oriented organization is forward-thinking, proactive, and adaptable, continuously scanning the external environment for opportunities and threats (Robertson et al., 2022; Bocar, Khan & Epoc, 2022). It emphasizes a deep understanding of its internal strengths and weaknesses, seeking to leverage competitive advantages while addressing limitations. SO fosters a culture of innovation, risk-taking, and learning, encouraging employees to collaborate and contribute to the organization's success (Khan et al., 2022b; Singh et al., 2020; Singh et al., 2019). By emphasizing long-term objectives and agility, SO enables the organization to navigate uncertainties, achieve sustainable growth, and maintain a competitive edge in a dynamic business landscape (Khan et al., 2019; Magd et al., 2022a). In this premise "strategic orientation" emerges as an important term that needs further investigation. For this purpose, also major keywords were extracted using the co-occurrence analysis in VOSviewer. For this purpose, 2454 documents were extracted from the Scopus database published between 2013 and 2023, minimum occurrence was kept at 35 to extract 30 keywords. The results of co-occurrence analysis are presented in Table 4.3; a network visualization map is also developed to represent the relationship between

TABLE 4.3
Co-occurrence Analysis for Strategic Orientation

Id	Keyword	Occurrences	Total Link Strength
1	Strategic orientations	718	864
2	Market orientation	186	328
3	Innovation	157	264
4	Entrepreneurial orientation	156	272
5	Performance	103	168
6	Strategic planning	93	180
7	Humans	85	171
8	Strategy	84	81
9	Sustainable development	75	174
10	Competition	71	172
11	Strategic approach	70	121
12	Firm performance	69	132
13	Sustainability	61	100
14	China	59	83
15	Strategic management	58	88
16	Commerce	57	170
17	SMEs	53	94
18	Decision-making	51	76
19	Knowledge management	50	88
20	Marketing	50	104
21	Article	49	90
22	Business strategy	48	76
23	Organizational performance	46	66
24	Industry	45	116
25	Business performance	43	87
26	Competitive advantage	42	94
27	Customer orientation	40	69
28	Technology orientation	39	102
29	Entrepreneurship	35	51
30	United States	33	49

these keywords (presented in Figure 4.2). Also, to understand the grouping of the terms cluster analysis was performed (as presented in Table 4.4).

A total of four clusters were extracted from the co-occurrence analysis for strategic performance. Cluster 1 contains keywords like "business performance" and "business strategy", which form the core components, emphasizing the importance of achieving objectives through well-defined plans. "Commerce" signifies the broader context of economic activities and transactions. "Competition" and "competitive advantage" highlight the significance of rivalry and differentiation for success. "Industry" refers to the specific sector in which businesses operate. "Knowledge management" underscores the

FIGURE 4.2 Network visualization map for Strategic Orientation.

(*Source*: extracted from Scopus database using VOSviewer).

TABLE 4.4
Cluster Analysis for Strategic Orientation

Cluster 1 (10 Items)	Cluster 2 (8 Items)	Cluster 1 (7 Items)	Cluster 1 (6 Items)
Business performance	China	Article	Entrepreneurial orientation
Business strategy	Customer orientation	Decision-making	Entrepreneurship
Commerce	Innovation	Humans	Firm performance
Competition	Organizational performance	Marketing	Market orientation
Competitive advantage	Strategic approach	Performance	SMEs
Industry	Sustainability	Strategy	Strategic orientations
Knowledge management	Sustainable development	United States	
Strategic management	Technology orientation		
Strategic orientations			
Strategic planning			

value of information and expertise in gaining a competitive edge. "Strategic management" and "strategic planning" encompass the systematic approach to decision-making and resource allocation. Lastly, "strategic orientations" emphasize the need for adaptive and forward-thinking perspectives. Together, these keywords showcase the multifaceted nature of strategic thinking, performance optimization, and sustainable growth in the business realm.

Cluster 2 revolves around various interconnected aspects of organizational success and development. China as a dominant keyword has emerged in this cluster in almost the same manner as it appeared in the cluster analysis of "strategy"; the keyword "China" has a strong relationship with keywords like "customer orientation", "innovation" and "technology orientation". "Customer orientation" emphasizes the significance of understanding and meeting customer needs. "Innovation" and "technology orientation" underscore the importance of staying ahead through technological advancements and creative problem-solving. "Organizational performance" suggests a focus on measuring and optimizing efficiency and effectiveness. "Strategic approach" signifies the need for a thoughtful, long-term plan to achieve objectives. "Sustainability" and "sustainable development" reflect the growing awareness of responsible business practices and environmental considerations. Together, these keywords form a cohesive picture of an organization that prioritizes customer-centricity, innovation, strategic planning, and sustainable growth in the dynamic landscape of a nation's (China in this case) market and beyond.

In Cluster 3 of the keywords, we can observe a connection between various elements linked to decision-making and performance in the context of marketing and strategy. "United States" has also emerged as a prominent keyword in this cluster, signifying that most of the studies within the context of the keywords in the cluster are performed on or within the United States. The keywords like "decision-making" and "humans" imply that the research may focus on human decision-making processes in marketing strategies. The term "marketing" signifies the domain of interest, where performance plays a crucial role. "Strategy" suggests the application of planned approaches to marketing efforts. "Article" suggests the presence of published research or academic studies, exploring the interplay of the other keywords. Although the inclusion of "United States" narrows the focus to the specific geographical context, but the way this cluster is formed indicates a potential research area concerning human decision-making and marketing strategy performance.

In Cluster 4, we observe a clear focus on the relationship between entrepreneurship, firm performance, and SOs. "Entrepreneurial orientation" and "entrepreneurship" suggest an emphasis on the mindset and activities associated with innovation, risk-taking, and proactiveness in creating and growing businesses. These entrepreneurial aspects are often critical for small and medium-sized enterprises (SMEs), denoted by the keyword "SMEs", as they strive to achieve superior "firm performance". Furthermore, the presence of "market orientation" indicates a concern for customer needs and market responsiveness, implying a customer-centric approach in formulating strategies. Overall, this cluster highlights the interplay between entrepreneurial behaviors, SOs, market focus, and firm performance, offering insights into how SMEs can thrive in competitive environments.

4.5 DISCUSSION

The concept of SO and its impact on OP has been the focus of significant research in the field of strategic management. Several studies have examined the relationship between SO and OP, as well as the factors that moderate this relationship (Kumar & Sharma, 2017; Evers et al., 2019; Danilwan & Dirhamsyah, 2022). One of the seminal studies on this topic was conducted by Narver and Slater (1990), who proposed the concept of market orientation as a key SO that can enhance OP. They defined market orientation as a culture that emphasizes customer orientation, competitor awareness, and inter-functional coordination. Their research found that companies with a strong market orientation outperformed those with weaker orientations, particularly in terms of sales growth and profitability.

Following Narver and Slater's work, several studies have investigated the impact of other SOs on OP. For example, Lumpkin and Dess (1996); Wales et al. (2020); Okoli et al. (2021), proposed the concept of EO, which emphasizes innovation, risk-taking, and proactiveness. Their research found that EO was positively related to new product development, sales growth, and profitability (Lumpkin and Dess, 1996; Wales et al., 2020; Okoli et al., 2021).

The above studies provide evidence for the positive impact of SO on OP; other research has explored the factors that moderate this relationship. For example, Li et al. (2008) and Nakos et al. (2019) found that the relationship between market orientation and performance was stronger for firms operating in highly competitive environments. Similarly, Soomro and Shah (2019) found that the relationship between EO and performance was stronger for firms with supportive organizational cultures. The term "entrepreneurship" itself has undergone various innovations and inclusions such as micro-entrepreneurship, self-help groups (SHGs), etc. (Gangwar & Khan, 2022).

Overall, the literature suggests that SO can significantly impact OP, but the nature of this relationship may vary depending on contextual factors such as industry environment and organizational culture. As such, organizations should carefully consider their approach to strategy development and implementation to optimize performance outcomes.

4.5.1 Strategic Orientation

There are several different types of SO that organizations can adopt. One common approach is market orientation, which involves a focus on understanding and meeting customer needs and preferences. Market-oriented companies tend to invest heavily in market research and use customer feedback to guide product development and marketing strategies. Another approach is EO, which emphasizes innovation, risk-taking, and proactivity. According to Magd et al. (2021a, 2021b), quality and knowledge management can lead to innovation. Entrepreneurial-oriented companies tend to be highly adaptable and responsive to changes in the business environment and are willing to take calculated risks to pursue new opportunities.

A third approach is learning orientation, which emphasizes knowledge acquisition, dissemination, and utilization. According to Magd et al. (2022b), self-efficacy and self-regulated approach are needed for a learning environment. Learning-oriented companies tend to have strong internal knowledge management systems and prioritize employee development and training to foster continuous improvement and innovation. The choice of SO depends on a range of factors, including the organization's goals and objectives, the nature of the business environment, and the organization's internal capabilities and resources. Ultimately, the goal of SO is to enable the organization to achieve sustainable competitive advantage and superior performance in the long run.

4.5.2 Technology Orientation

A technology-oriented organization tends to view technology as a key enabler of innovation, productivity, and growth (Ansari, 2022a, 2022b). Such organizations typically invest heavily in technology research and development, as well as in the adoption of cutting-edge technologies to support their business operations (Ansari et al., 2016a, 2016b). In a technology-oriented organization, technology is often seen as a strategic asset that can help the organization achieve its goals and objectives more effectively (Khan & Magd, 2023; Sonmez Cakir et al., 2023). This may involve the development of proprietary technologies that provide a competitive advantage or the adoption of existing technologies to improve operational efficiency and reduce costs.

However, it is important to note that technology orientation is not a one-size-fits-all approach. The adoption of technology must be aligned with the organization's overall goals and objectives, as well as with the needs of its customers and other stakeholders (El Khatib et al., 2022; Sonmez Cakir et al., 2023). Additionally, technology adoption and implementation require significant investment in terms of financial and human resources, as well as a strong culture of innovation and experimentation (Khatri et al., 2023; Zareen & Khan, 2023).

Overall, technology orientation can be a powerful driver of OP and competitiveness, particularly in industries where technology plays a critical role. However, organizations must carefully consider the costs and benefits of technology adoption and ensure that it is aligned with their overall strategic goals and objectives.

4.5.3 Market, Customer, and Competitor Orientation

Market, customer, and competitor orientation are all different SOs that organizations can adopt to achieve competitive advantage and superior performance. Each of these orientations focuses on different aspects of the external business environment and involves a unique set of strategies and practices (Crick, 2021; Gotteland et al., 2020).

Market orientation involves a focus on understanding and responding to the needs and preferences of the overall market. This approach involves extensive market research to identify customer needs and preferences, and the development of products and services that meet those needs (Kamal et al., 2022; Khan et al., 2021). Market-oriented companies also tend to monitor competitors closely, to identify emerging trends and opportunities. According to Ansari (2021), company's internal performance and support to develop relationships with customers and suppliers.

Customer orientation, on the other hand, involves a more specific focus on understanding and meeting the needs and preferences of individual customers (Kopalle et al.,

2020). This approach involves building strong relationships with customers and using customer feedback to inform product development and marketing strategies (Habel et al., 2020). Customer-oriented companies tend to prioritize customer satisfaction and loyalty and may invest heavily in customer service and support.

Competitor orientation involves a focus on understanding and responding to the actions and strategies of competitors in the market (Na et al., 2019). This approach involves analyzing the strengths and weaknesses of competitors, as well as their strategies for competing in the market (Ranjan & Foropon, 2021). Competitor-oriented companies may focus on developing products or services that offer a unique value proposition relative to competitors, or on finding ways to outcompete competitors on price, quality, or other factors. According to Magd, Ansari, and Negi (2021a) and Magd, Negi, and Ansari (2021b), quality practices have a strong linkage to OP.

Overall, each of these orientations can be effective in different contexts, depending on the nature of the business environment and the organization's goals and objectives. However, all these orientations share a common emphasis on understanding and responding to external market conditions, and on using that knowledge to drive strategic decision-making and performance improvement. According to Magd, Ansari, and Negi (2021a) and Magd, Negi, and Ansari (2021b), quality practices can effectively improve OP.

4.5.4 Entrepreneurial Orientation

Entrepreneurial orientation is a strategic approach that emphasizes innovation, risk-taking, and proactivity in pursuit of opportunities (Khan et al., 2019; Magd, Khan, and Bhuyan, 2022a). This approach involves a willingness to take calculated risks and a focus on identifying and exploiting new business opportunities (Tenhiälä & Laamanen, 2018). An entrepreneurial-oriented organization tends to be highly adaptable and responsive to changes in the business environment. Such organizations often have a flat organizational structure that enables rapid decision-making and flexibility and prioritize creativity and experimentation in pursuit of innovation. EO involves several key dimensions (Lumpkin & Dess, 2015), including innovativeness, competitive aggressiveness, proactiveness, autonomy, and risk-taking. EO can be an effective approach for organizations that operate in dynamic and uncertain environments, where new opportunities are constantly emerging. However, it requires a culture of innovation, experimentation, and risk-taking, and may be challenging to implement in more traditional, risk-averse organizations (Kasemsap, 2017). Overall, EO is an important SO that can help organizations to achieve competitive advantage and superior performance by identifying and exploiting new opportunities and staying ahead of the competition.

4.5.5 Innovative Climate and Learning Culture

An innovative climate is a work environment that fosters and supports creativity, experimentation, and innovation among employees (Malibari & Bajaba, 2022). Such a climate involves a culture that values and encourages latest ideas and provides the resources and support necessary for employees to explore and develop those ideas (Malibari & Bajaba, 2022; Chaubey & Sahoo, 2022). Employee motivation positively reflects on the innovation activities that the businesses aim to perform (Wang & Chang, 2017). In an innovative climate, employees are encouraged to take risks and think outside the box and are given the freedom and autonomy to explore new ideas and approaches. This can be highly motivating for employees, as it provides a sense of purpose and meaning, and can lead to a greater sense of engagement and satisfaction in their work. In addition to this, an innovative climate can lead to work motivation by providing employees with opportunities for growth and development. By encouraging employees to learn and develop new skills, an innovative climate can help to foster a sense of mastery and accomplishment, which can be highly motivating. Research has shown that there is a strong relationship between an innovative climate and work motivation. Overall, an innovative climate can be a powerful motivator for employees, providing a sense of purpose and meaning, as well as opportunities for growth and development. By fostering a culture of creativity, experimentation, and innovation, organizations can create a work environment that is both motivating and rewarding for employees.

On the other hand, learning culture is a culture that values and encourages continuous learning and development among employees for improved satisfaction. According to Ansari (2020), employee satisfaction provides the major impetus for businesses to remain profitable. This approach involves creating a supportive environment for learning, where employees are supported to take risks, experiment, and learn from their achievements and failures. According to Chen and Huang (2009), learning orientation as a culture that values knowledge acquisition, dissemination, and utilization. In a contingency approach, the effectiveness of a management strategy or approach depends on the unique characteristics of the organization and the environment in which it operates. Similarly, a learning culture involves adapting to the specific needs and challenges of the organization and encouraging employees to learn and develop in response to those challenges. According to Ansari, Farooquie, and Gattoufi (2016b) and Ansari, Farooqui, and Gattoufi (2018), a learning culture engages employees and enhances internal performance, fostering stronger relationships with customers.

In a learning culture, employees are encouraged to be proactive in their learning and development, and to take ownership of their own learning. This can help organizations to adapt to changing circumstances and to stay ahead of the competition by fostering innovation and creativity among employees. In short, motivated employees can lead to a significant increase in business performance (Garg, 2017).

Overall, the combination of contingency theory and a learning culture can be a powerful approach to organizational management, enabling organizations to adapt to changing circumstances and to foster a culture of innovation and continuous learning. The significance of proficiently and productively handling these valuable human resources is continuously on the rise (Steinbauer et al., 2018). By recognizing the unique characteristics of the organization and the environment in which it operates, and by creating a supportive environment for learning and development, organizations can achieve greater effectiveness and long-term success.

4.5.6 Organizational Performance

OP, also known as business performance, refers to the ability of an organization to achieve its goals and objectives (Alrowwad et al., 2020). Business performance can be measured in a variety of ways, including financial performance, operational efficiency, customer satisfaction, employee engagement and satisfaction, and market share (Chen et al., 2021).

Financial performance is a common measure of OP and includes metrics such as revenue, profitability, return on investment (ROI), and shareholder value. Operational efficiency, on the other hand, refers to how well an organization manages its resources to achieve its objectives and includes metrics, such as productivity, quality, and efficiency. According to Ansari (2020), quality can be maintained with the right environment with committed leadership and supportive infrastructure.

Customer satisfaction is another important measure of OP, as it reflects the ability of an organization to meet the needs and expectations of its customers. This includes metrics such as customer retention, customer loyalty, and customer satisfaction ratings. Employee engagement and satisfaction are also important indicators of OP, as they reflect the ability of an organization to create a positive work environment and to motivate and retain its employees. According to Nadhar et al. (2017), work motivation has a positive effect on OP. This includes metrics such as employee turnover rates, employee satisfaction ratings, and employee engagement surveys.

Finally, market share is an important measure of OP, as it reflects the organization's ability to compete effectively in the marketplace and to capture a share of the available market. Overall, OP is a multidimensional concept that encompasses a variety of measures of organizational success. By measuring and tracking these metrics, organizations can assess their performance, identify areas for improvement, and take action to achieve their goals and objectives.

4.6 CONCLUSIONS

Based on the discussion and analysis of this study, the conceptual framework for SO is developed and presented in Figure 4.3. The model proposes that there are four distinct SOs: market orientation, EO, technological orientation, and learning orientation. Market orientation focuses on understanding and meeting customer needs, while EO emphasizes innovation and risk-taking. Technology orientation covers the aspects of digital technology, process management, and sustainability. Learning orientation prioritizes knowledge acquisition and development.

4.7 FUTURE DIRECTIONS

An empirical study on this model needs to be performed to check the relationship between the SO factors and the OP. To measure OP, various key performance indicators (KPIs) can be utilized such as financial metrics (e.g., revenue, profitability), operational efficiency (e.g., productivity, customer satisfaction), and employee engagement (e.g., turnover rates, job satisfaction). Additionally, performance measurement may

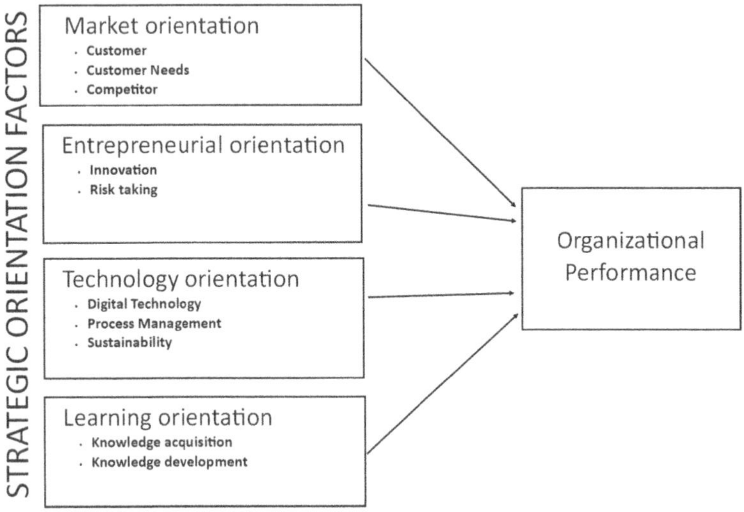

FIGURE 4.3 Proposed framework.

(*Source*: authors' proposal).

encompass qualitative factors, like innovation, corporate reputation, and sustainability practices. By analyzing and interpreting these data points, organizations can gain valuable insights into their overall effectiveness and identify areas for improvement, facilitating informed decision-making and fostering continuous growth and success.

REFERENCES

Agyepong, E., Cherdantseva, Y., Reinecke, P., & Burnap, P. (2020). Challenges and performance metrics for security operations center analysts: A systematic review. *Journal of Cyber Security Technology*, *4*(3), 125–152.

Al-Ansaari, Y., Bederr, H., & Chen, C. (2015). Strategic orientation and business performance: An empirical study in the UAE context. *Management Decision*, *53*(10), 2287–2302.

Ali, I., & Gölgeci, I. (2019). Where is supply chain resilience research heading? A systematic and co-occurrence analysis. *International Journal of Physical Distribution & Logistics Management*, *49*(8), 793–815.

Alrowwad, A. A., Abualoush, S. H., & Masa'deh, R. E. (2020). Innovation and intellectual capital as intermediary variables among transformational leadership, transactional leadership, and organizational performance. *Journal of Management Development*, *39*(2), 196–222.

Al-Shammari, M. M. (2021). A strategic framework for designing knowledge-based customer-centric organizations. *International Journal of eBusiness and eGovernment Studies*, *13*(2), 1–16.

Ansari, M. S. A. (2020). Extended service profit chain in telecom service industry in Oman–An empirical validation. *Sustainable Futures*, *2*, 100032.

Ansari, M. S. A. (2021). An innovative approach of integrating service quality, employee loyalty and profitability with service profit chain in telecom service industry: An empirical validation. *Proceedings on Engineering*, *3*(1), 1–12.

Ansari, M. S. A. (2022a). TQM framework for healthcare sectors: Barriers to implementation. *Quality Innovation Prosperity*, *26*(1), 1–23. https://doi.org/10.12776/qip.v26i1.1611

Ansari, M. S. A., Farooqui, J. A., & Gattoufi, S. M. (2018). Emotional intelligence and extended service profit chain in telecom industry in Oman–An empirical validation. *International Business Research*, *11*(3), 133–148.

Ansari, M. S. A., Farooquie, J. A., & Gattoufi, S. M. (2016a). Assessing the Impact of Service Quality on Customers and Operators: Empirical Study. *International Journal of Business and Management*, *11*(9), 207–217.

Ansari, M. S. A., Farooquie, J. A., & Gattoufi, S. M. (2016b). Does emotional intelligence influence employees, customers and operational efficiency? An empirical validation. *International Journal of Marketing Studies*, *8*(6), 77–88.

Ansari, M.S.A. (2022b), Lean six sigma in healthcare: Some Sobering thoughts on implementation. *Proceedings on Engineering Sciences*, *4* (4), 457–468. https://doi.org/10.24874/PES04.04.007

Babu, M. M., Dey, B. L., Rahman, M., Roy, S. K., Alwi, S. F. S., & Kamal, M. M. (2020). Value co-creation through social innovation: A study of sustainable strategic alliance in telecommunication and financial services sectors in Bangladesh. *Industrial Marketing Management*, *89*, 13–27.

Barbosa, M., Castañeda-Ayarza, J. A., & Ferreira, D. H. L. (2020). Sustainable strategic management (GES): Sustainability in small business. *Journal of Cleaner Production*, *258*, 120880.

Bocar, A.C., Khan, S.A., Epoc, F. (2022). COVID-19 Work from Home Stressors and the Degree of its Impact: Employers and Employees Actions. *International Journal of Technology Transfer and Commercialisation*, *19* (2), 270–291. https://doi.org/10.1504/ijttc.2022.124349

Bryson, J., & George, B. (2020). Strategic management in public administration. In *Oxford Research Encyclopedia of Politics*. Oxford University Press.

Chaubey, A., & Sahoo, C. K. (2022). The drivers of employee creativity and organizational innovation: A dynamic capability view. *Benchmarking: An International Journal*, *29*(8), 2417–2449.

Chen, C. J., & Huang, J. W. (2009). Strategic human resource practices and innovation performance—The mediating role of knowledge management capacity. *Journal of Business Research*, *62*(1), 104–114.

Chen, X., You, X., & Chang, V. (2021). FinTech and commercial banks' performance in China: A leap forward or survival of the fittest?. *Technological Forecasting and Social Change*, *166*, 120645.

Chevrollier, N., & Kuijf, F. (2023). Sensing and seizing in the apparel industry: the role of dynamic capabilities in fostering sustainable strategic orientations. *International Journal of Organizational Analysis*, *31*(3), 605–623.

Crick, J. M. (2021). The dimensionality of the market orientation construct. *Journal of Strategic Marketing*, *29*(4), 281–300.

Danilwan, Y., & Dirhamsyah, I. P. (2022). The impact of the human resource practices on the organizational performance: does ethical climate matter?. *Journal of Positive School Psychology*, *6*(3), 1–16.

Day, G. S. (2011). Closing the marketing capabilities gap. *Journal of Marketing*, *75*(4), 183–195.

El Khatib, M., Al Mulla, A., & Al Ketbi, W. (2022). The role of Blockchain in E-Governance and decision-making in project and program management. *Advances in Internet of Things*, *12*(3), 88–109.

Ennis, S., & Ennis, S. (2020). Globalisation of the sports product. *Sports Marketing: A Global Approach to Theory and Practice*, 213–243, Springer Nature.

Epstein, M. J., & Roy, M. J. (2003). Improving sustainability performance: specifying, implementing and measuring key principles. *Journal of General Management*, *29*(1), 15–31.

Evans, S., & Bahrami, H. (2020). Super-flexibility in practice: Insights from a crisis. *Global Journal of Flexible Systems Management*, *21*, 207–214.

Evers, N., Gliga, G., & Rialp-Criado, A. (2019). Strategic orientation pathways in international new ventures and born global firms—Towards a research agenda. *Journal of International Entrepreneurship*, *17*, 287–304.

Gangwar, V. P. & Khan, S. A. (2022). Analyzing the role of micro-entrepreneurship and self-help groups (SHGs) in women empowerment and development: A bottom-of-pyramid perspective. In M. Arafat, I. Saleem, J. Ali, A. Khan, & H. Balhareth (Eds.), *Driving Factors for Venture Creation and Success in Agricultural Entrepreneurship* (pp. 213–226). IGI Global. https://doi.org/10.4018/978-1-6684-2349-3.ch011

Garg, N. (2017). Workplace spirituality and organizational performance in Indian context: Mediating effect of organizational commitment, work motivation and employee engagement. *South Asian Journal of Human Resources Management*, *4*(2), 191–211.

Gotteland, D., Shock, J., & Sarin, S. (2020). Strategic orientations, marketing proactivity and firm market performance. *Industrial Marketing Management*, *91*, 610–620.

Habel, J., Kassemeier, R., Alavi, S., Haaf, P., Schmitz, C., & Wieseke, J. (2020). When do customers perceive customer centricity? The role of a firm's and salespeople's customer orientation. *Journal of Personal Selling & Sales Management*, *40*(1), 25–42.

Hanifah, H., Abdul Halim, H., Ahmad, N. H., & Vafaei-Zadeh, A. (2019). Emanating the key factors of innovation performance: leveraging on the innovation culture among SMEs in Malaysia. *Journal of Asia Business Studies*, *13*(4), 559–587.

He, Q., Meadows, M., Angwin, D., Gomes, E., & Child, J. (2020). Strategic alliance research in the era of digital transformation: Perspectives on future research. *British Journal of Management*, *31*(3), 589–617.

Ikram, M., Zhang, Q., & Sroufe, R. (2020). Developing integrated management systems using an AHP-Fuzzy VIKOR approach. *Business Strategy and the Environment*, *29*(6), 2265–2283.

Kamal, S., Naim, A., Magd, H., Khan, S. A., & Khan, F. M. (2022). The relationship between E-service quality, ease of use, and E-CRM performance referred by brand image. In A. Naim & S. Kautish (Eds.), *Building a Brand Image Through Electronic Customer Relationship Management* (pp. 84–108). IGI Global. https://doi.org/10.4018/978-1-6684-5386-5.ch005

Kasemsap, K. (2017). Strategic innovation management: An integrative framework and causal model of knowledge management, strategic orientation, organizational innovation, and organizational performance. In *Organizational culture and behavior: Concepts, methodologies, tools, and applications* (pp. 86–101). IGI Global.

Khan, S. A. & Rena, R. (2023). Emerging green practices, internet of things, and digital marketing: A response to the global economic and climate crises. In A. Naim & V. Devi (Eds.), *Global applications of the internet of things in digital marketing* (pp. 1–16). IGI Global. https://doi.org/10.4018/978-1-6684-8166-0.ch001

Khan, S. A. (2023). E-marketing, E-commerce, E-business, and Internet of Things: An overview of terms in the context of small and medium enterprises (SMEs). In A. Naim & V. Devi (Eds.), *Global applications of the internet of things in digital marketing* (pp. 332–348). IGI Global. https://doi.org/10.4018/978-1-6684-8166-0.ch017

Khan, S. A., & Magd, H. A. E. (2023). New technology anxiety and acceptance of technology. *Advances in Distance Learning in Times of Pandemic* (pp. 105–133). CRC Press.

Khan, S. A., Magd, H., Al Shamsi, I. R., & Masoom, K. (2022a). Social entrepreneurship through innovations in agriculture. In H. Magd, D. Singh, R. Syed, & D. Spicer (Eds.), *International perspectives on value creation and sustainability through social entrepreneurship* (pp. 209–222). IGI Global. https://doi.org/10.4018/978-1-6684-4666-9.ch010

Khan, S. A., Magd, H., Khatri, B., Arora, S., & Sharma, N. (2023). Critical success factors of internet of things and digital marketing. In A. Naim & V. Devi (Eds.), *Global applications of the internet of things in digital marketing* (pp. 233–253). IGI Global. https://doi.org/10.4018/978-1-6684-8166-0.ch012

Khan, S. A., Magd. H., & Epoc, F. (2022b). Application of data management system in business to business electronic commerce. In Naim, A., & Malik, P.K. (Ed.), *Competitive trends and technologies in business management.* (pp. 109–124). Nova Science Publishers, USA.

Khan, S. A., Sharma, P.P., & Thoudam, P. (2019). Role of attitude and entrepreneurship education towards entrepreneurial orientation among business students of Bhutan. *International Journal of Recent Technology and Engineering (IJRTE).* 10(10), 335–342 September 2019. https://doi.org/10.35940/ijrte.C1072.1083S19

Khan, S.A., Epoc, F., Gangwar, V.P., Ligori T.A.A., Ansari, Z.A. (2021). Will Online banking sustain in Bhutan post Covid – 19? A quantitative analysis of the customer e-satisfaction and e-loyalty in the Kingdom of Bhutan, *Transnational Marketing Journal*, *9*(3), 607–624 https://doi.org/10.33182/tmj.v9i3.1288

Khan, S.A., & Gurung, M., (2019). Green public procurement through lens of practicality and policies: A study on Royal University of Bhutan. *Delhi Business Review*, *20*(1), 23–32.

Khan, S.A., Devi, T.P., Ligori, T.A.A., & Saleem, M. (2020). Customer satisfaction and customer loyalty in online shopping: A study on university students of Bhutan. *Delhi Business Review*. *21*(2). https://doi.org/10.51768/dbr.v21i2.212202002

Khan. S.A. & Magd. H. (2021). Empirical examination of MS teams in conducting webinar: Evidence from international online program conducted in Oman. *Journal of Content, Community and Communication*, *14*, 159–175. https://doi.org/10.31620/JCCC.12.21/13

Khatri, B., Arora, S., Magd, H., & Khan, S. A. K. (2022). Bibliometric analysis of social entrepreneurship. In H. Magd, D. Singh, R. Syed, & D. Spicer (Eds.), *International perspectives on value creation and sustainability through social entrepreneurship* (pp. 46–60). IGI Global. https://doi.org/10.4018/978-1-6684-4666-9.ch003

Khatri, B., Shrimali, H., Khan, S. A., & Naim, A. (2023). Role of HR analytics in ensuring psychological wellbeing and job security: Learnings from COVID-19. In *HR analytics in an era of rapid automation* (pp. 36–53). IGI Global.

Kopalle, P. K., Kumar, V., & Subramaniam, M. (2020). How legacy firms can embrace the digital ecosystem via digital customer orientation. *Journal of the Academy of Marketing Science*, *48*, 114–131.

Kumar, V., & Sharma, R. R. K. (2017). An empirical investigation of critical success factors influencing the successful TQM implementation for firms with different strategic orientation. *International Journal of Quality & Reliability Management*, *34*(9), 1530–1550.

Li, Y., Zhao, Y., Tan, J., & Liu, Y. (2008). Moderating effects of entrepreneurial orientation on market orientation-performance linkage: Evidence from Chinese small firms. *Journal of Small Business Management*, *46*(1), 113–133.

Lumpkin, G. T., & Dess, G. G. (1996). Clarifying the entrepreneurial orientation construct and linking it to performance. *Academy of Management Review*, *21*(1), 135–172.

Lumpkin, G. T., & Dess, G. G. (2015). Entrepreneurial orientation. *Wiley Encyclopedia of Management* (pp. 1–4). John Wiley & Sons.

Magd, H. & Khan, S. A. (2022a). Strategic framework for entrepreneurship education in promoting social entrepreneurship in GCC countries during and post COVID-19. In H. Magd, D. Singh, R. Syed, & D. Spicer (Eds.), *International perspectives on value creation and sustainability through social entrepreneurship* (pp. 61–75). IGI Global. https://doi.org/10.4018/978-1-6684-4666-9.ch004

Magd, H., Ansari, M., & Negi, S. (2021a). The relationship between TQM, knowledge management, and innovation: A framework to achieve organizational excellence in service industry. *Global Business & Management Research*, *13*(3), 283–296.

Magd, H., Khan, S. A. K., & Bhuyan, U. (2022a). Social Entrepreneurship Intentions Among Business Students in Oman. In H. Magd, D. Singh, R. Syed, & D. Spicer (Eds.), *International Perspectives on Value Creation and Sustainability Through Social Entrepreneurship* (pp. 76–93). IGI Global. https://doi.org/10.4018/978-1-6684-4666-9.ch005

Magd, H., Khan, S. A., Khatri, B., Sharma, N., & Arora, S. (2023). Understanding the relationship between IoT and digital marketing: A bibliometric analysis. In A. Naim & V. Devi (Eds.), *Global applications of the internet of things in digital marketing* (pp. 123–140). IGI Global. https://doi.org/10.4018/978-1-6684-8166-0.ch007

Magd, H., Negi, S., & Ansari, M. S. A. (2021b). Effective TQM implementation in the service industry: a proposed framework. *Quality Innovation Prosperity*, *25*(2), 95–129.

Magd, H., Nzomkunda, A., Negi, S., & Ansari, M. (2022b). Critical success factors of E-learning implementation in higher education institutions: A proposed framework for success. *Global Business & Management Research*, *14* (2), 20–38.

Magd, H., & Khan, S.A. (2022). Effectiveness of using online teaching platforms as communication tools in higher education institutions in Oman: Stakeholders perspectives. *Journal of Content, Community and Communication*, *16*, 148–160. https://doi.org/10.31620/JCCC.12.22/13

Malibari, M. A., & Bajaba, S. (2022). Entrepreneurial leadership and employees' innovative behavior: A sequential mediation analysis of innovation climate and employees' intellectual agility. *Journal of Innovation & Knowledge*, *7*(4), 100255.

Na, Y. K., Kang, S., & Jeong, H. Y. (2019). The effect of market orientation on performance of sharing economy business: Focusing on marketing innovation and sustainable competitive advantage. *Sustainability*, *11*(3), 729.

Nadhar, M., Tawe, A. and Parawansa, D.A. (2017). The effect of work motivation and entrepreneurship orientation on business performance through entrepreneurial commitments of coffee shops in Makassar. *International Review of Management and Marketing*, *7*(1), 470–474.

Naim, A. & Khan, S. A. (2023). Impact and assessment of electronic commerce on consumer buying behaviour. In A. Naim & V. Devi (Eds.), *Global applications of the internet of things in digital marketing* (pp. 264–289). IGI Global. https://doi.org/10.4018/978-1-6684-8166-0.ch014

Nakos, G., Dimitratos, P., & Elbanna, S. (2019). The mediating role of alliances in the international market orientation-performance relationship of SMEs. *International Business Review*, *28*(3), 603–612.

Narver, J. C., & Slater, S. F. (1990). The effect of a market orientation on business profitability. *Journal of Marketing*, *54*(4), 20–35.

Newman, S. A., & Ford, R. C. (2021). Five steps to leading your team in the virtual COVID-19 workplace. *Organizational Dynamics*, *50*(1), 100802.

Okoli, I. E. N., Nwosu, K. C., & Okechukwu, M. E. (2021). Entrepreneurial orientation and performance of selected SMEs in Southeast, Nigeria. *European Journal of Business and Management Research*, *6*(4), 108–115.

Ramírez-Solis, E. R., Llonch-Andreu, J., & Malpica-Romero, A. D. (2022). Relational capital and strategic orientations as antecedents of innovation: Evidence from Mexican SMEs. *Journal of Innovation and Entrepreneurship*, *11*(1), 42.

Ranjan, J., & Foropon, C. (2021). Big data analytics in building the competitive intelligence of organizations. *International Journal of Information Management*, *56*, 102231.

Robertson, J., Botha, E., Walker, B., Wordsworth, R., & Balzarova, M. (2022). Fortune favours the digitally mature: The impact of digital maturity on the organisational resilience of SME retailers during COVID-19. *International Journal of Retail & Distribution Management*, *50*(8/9), 1182–1204.

Sadegh Sharifirad, M., & Ataei, V. (2012). Organizational culture and innovation culture: Exploring the relationships between constructs. *Leadership & Organization Development Journal*, *33*(5), 494–517.

Saleem, M., Khan, S. A., & Magd, H. (2022). Content marketing framework for building brand image: A case study of Sohar International School, Oman. In A. Naim & S. Kautish (Eds.), *Building a brand image through electronic customer relationship management* (pp. 64–83). IGI Global. https://doi.org/10.4018/978-1-6684-5386-5.ch004

Saleem, M., Khan, S. A., Al Shamsi, I. R., & Magd, H. (2023). Digital marketing through social media best practices: A case study of HEIs in the GCC region. In A. Naim & V. Devi (Eds.), *Global Applications of the Internet of Things in Digital Marketing* (pp. 17–30). IGI Global. https://doi.org/10.4018/978-1-6684-8166-0.ch002

Shafi, M., Lei, Z., Song, X., & Sarker, M. N. I. (2020). The effects of transformational leadership on employee creativity: Moderating role of intrinsic motivation. *Asia Pacific Management Review*, *25*(3), 166–176.

Sharma, N., Khatri, B., & Khan, S. A. (2023a). Do e-WOM persuade travelers destination visit intentions? An investigation on how travelers adopt the information from the social media channels. *Journal of Content, Community and Communication*, *17*, 147–161.

Sharma, N., Khatri, B., Khan, S. A., & Shamsi, M. S. (2023b). Extending the UTAUT model to examine the influence of social media on Tourists' Destination Selection. *Indian Journal of Marketing*, *53*(4), 47–64. https://doi.org/10.17010/ijom/2023/v53/i4/172689

Singh, E.A., Khan, S.A., Thoudam, P., & Sharma, P.P. (2019). Factor affecting the choice of Cheese in Bhutan: A choice architecture perspective. *International Journal of Engineering and Advanced Technology (IJEAT)*, *8*(5), 1880–1888. (ISSN: 2249-8958)

Singh, E.H., Wangda, S., Khan, S., Khan, S.A. (2020). Exploring the obstacles for start-ups in Bhutan: From a prevented entrepreneur perspective. *International Journal of Innovation, Creativity and Change*. *11*(4), 70–87.

Sonmez Cakir, F., Kalaycioglu, O., & Adiguzel, Z. (2023). Examination the effects of organizational innovation and knowledge management strategy in information technology companies in R&D departments on service quality and product innovation. *Information Technology & People*. https://doi.org/10.1108/ITP-03-2022-0196

Soomro, B. A., & Shah, N. (2019). Determining the impact of entrepreneurial orientation and organizational culture on job satisfaction, organizational commitment, and employee's performance. *South Asian Journal of Business Studies*, *8*(3), 266–282.

Steinbauer, R., Renn, R. W., Chen, H. S., & Rhew, N. (2018). Workplace ostracism, self-regulation, and job performance: Moderating role of intrinsic work motivation. *The Journal of Social Psychology*, *158*(6), 767–783.

Taylor, M. (2016). Risky ventures: Financial inclusion, risk management and the uncertain rise of index-based insurance. In *Risking capitalism* (Vol. 31, pp. 237–266). Emerald Group Publishing Limited.

Tenhiälä, A., & Laamanen, T. (2018). Right on the money? The contingent effects of strategic orientation and pay system design on firm performance. *Strategic Management Journal*, *39*(13), 3408–3433.

Tirwa, I.P., & Khan, S.A. (2022). Perception of business owners about the factors influencing the use of social media as a tool of marketing by the business community of Bhutan. *Anusandhan-NDIM's Journal of Business and Management Research*, 4(2), 1–11. https://doi.org/10.56411/anusandhan.2022.v4i2.1-11

Tirwa, I.P., Bhuyan, U., Magd, H., & Khan, S.A. (2022). Implications of liquidity risk and credit risk on the Bank of Bhutan's Financial Performance (BOB). *Global Business and Management Research: An International Journal*, *14* (2), 105–117.

Wales, W. J., Covin, J. G., & Monsen, E. (2020). Entrepreneurial orientation: The necessity of a multilevel conceptualization. *Strategic Entrepreneurship Journal*, *14*(4), 639–660.

Wang, H., & Chang, Y. (2017). The influence of organizational creative climate and work motivation on employee's creative behavior. *Journal of Management Science*, *30*(3), 51–62.

Wickert, C. (2021). Corporate social responsibility research in the Journal of Management Studies: A shift from a business-centric to a society-centric focus. *Journal of Management Studies*, *58*(8), E1–E17.

Zareen, S. & Khan, S. A. (2023). Exploring dependence of Human Resource Management (HRM) on Internet of Things (IoT) and digital marketing in the Digital Era. In A. Naim & V. Devi (Eds.), *Global applications of the internet of things in digital marketing* (pp. 51–66). IGI Global. https://doi.org/10.4018/978-1-6684-8166-0.ch004

5 Critical Factors for Relationship between Strategic Orientation, Corporate Success, and Innovation

Shad Ahmad Khan
University of Buraimi, Oman

Mohammad Sultan Ahmad Ansari
Modern College of Business and Science

5.1 INTRODUCTION

The correlation between strategic management, implementation, evaluation, and feedback is essential for corporate success (CS) irrespective of company size. The contribution of large corporations, small-to-medium-sized enterprises (SMEs), or even the cottage industries cannot be ignored. Particularly, SMEs are substantial for a country's economic development as they generate a variety of jobs. It competently utilizes the resources in developing countries' gross domestic product (GDP), which contributes to building national income and overcoming poverty. According to Nassani and Aldakhil (2023), the association between strategic orientation (SO) and strategic alignment is stronger when SMEs are strategically flexible. Therefore, SMEs are considered key drivers of national economic advancement (Singh et al., 2020). However, it is essential for SMEs to accelerate their development, prove strategic direction, and create inventions for achieving goals (Hsieh & Wu, 2019).

Strategy reflects procedures, processes, attitudes, applications, and decision-making techniques that govern the organization's operations. It induces the desired behavior for improving organizational performance, growth, and sustainability. According to Ansari et al. (2016), findings suggest that technological support would improve service delivery system and service system, which can be a critical factor for success. The corporation cannot achieve success until the growth and pace of modern technology are considered, which will lead to innovation and differentiation in terms of product and service, which is a critical success factor (CSF) for any firm (Khan et al., 2023). In the emergence of successful innovation, the existence of a supportive organizational environment is very important (Lee & Trimi, 2021; Adiguzel & Sonmez Cakir, 2022).

DOI: 10.1201/9781032616810-5

Strategic innovation, corporate development, and customer satisfaction form the strategic posture for organizational success. It supports organizational structure with presence of administrative systems that can adapt to unexpected external environments steering, which remains beyond the corporate control. It becomes challenging in underdeveloped and developing countries due to a scarcity of competent resources. According to prevailing literature, customers, competitors, organizational learning, entrepreneurship, and technological orientation are all important factors of CS (Xue et al., 2023; Ferreira et al., 2021). In which dependable competencies must be established to drive organizational excellence. There are monetary or non-monetary variables that can support in strategy development, strategy implementation, monitoring, and continuous feedback for achieving SO (He et al., 2020; Kabeyi, 2019; Khatri et al., 2023). SO reflects an outward-looking view of the business's strategic choices and environmental compatibility (He et al., 2020).

SO refers to the deliberate and purposeful direction that organizations adopt to align their resources and capabilities with market demands and competitive forces (Chaganti & Sambharya, 1987; Zhou et al., 2005). A well-defined SO provides a roadmap for decision-making, resource allocation, and goal setting, enabling companies to navigate complexities and uncertainties in their industries (Enholm et al., 2022; Kamal et al., 2022). However, the actual realization of CS hinges on the effective implementation of innovative practices within the strategic framework (Magd & Khan, 2022). Innovation acts as a powerful catalyst, converting strategic intent into tangible outcomes by fostering creative thinking, generating new ideas, and translating them into products, services, or processes that address market needs and differentiate the organization from its competitors (Begum et al., 2022; Khan & Rena, 2023; Khan et al., 2022b). Embracing INN as an integral part of SO empowers companies to seize emerging opportunities, respond to dynamic market changes, and create sustainable competitive advantages, ultimately propelling them toward greater achievements and long-term success (Adams et al., 2019; Keiningham et al., 2020; Khan, 2023). Critical factors are essential for researchers and practitioners as they help them to understand the domain in a better way. One of the ways of identifying the critical factors is by identifying the prominent keywords in the domain (Magd et al., 2023b; Khan et al., 2023). Further, if one of the domains in the research items is less researched, it becomes even more important to identify the critical factors that may be associated with such less-researched domain (Khan et al., 2023; Jamil et al., 2023; Moeuf et al., 2020). As CS is a less researched area, only one document has been published on this domain by Jamil et al. (2023). It was important to understand the relationship between CS, innovation, and SO through a valid tool.

5.2 METHODOLOGY

The methodology for this article employs a quantitative approach utilizing bibliometric analysis to investigate the relationship between SO and CS, with a focus on the intervening role of innovation. For this purpose, the keywords "Innovation"; "Strategic Orientation" AND "Innovation"; "Corporate Success" AND "Innovation"; and "Corporate Success" were used as a prompt in the Scopus database (Magd et al., 2023b; Khatri et al., 2022; Khan et al., 2022a) for which

12,628; 547; 63; and 01 documents, respectively, were extracted that were published between 2013 and 2023. VOSviewer, a standard tool for analysis of bibliometric data, was utilized, and mainly the co-occurrence analysis was performed as the intention was to understand the three keywords, that is, innovation, CS, and SO.

5.3 BIBLIOMETRIC ANALYSIS

5.3.1 Exploring Innovation

Innovation, the process of creating and applying new ideas, products, or services, lies at the heart of human progress (Bledow et al., 2009; Khan et al., 2023). Throughout history, it has been the driving force behind advancements in technology, science, and society, transforming the way we live, work, and interact (Magd & Khan, 2022; Khan & Magd, 2023; Van Veldhoven & Vanthienen, 2022; Alqahtani & Rajkhan 2020). From the wheel to the internet, from the printing press to space exploration, every major leap in human civilization owes its existence to the spirit of innovation (Van Veldhoven & Vanthienen, 2022; Kamal et al., 2022; Magd et al., 2023a).

Innovation is a broad concept encompassing a variety of forms (Silvestre & Ţîrcă, 2019). It includes technological INN, where novel advancements in products, processes, or services result in improved efficiency or effectiveness (Chiarini et al., 2020). Process innovation involves optimizing existing systems and workflows to enhance productivity and reduce costs (Lim et al., 2020). Additionally, organizational innovation focuses on creating a dynamic and adaptable work culture that fosters creativity and collaboration (Azeem et al., 2021). Social innovation seeks to address societal challenges, such as poverty, education, and healthcare, by introducing novel solutions and approaches (Gupta et al., 2020). It is an important consideration to understand what exactly is happening in the field of innovation, thus, bibliometric analysis is employed to identify the trends, keywords, and keyword clusters in the field of innovation. In order to learn about innovation, bibliometric data extracted from the Scopus database was utilized. A total of 12,628 documents were found in the domain of innovation. Upon running the co-occurrence analysis on this data, 25 keywords were identified with a minimum of 318 occurrences. These occurrences with their respective link strength are presented in Table 5.1. A bibliometric graph is also presented for these keywords in Figure 5.1.

As visible from Table 5.1, innovation as a keyword has the highest occurrence for obvious reasons. However, the other 24 keywords for innovation reveal several key themes and trends in the field. First, sustainability and sustainable development emerge as critical factors, highlighting the growing emphasis on environmentally responsible and socially conscious innovation practices. China appears prominently, indicating its significant role in driving innovation globally. Second, technological development and research and development (R&D) underline the importance of technological advancements and continuous research efforts in fostering innovation. Third, manufacturing signifies the relevance of innovation in the production sector, while governance approaches suggest the importance of regulatory frameworks in promoting innovative practices. The presence of "entrepreneur" and "entrepreneurship" implies the crucial role of startups and small businesses in driving innovative

TABLE 5.1
Co-occurrence Analysis for Innovation

S. No	Keyword	Occurrences	Total Link Strength
1	Innovation	12,594	12,691
2	Sustainability	1,715	4,024
3	China	1,212	2,716
4	Sustainable development	1,109	2,852
5	Technological development	716	1,693
6	Knowledge	612	1,412
7	Research and development	509	1,120
8	Manufacturing	469	1,172
9	Governance approach	436	1,070
10	Entrepreneur	431	1,088
11	Empirical analysis	424	1,073
12	Strategic approach	413	1,111
13	Stakeholder	411	1,027
14	Entrepreneurship	395	677
15	Europe	381	851
16	Business	368	925
17	Technology adoption	365	766
18	Decision making	359	830
19	Conceptual framework	357	885
20	Economic growth	344	888
21	Learning	342	743
22	Performance assessment	341	878
23	Economic development	335	882
24	United Kingdom	329	702
25	Research work	318	856

solutions. Empirical analysis and strategic approach indicate a growing focus on evidence-based decision-making and a structured approach to innovation. Stakeholder involvement reflects the recognition of diverse perspectives and collaborations in the innovation process. Europe, the United Kingdom, and China's prominence suggests regional clusters of innovation activities. "Technology adoption" points to the adoption and integration of new technologies as a catalyst for innovation. Lastly, "performance assessment" and "economic growth" highlight the need to measure the impact of innovation on economic development and organizational success, while "knowledge," "decision making," and "learning" underscore the cognitive aspects of innovation, emphasizing the importance of knowledge acquisition, informed decision-making, and continuous learning to drive successful innovation processes. This analysis offers valuable insights into the multidimensional nature of innovation and its intersection with sustainability, technology, and economic development. In order to understand the ways these keywords behave in group, cluster analysis is performed for this co-occurrence analysis presented in Table 5.2.

As presented in Table 5.2, Cluster 1 of keywords encompasses several critical aspects of organizational dynamics and innovation within the context of Europe, specifically focusing on the United Kingdom (UK). The inclusion of "decision making"

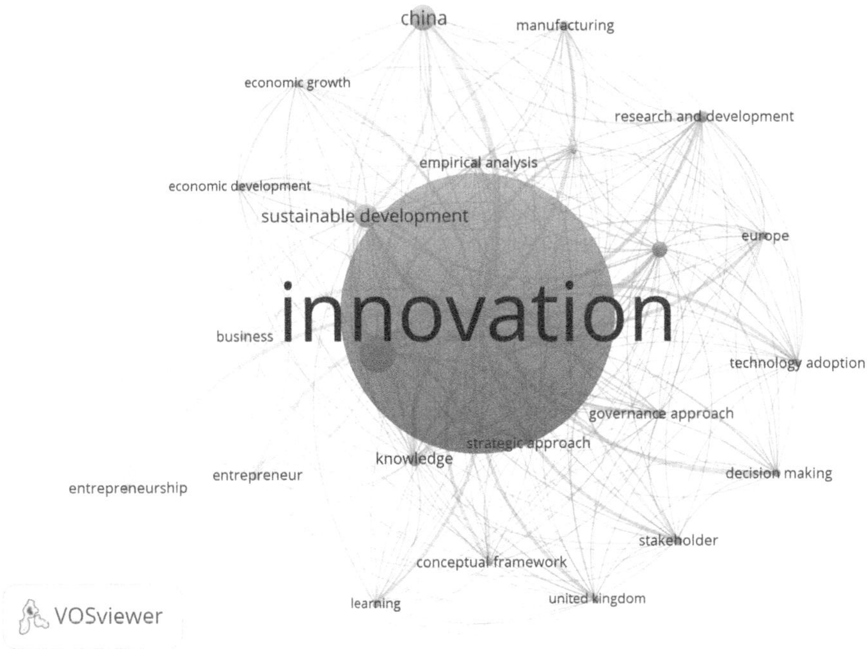

FIGURE 5.1 Bibliometric graph for co-occurrence of keywords for Innovation.

(***Source***: **extracted from Scopus database using VOSviewer).**

TABLE 5.2
Cluster Analysis on Co-occurrence Analysis of Innovation

Cluster 1 (9 Items)	Cluster 2 (4 Items)	Cluster 3 (4 Items)	Cluster 4 (3 Items)	Cluster 5 (3 Items)	Cluster 5 (3 Items)
Decision making	China	Conceptual framework	Business	Empirical analysis	Strategic approach
Europe	Economic development	Knowledge	Entrepreneur	Manufacturing	Sustainability
Governance approach	Economic growth	Learning	Entrepreneurship	Performance assessment	
Innovation	Sustainable development	Research work			
Research and development					
Stakeholder					
technological development					
Technology adoption					
United Kingdom					

and "governance approach" indicates the significance of strategic and managerial choices in the region's businesses and institutions. The presence of "innovation," "research and development," and "technological development" underscores the growing importance of technological advancements and their potential impact on economic growth and competitiveness. Furthermore, "technology adoption" highlights the relevance of understanding how new technologies are embraced and integrated into existing systems. The term "stakeholder" suggests a focus on inclusive and collaborative approaches to decision-making, acknowledging the importance of various stakeholders' perspectives and interests. Overall, this keyword cluster signals the intricate relationship between decision-making processes, innovation, technology, and stakeholder engagement, shaping the business landscape and future prospects in the European region, particularly in the UK. This cluster can be utilized as a case study of the UK and European region for any nation or organization that wishes to understand organizational dynamics and innovation.

Cluster 2 also brings to the surface the keyword that is the name of a country, that is, "China." This cluster signifies the intricate and multifaceted relationship between China's economic progress and its pursuit of sustainability. Over the past few decades, China has experienced unprecedented economic growth, transforming into a major global economic powerhouse (Zhang et al., 2021). This rapid development, however, has also raised concerns about its environmental and social sustainability. China's pursuit of sustainable development aims to balance economic growth with environmental conservation, social equity, and resource efficiency. Initiatives such as the Belt and Road Initiative and the shift toward renewable energy sources demonstrate China's commitment to sustainable development on a global scale (Zhang et al., 2021; Wu et al., 2023). Nevertheless, challenges remain, including air and water pollution, income inequality, and the conservation of natural resources. As China continues to address these challenges, it becomes crucial for policymakers and stakeholders to strike a harmonious balance between economic development and sustainable practices to ensure a prosperous and environmentally responsible future for the nation and the world at large (Zhang et al., 2021).

Cluster 3 of keywords "conceptual framework, knowledge, learning, research work" denotes a cohesive and interrelated set of concepts within the realm of academic inquiry and intellectual advancement. A conceptual framework provides the structural basis for organizing knowledge and understanding complex phenomena, guiding research work and exploration. Knowledge represents the accumulation of information, insights, and expertise gained through learning processes, both formal and informal. Learning, as a fundamental process, facilitates the acquisition and assimilation of knowledge, fostering personal and organizational growth. Research work embodies the systematic investigation and inquiry into specific topics, informed by conceptual frameworks, to expand knowledge and contribute to the collective understanding of various disciplines. This cluster highlights the essential elements and dynamics underpinning the advancement of human understanding and highlights the importance of robust conceptual frameworks and continuous learning in conducting meaningful research and driving intellectual progress.

Cluster 4 of keywords "business, entrepreneur, entrepreneurship" indicates a close thematic relationship centered around entrepreneurship within the business context. As combined, these keywords suggest a focus on the study of individuals or entities involved in entrepreneurial endeavors, exploring their innovative approaches, strategies, challenges, and contributions to the overall economic and societal landscape. The analysis of this keyword cluster is essential for gaining insights into the world of entrepreneurship, fostering entrepreneurial ecosystems, and understanding the factors that drive entrepreneurial success and its impact on economic growth and development.

Cluster 5 of keywords points toward a research context focused on examining and evaluating the performance of manufacturing firms through an empirical approach. The research is likely to involve gathering real-world data and conducting statistical analysis to understand the factors influencing manufacturing performance. The empirical study also signifies the utilization of quantitative techniques to assess the performance. This is crucial to enhance the productivity and efficiency of the organizations.

Cluster 6 of keywords "strategic approach" and "sustainability" signifies a significant and timely topic in the business and organizational landscape. A strategic approach to sustainability involves the integration of environmental, social, and economic considerations into the core decision-making processes of a company. Sustainability as a term has a bigger orientation toward environmental sensitivity, although it includes terms like organizational and business sustainability as well (in terms of economics). This cluster suggests a growing recognition among businesses of the need to align their strategies with sustainability principles to address global challenges such as climate change, resource depletion, and social inequality. It also highlights the increasing importance of sustainability as a competitive advantage, as consumers, investors, and other stakeholders demand ethical and responsible practices.

5.4 CORPORATE SUCCESS AND INNOVATION

The relationship between CS and innovation is deeply connected, as innovation plays a critical role in driving sustainable growth and competitive advantage for organizations (Zhou et al., 2020). In today's dynamic business landscape, companies that prioritize and foster a culture of innovation are better positioned to adapt to changing market demands, exploit new opportunities, and stay ahead of their competitors (Magd et al., 2023b). Innovative products, services, and processes not only attract customers but also enable cost efficiency and operational improvements (Ansari & Khan, 2023; Kohli & Melville, 2019; Singh et al., 2020). Additionally, innovation enhances employee engagement and creativity, leading to higher productivity and a more resilient workforce (Zareen & Khan, 2023; Aldabbas et al., 2023). As organizations embrace innovation, they can effectively address challenges, disrupt traditional industries, and explore new markets, ultimately contributing to enhanced financial performance and long-term viability (Zalan & Toufaily, 2017). Thus, CS and innovation share a symbiotic relationship, where innovation serves as a driving force behind achieving sustainable growth and prosperity for businesses in an ever-evolving world (Moore, 2015; Fan et al., 2023; Khan et al., 2021).

As this relationship is crucial, the line of discussion on these two terms combined, that is, "Corporate Success" AND "Innovation," was checked in the Scopus database.

A total of 63 documents were extracted from the period 2013 to 2023. A minimum occurrence of 3 was kept extracting 13 keywords as presented in Table 5.3, and bibliometric graph for this relation is presented in Figure 5.2. Cluster analysis was also performed on these 13 keywords, a total of 3 clusters were extracted as presented in Table 5.4.

As visible from Table 5.3 and Figure 5.2, "innovation" stands as the central theme, emphasizing the paramount importance of novel ideas, products, and processes in today's highly competitive markets. The concept of "competitive advantage" highlights the pursuit of unique strengths that differentiate businesses from their rivals, while "creativity" emerges as a critical driver for generating innovative solutions. Additionally, the presence of "competition" and "competitiveness" underscores the significance of strategic positioning and continuous improvement to stay ahead in the market. "Corporate social responsibility" signifies a growing trend of integrating ethical and sustainable practices into business operations, complementing the focus on "sustainable development." "Customer satisfaction" reflects the customer-centric approach that fosters innovation by understanding and addressing consumer needs. Meanwhile, "leadership" and "management" play pivotal roles in fostering innovation culture and facilitating the implementation of innovative ideas. Moreover, "organizational culture" highlights how a supportive and adaptive culture nurtures and sustains innovation efforts within a company. "Product design" represents the creative process of developing new offerings that align with market demands, while "project management" ensures efficient execution of innovation initiatives. In conclusion, these keywords collectively depict a multifaceted perspective on innovation and its multifarious implications for contemporary businesses striving to succeed in dynamic and ever-evolving markets.

Cluster 1 highlights the significance of the market landscape and the constant drive to outperform rivals. Companies strive for a "competitive advantage" by leveraging unique strengths and capabilities that differentiate them from competitors,

TABLE 5.3
Co-occurrence Analysis for Corporate Success and Innovation

S. No	Keyword	Occurrences	Total Link Strength
1	Innovation	22	12
2	Competitive advantage	4	8
3	Creativity	4	11
4	Competition	3	7
5	Competitiveness	3	4
6	Corporate social responsibility	3	3
7	Customer satisfaction	3	3
8	Leadership	3	3
9	Management	3	8
10	Organizational culture	3	9
11	Product design	3	5
12	Project management	3	3
13	Sustainable development	3	0

FIGURE 5.2 Bibliometric graph for co-occurrence of keywords for corporate success and Innovation.

(*Source*: extracted from Scopus database using VOSviewer).

TABLE 5.4
Cluster Analysis on Co-occurrence Analysis of Corporate Success and Innovation

Cluster 1 (5 Items)	Cluster 2 (3 Items)	Cluster 3 (3 Items)
Competition	Customer satisfaction	Creativity
Competitive advantage	Leadership	Innovation
Competitiveness	Project management	Organizational culture
Corporate social responsibility		
Product design		

thereby ensuring long-term success. In parallel, "corporate social responsibility" emphasizes the increasing importance of ethical and sustainable practices in the business world, acknowledging the role of responsible actions in enhancing a company's reputation and competitiveness. Additionally, "product design" signifies the creative process of crafting offerings that meet consumer needs, staying ahead of competitors, and maintaining relevance in the market. Together, this cluster accentuates the intricate interplay between strategic positioning, ethical considerations, and innovation in shaping the success and sustainability of businesses in a competitive landscape. Organizations must strike a balance between these elements to achieve a meaningful and impactful presence in the market while being mindful of their societal and environmental responsibilities.

Cluster 2 signifies a critical intersection of factors crucial for organizational success. "Customer satisfaction" is a pivotal metric that measures the level of contentment and loyalty among customers, directly impacting business performance and competitiveness. Effective "leadership" plays a central role in driving innovation, fostering a culture of continuous improvement and aligning organizational goals with customer needs. Leaders who prioritize customer satisfaction and champion a customer-centric approach can significantly enhance innovation efforts. Moreover, "project management" acts as the bridge that translates innovative ideas into tangible outcomes, ensuring efficient planning, execution, and control of innovation initiatives. The successful implementation of innovation projects directly influences customer satisfaction, as it leads to the development of new products, services, and solutions that meet or exceed customer expectations. Thus, this cluster of keywords emphasizes the crucial link between customer satisfaction, effective leadership, and proficient project management as key drivers for achieving sustained innovation and organizational excellence.

Cluster 3 illuminates the interplay between individual and collective factors that drive organizational success and competitive advantage. "Creativity" serves as the foundational element, representing the generation of novel and original ideas, a crucial precursor to innovation. "Innovation," on the other hand, emphasizes the practical application and implementation of creative ideas to create value, achieve growth, and stay relevant in the market. The presence of "organizational culture" underscores the significance of fostering an environment that nurtures and encourages creativity and innovation.

5.5 STRATEGIC ORIENTATION AND INNOVATION

SO and innovation represent two interconnected pillars that drive the growth and success of organizations in a rapidly evolving business landscape (Grawe et al., 2009; Kasemsap, 2017). SO refers to the deliberate and forward-thinking approach that companies adopt to align their resources, capabilities, and objectives with the ever-changing market dynamics and customer demands (Chaganti & Sambharya, 1987; Zhou et al., 2005). It involves a thorough understanding of market trends, competitive forces, and internal strengths, enabling organizations to make informed decisions and anticipate future challenges and opportunities (Enholm et al., 2022; Kamal et al., 2022). On the other hand, innovation entails the continuous pursuit of creative and novel solutions that address emerging needs, enhance products, services, or processes, and create a sustainable competitive advantage (Begum et al., 2022; Khan & Rena, 2023). By fostering a culture of innovation and encouraging experimentation, organizations can adapt to shifting market demands, differentiate themselves from competitors, and maintain relevance over time (Palacios-Marqués et al., 2021). The synergy between SO and innovation enables companies to proactively explore uncharted territories, capitalize on disruptive technologies, and stay at the forefront of their industries, ultimately leading to long-term growth and prosperity (Adams et al., 2019; Keiningham et al., 2020; Khan, 2023).

As presented in Table 5.5, the keywords revolve around SO and innovation, encompassing various facets of business and organizational development. SO

TABLE 5.5
Co-occurrence Analysis for Strategic Orientation and Innovation

S. No	Keyword	Occurrences	Total Link Strength
1	Strategic Orientation	208	312
2	Innovation	157	226
3	Market Orientation	53	111
4	Entrepreneurial Orientation	44	82
5	Competition	29	70
6	Performance	24	38
7	Commerce	23	61
8	Innovation Performance	23	52
9	Sustainable Development	23	57
10	Industry	21	59
11	Strategic Planning	20	48
12	China	19	34
13	Firm Performance	19	44
14	Product Development	19	47
15	SMEs	18	32
16	Competitive Advantage	17	37
17	Customer Orientation	17	31
18	Knowledge Management	17	27
19	Sustainability	17	24

emphasizes the long-term planning and direction of a company, taking into account factors like market orientation, competition, and strategic planning. Innovation is a key driver for sustainable development and growth, involving product development and knowledge management (KM) to achieve a competitive advantage. The keywords also highlight the significance of performance and firm performance, both in terms of financial performance and innovation performance. Furthermore, the inclusion of SMEs indicates a focus on fostering entrepreneurship and customer orientation, with an eye toward sustainability and industry advancements. Additionally, China's presence in the keywords points to the country's role as a major player in the global business landscape, further emphasizing the importance of SO and innovation for success in diverse markets. Overall, these keywords showcase the interplay between strategic planning, innovation, and business performance to drive sustainable growth and success in a competitive environment (Figure 5.3).

As presented in Table 5.6, cluster 1 of keywords underscores the interconnectedness of commerce, SO, innovation, and customer-centricity in achieving success in the business world. By effectively aligning strategic planning, fostering innovation, and prioritizing customer needs, organizations can position themselves for growth, profitability, and long-term sustainability in their respective industries. Commerce serves as the foundation for business transactions and market interactions, highlighting the importance of a well-functioning economic ecosystem (Iheanachor et al., 2023; Cao & Shi, 2021). SO and strategic planning indicate the significance of long-term vision and deliberate decision-making in achieving organizational goals and staying ahead in the competitive landscape. Industry signifies the specific sector or

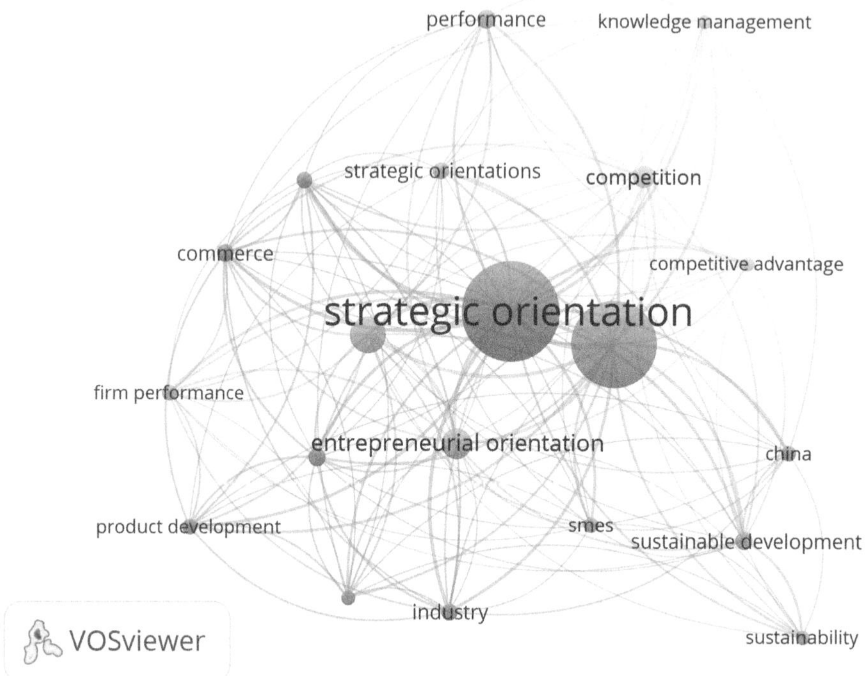

FIGURE 5.3 Bibliometric graph for co-occurrence of keywords for strategic orientation and innovation.

(*Source*: extracted from Scopus database using VOSviewer).

TABLE 5.6
Cluster Analysis on Co-occurrence Analysis of Strategic Orientation and Innovation

Cluster 1	Cluster 2	Cluster 3	Cluster 4
Commerce	Entrepreneurial orientation	China	Competition
Customer orientation	Firm performance	Innovation	Competitive advantage
Industry	Market orientation	SMEs	Knowledge management
Innovation performance	Performance	Sustainability	
Product development		Sustainable development	
Strategic orientation			
Strategic planning			

market in which the business operates, emphasizing the need to align strategies with industry-specific dynamics and trends (Arian et al., 2023). Innovation performance and product development demonstrate a commitment to continuous improvement and the creation of novel solutions to meet customer needs and demands. These elements are crucial for maintaining a competitive advantage and driving sustainable growth. Additionally, the inclusion of customer orientation reinforces the importance of understanding and meeting customer preferences and expectations, which ultimately leads to increased customer satisfaction and loyalty.

Cluster 2 of keywords suggests a strong emphasis on the relationship between entrepreneurship, firm performance, and market orientation. Entrepreneurial orientation refers to a company's willingness and ability to take risks, innovate, and pursue opportunities to achieve a competitive advantage. This entrepreneurial mindset is likely to influence firm performance, encompassing various aspects such as financial success, growth, and overall efficiency. Market orientation, on the other hand, highlights a company's focus on understanding and meeting customer needs, thereby aligning its strategies with market demands and trends. This customer-centric approach can significantly impact firm performance by enhancing customer satisfaction and loyalty. The presence of "performance" as a keyword further reinforces the significance of measuring and improving business outcomes, which could be influenced by entrepreneurial and market orientations. Together, these keywords indicate a comprehensive perspective on how entrepreneurial spirit, market understanding, and firm performance are interconnected and pivotal for sustained success in competitive business environments. Organizations that embrace entrepreneurial and market-oriented approaches are more likely to enhance their performance and achieve favorable results in the marketplace.

Cluster 3 of keywords suggests that China is positioning itself as a country that values innovation, supports the growth of small businesses, and aims to achieve sustainable development while staying competitive in the global market. It highlights China's SO toward becoming a leader in innovation and sustainability, potentially serving as a role model for other economies aiming for similar goals. Apart from China, it can be said that any country that needs to grow in economic terms needs to include innovation that resembles advancements, research, and development. Further, this cluster highlights the importance of SMEs for an economy. Further, "sustainability" and "sustainable development" point to countries' recognition of the importance of environmentally and socially responsible practices. This indicates a desire to achieve long-term economic growth without compromising future generations' ability to meet their needs.

Keywords in Cluster 4 revolve around the fundamental aspects of gaining a competitive edge in the business landscape. Competition refers to the rivalry between organizations striving to capture market share and achieve success in their respective industries. To thrive amid competition, businesses seek to establish a competitive advantage. Knowledge management plays a crucial role in achieving and sustaining a competitive advantage (Azeem et al., 2021; Arsawan et al., 2022). It involves the systematic process of identifying, capturing, organizing, and utilizing knowledge within an organization (Oktari et al., 2020). By effectively managing knowledge, companies can make better-informed decisions, innovate, and optimize their processes, leading to improved performance and a competitive edge.

5.6 DISCUSSION

Based on the bibliometric analysis, certain conclusions can be drawn. Since internal resources allow businesses to gain a competitive advantage by developing greater performance capabilities. Firm-level resources support the ability to innovate in products, processes, and procedures. For long-term business performance that allows for market explanation and understanding, it is beneficial to explain the firm's level of orientation toward innovative attitude for products and service process levels. SO evaluates the ways in which businesses acquire, allocate, and use resources to create dynamic capabilities (Kero & Sogbossi, 2017). It worked hard to maintain its competitiveness by implementing essential products or process improvements. At the same time, customer-orientated attitude allows organizations to engage with customers, understand their requirements, and develop products and services to meet and exceed customers' expectations for ultimately achieving improved OS. KM is another factor to consider. KM principles have been studied and utilized in every organizational field and profession (Ganguly et al., 2019; Swanson et al., 2020). Academics and practitioners have underlined the significance of knowledge as the foundation of success. Due to its tacit nature, developing organizational skills is essential, as knowledge often remains personal and non-transferable. However, simply knowing something does not guarantee a competitive advantage. To achieve personal and professional success, knowledge must be utilized, controlled, and researched. According to Magd et al. (2021), KM has a reciprocal relationship with innovation that complements each other. The firms that generate new knowledge and apply it effectively and efficiently will be able to develop long-term competitive advantages. As a result, firms shall strive to actively manage their professional, intellectual capital, expertise, and skills. Nonetheless, many KM systems have failed for a variety of reasons, such as an over-focus on IT, due to poor KM approaches, lack of understanding and translation of KM consequences. Recent developments in information-sharing technologies, which have crossed multiple trajectories and are heading toward maturity, can be used by academics and practitioners to reflect on the factors that contributed to their success. According to Magd et al. (2022), the dominant CSFs for e-learning are recognized as teachers' and students' commitment to pedagogy, technology, and social interaction. Furthermore, there is already a divide between practitioner and academic perspectives on KM perception.

Further, an organization's strategy remains one of the most important pillars for its success. It has substantial implications on organization's structure, operations management, market relationships, investment portfolios, and overall company performance. A strategy enables organizations and managers to gather specific resources and identify opportunities for providing valuable products and services. While businesses have many assets, including human resources, capital, finance, infrastructure, and talents, the most prominent element among them is the human factor (Sara et al., 2021). The strategy works out the product and service placement through strategic marketing for better profits. It also supports organizations in working out solutions to developing new capabilities for improving business performance. SO has a positive effect on firm performance and efficiency in service and productivity, while frugality in production defines performance (Kerdpitak & Boonrattanakittibhumi, 2020).

Market orientation is also a key strategic issue in the current environment. It is regarded as a valuable practice or capability that advantages organizations to stay competitive in current unstable business environment. Market orientation is part of strategic management practice and a great marketing concept for expansion of the market presence, market share, and market footfall (Gul et al., 2021; Kurniawan et al., 2020). According to various practitioners, researchers, and scholars, market orientation supports the development of marketing knowledge, competitive advantage, and superior performance. Scholars have explained market orientation differently. One of the understandings of market orientation is considered part of managerial caprice and whim toward organizational commitment through a marketing approach. Depending on the market situation, each organization must use its resources strategically to increase its competitive position (Al-Ansaari et al., 2015). It is an organization-wide approach for collecting market intelligence on the current and future expectations from the customer's point of view. The dissemination and spreading of market intelligence across departments remains corporate responsiveness. Market orientation mainly consists of three dimensions, such as competitor orientation, customer orientation, and organization sensitivity. According to Ansari (2022), improved customer and employee satisfaction and loyalty increase profitability and shareholder values. One of the tasks of corporate houses is to monitor the changing external environment, which remains out of control; changing consumer needs and wants for improved customer satisfaction and loyalty; and enhanced product innovation to gain a competitive edge. Improving customer orientation, competitor orientation, and inter-functional cooperation can be part of market orientation. According to Adiguzel and Sonmez Cakir (2022), the importance of SO and work motivation is essential for an organization's success. Businesses can identify warning or opportunity signals, process them, and transform them into actions through inter-functional coordination.

Further, innovation is not a "black box" that can provide all the data, but innovation can be dependent on the resources of the firm and how these resources can be put together for organizational success (Asiaei et al., 2020; Hofmann & Jaeger-Erben, 2020). A successful organization must combine resources such as skilled manpower, marketing expertise, organizational capabilities, and technological abilities to thrive. The most critical resource of any organization is its intangible resources such as brand loyalty, trademark, goodwill, patent, and intellectual property of the company (Asiaei et al., 2020). These intangible resources could be great assets for organizations gaining competitive advantage through innovation. In the emergence of successful innovation, the existence of a supportive organizational environment is very important (Lee & Trimi, 2021; Adiguzel & Sonmez Cakir, 2022). Innovation remains an internal factor and one of the main planks of innovation is to investigate the customer expectations that could support in integrating the relationship between innovation and organizational performance. The final goal of adopting innovation is to improve organizational performance in terms of improved market share.

In addition to this, the orientation toward entrepreneurship, sustainability, and creativity is also found to be an important consideration. Entrepreneurial orientation is linked to the SO that dealt with the firm's strategy on how an entrepreneurial activity will be taken up to outperform the competitor. According to Lee et al. (2017), SO is adopted within an intensely competitive and technological environment with unpredictable development (Lee et al., 2017). Advances in innovation open doors to new products or services, offering opportunities in constantly evolving markets. Implementing entrepreneurial characteristics within an organization's policies, procedures, practices, and decision-making attitudes is essential. Entrepreneurial orientation relates to managers' capabilities in exploring new avenues and elevating the company to higher levels. According to the entrepreneurial orientation concept, companies should be able to explore new avenues for achieving outstanding performance. In other words, the companies should carve out concrete strategies, explore behaviors like innovation, and create risk awareness and backup plans for any unforeseen. At the same time, the outright support of top management is essential. Entrepreneurship is also characterized by the company's position as a first mover in launching products and services, achieved through innovative exploration while strategically avoiding excessively risky ventures (Singh et al., 2019). According to Ansari and Khan (2023), entrepreneurship is essential for business success. Entrepreneurial orientation is represented in this study by innovativeness, risk-taking, and pro-activeness. A company can have an innovative inclination toward creativity and experimentation in the introduction of new products/services. The fact that organizations take risks in innovation and exhibit an active strategic stance may affect their performance in a competitive environment (Kasemsap, 2017). The new product can be successful with the most modern research and development (R&D) facilities that would give technological leadership an advantage in evolving new processes. It will improve proactivity that will impact on revised market opportunities by exploring expected market demands and opportunities.

In modern times, issues like climate change, ecological imbalances, and planetary ecosystems should gain priority; thus, sustainable business is one that not only considers the sustainability of the profits and business growth but also takes measures to enhance the quality of life and people.

5.7 CONCLUSIONS

The findings from the bibliometric analysis offer valuable insights into the multifaceted factors that contribute to organizational success and long-term competitive advantage. It is evident that a firm's ability to harness its internal resources plays a pivotal role in shaping its performance capabilities. This includes a firm's orientation toward innovation, as highlighted by the development of innovative products, processes, and procedures. Strategic resource allocation and utilization, as well as effective KM, are essential components of this equation. Moreover, the reciprocal relationship between KM and innovation underscores the significance of actively managing intellectual capital and expertise within organizations. While knowledge is undoubtedly a valuable asset, it must be harnessed and applied effectively to confer a competitive edge. Nevertheless, the failure of some KM systems serves as a reminder of the importance of holistic approaches and avoiding an overemphasis on technology.

Furthermore, an organization's strategy remains a cornerstone of its success, impacting various facets of its operations and overall performance. Human resources are singled out as a prominent element among an organization's assets, highlighting the importance of strategic resource acquisition. Market orientation, characterized by customer-centricity, competitor awareness, and organizational sensitivity, is a strategic imperative for navigating today's volatile business landscape and achieving improved customer satisfaction and loyalty. Innovation, a key driver of organizational success, necessitates the careful integration of various resources, including skilled manpower, technological capabilities, and intangible assets such as brand loyalty and intellectual property. Supportive organizational environments are crucial for nurturing innovation, ultimately contributing to improved market share.

In addition to innovation, an entrepreneurial orientation, marked by innovativeness, risk-taking, and pro-activeness, is instrumental in thriving in competitive and rapidly evolving markets. This orientation empowers organizations to explore new avenues, embrace creativity, and respond proactively to market demands and opportunities. Lastly, the imperative of sustainability is recognized as a critical consideration in modern business practices. Sustainable businesses not only focus on profitability and growth but also prioritize the enhancement of quality of life and environmental stewardship. In a world facing challenges like climate change and ecological imbalances, sustainable practices are essential for long-term success and responsible corporate citizenship.

In essence, the multifaceted nature of organizational success is underscored by these findings, emphasizing the need for strategic resource management, innovation, knowledge utilization, market orientation, entrepreneurship, and sustainability as interconnected elements in the pursuit of enduring competitive advantage and holistic business excellence.

5.8 LIMITATION

Being based on bibliometric data, this study also faces the limitations of a bibliometric study. Bibliometric analysis relies heavily on published literature, which may suffer from publication bias. This means that studies with statistically significant or positive findings are more likely to be published, potentially skewing the analysis toward certain trends or results. Further, the data is highly generalized and may not be suitable for specific contexts. Further, the bibliometric analysis may not fully capture qualitative aspects of organizational success, such as cultural factors, leadership styles, and ethical considerations. Also, the flair of industry experience in the context of this study is missing, as only the academic published data has been taken for the bibliometric analysis.

5.9 FUTURE DIRECTIONS

Based on the outcomes of this study, empirical investigation is recommended for future researchers where innovation can be considered as a mediating variable between CS and SO. Also, systematic literature review of top researches is also advisable so as to bring more details to the surface.

REFERENCES

Adams, P., Freitas, I. M. B., & Fontana, R. (2019). Strategic orientation, innovation performance and the moderating influence of marketing management. *Journal of Business Research*, *97*, 129–140.

Adiguzel, Z., & Sonmez Cakir, F. (2022). Examining the effects of strategic orientation and motivation on performance and innovation in the production sector of automobile spare parts. *European Journal of Management Studies*, *27*(2), 131–153. https://doi.org/10.1108/EJMS-01-2022-0007

Al-Ansaari, Y., Bederr, H., & Chen, C. (2015). Strategic orientation and business performance: An empirical study in the UAE context. *Management Decision*, *53*(10), 2287–2302.

Aldabbas, H., Pinnington, A., & Lahrech, A. (2023). The influence of perceived organizational support on employee creativity: The mediating role of work engagement. *Current Psychology*, *42*(8), 6501–6515.

Alqahtani, A. Y., & Rajkhan, A. A. (2020). E-learning critical success factors during the covid-19 pandemic: A comprehensive analysis of e-learning managerial perspectives. *Education Sciences*, *10*(9), 216.

Ansari, M. S. A. (2022). TQM Framework for Healthcare Sectors: Barriers to Implementation. *Quality Innovation Prosperity*, *26*(1), 1–23. https://doi.org/10.12776/qip.v26i1.1611

Ansari, M. S. A., Farooquie, J. A., & Gattoufi, S. M. (2016). Assessing the impact of service quality on customers and operators: Empirical study. *International Journal of Business and Management*, *11*(9), 207–217.

Ansari, M. S. A., & Khan, S. A. (2023). FDI, disinvestment and growth: An appraisal of Bhutanese economy. *Journal of Chinese Economic and Foreign Trade Studies*, 16(1), 64–82. https://doi.org/10.1108/JCEFTS-05-2022-0031

Arian, A., Sands, J., & Tooley, S. (2023). Industry and stakeholder impacts on corporate social responsibility (CSR) and financial performance: Consumer vs. Industrial Sectors. *Sustainability*, *15*(16), 12254.

Arsawan, I. W. E., Koval, V., Rajiani, I., Rustiarini, N. W., Supartha, W. G., & Suryantini, N. P. S. (2022). Leveraging knowledge sharing and innovation culture into SMEs sustainable competitive advantage. *International Journal of Productivity and Performance Management*, *71*(2), 405–428.

Asiaei, K., Barani, O., Bontis, N., & Arabahmadi, M. (2020). Unpacking the black box: How intrapreneurship intervenes in the intellectual capital-performance relationship? *Journal of Intellectual Capital*, *21*(6), 809–834.

Azeem, M., Ahmed, M., Haider, S., & Sajjad, M. (2021). Expanding competitive advantage through organizational culture, knowledge sharing and organizational innovation. *Technology in Society*, *66*, 101635.

Begum, S., Ashfaq, M., Xia, E., & Awan, U. (2022). Does green transformational leadership lead to green innovation? The role of green thinking and creative process engagement. *Business Strategy and the Environment*, *31*(1), 580–597.

Bledow, R., Frese, M., Anderson, N., Erez, M., & Farr, J. (2009). A dialectic perspective on innovation: Conflicting demands, multiple pathways, and ambidexterity. *Industrial and Organizational Psychology*, *2*(3), 305–337.

Cao, Z., & Shi, X. (2021). A systematic literature review of entrepreneurial ecosystems in advanced and emerging economies. *Small Business Economics*, *57*, 75–110.

Chaganti, R., & Sambharya, R. (1987). Strategic orientation and characteristics of upper management. *Strategic Management Journal*, *8*(4), 393–401.

Chiarini, A., Belvedere, V., & Grando, A. (2020). Industry 4.0 strategies and technological developments. An exploratory research from Italian manufacturing companies. *Production Planning & Control*, *31*(16), 1385–1398.

Enholm, I. M., Papagiannidis, E., Mikalef, P., & Krogstie, J. (2022). Artificial intelligence and business value: A literature review. *Information Systems Frontiers*, *24*(5), 1709–1734.

Fan, Q., Abbas, J., Zhong, Y., Pawar, P. S., Adam, N. A., & Alarif, G. B. (2023). Role of organizational and environmental factors in firm green innovation and sustainable development: Moderating role of knowledge absorptive capacity. *Journal of Cleaner Production*, *411*, 137262.

Ferreira, J., Cardim, S., & Coelho, A. (2021). Dynamic capabilities and mediating effects of innovation on the competitive advantage and firm's performance: The moderating role of organizational learning capability. *Journal of the Knowledge Economy*, *12*, 620–644.

Ganguly, A., Talukdar, A., & Chatterjee, D. (2019). Evaluating the role of social capital, tacit knowledge sharing, knowledge quality and reciprocity in determining innovation capability of an organization. *Journal of Knowledge Management*, *23*(6), 1105–1135.

Grawe, S. J., Chen, H., & Daugherty, P. J. (2009). The relationship between strategic orientation, service innovation, and performance. *International Journal of Physical Distribution & Logistics Management*, *39*(4), 282–300.

Gul, R. F., Liu, D., Jamil, K., Baig, S. A., Awan, F. H., & Liu, M. (2021). Linkages between market orientation and brand performance with positioning strategies of significant fashion apparels in Pakistan. *Fashion and Textiles*, *8*(1), 1–19.

Gupta, S., Kumar, V., & Karam, E. (2020). New-age technologies-driven social innovation: What, how, where, and why? *Industrial Marketing Management*, *89*, 499–516.

He, J., Chen, H., & Tsai, F. S. (2020). Strategy orientation, innovation capacity endowment, and international R&D intensity of listed companies in China. *Sustainability*, *12*(1), 344.

Hofmann, F., & Jaeger-Erben, M. (2020). Organizational transition management of circular business model innovations. *Business Strategy and the Environment*, *29*(6), 2770–2788.

Hsieh, Y. J., & Wu, Y. J. (2019). Entrepreneurship through the platform strategy in the digital era: Insights and research opportunities. *Computers in Human Behavior*, *95*, 315–323.

Iheanachor, N., Umukoro, I., & Aránega, A. Y. (2023). Ecosystem emergence in emerging markets: Evidence from the Nigerian digital financial services ecosystem. *Technological Forecasting and Social Change*, *190*, 122426.

Jamil, M., He, N., Wei, Z., Gupta, M. K., & Khan, A. M. (2023). A novel quantifiable approach of estimating energy consumption, carbon emissions and cost factors in manufacturing of bearing steel based on triple bottom-line approach. *Sustainable Materials and Technologies*, *36*, e00593.

Kabeyi, M. (2019). Organizational strategic planning, implementation and evaluation with analysis of challenges and benefits. *International Journal of Applied Research and Studies*, *5*(6), 27–32.

Kamal, S., Naim, A., Magd, H., Khan, S. A., & Khan, F. M. (2022). The Relationship Between E-Service Quality, Ease of Use, and E-CRM Performance Referred by Brand Image. In A. Naim & S. Kautish (Eds.), *Building a Brand Image Through Electronic Customer Relationship Management* (pp. 84–108). IGI Global. https://doi.org/10.4018/978-1-6684-5386-5.ch005

Kasemsap, K. (2017). Strategic Innovation Management: An Integrative Framework and Causal Model of Knowledge Management, Strategic Orientation, Organizational Innovation, and Organizational Performance. In I. Management Association (Ed.),*Organizational Culture and Behavior: Concepts, Methodologies, Tools, and Applications* (pp. 86–101). IGI Global.

Keiningham, T., Aksoy, L., Bruce, H. L., Cadet, F., Clennell, N., Hodgkinson, I. R., & Kearney, T. (2020). Customer experience driven business model innovation. *Journal of Business Research*, *116*, 431–440.

Kerdpitak, C., & Boonrattanakittibhumi, C. (2020). Effect of strategic orientation and organizational culture on firm's performance. Role of organizational commitments. *Journal of Security & Sustainability Issues*, *9*, 70–80.

Kero, C. A., & Sogbossi, B. B. (2017). Competitive strategy orientation and innovative success: Mediating market orientation a study of small-medium enterprises. *Global Journal of Management and Business Research*, *17*, 75–89.

Khan, S. A. & Rena, R. (2023). Emerging Green Practices, Internet of Things, and Digital Marketing: A Response to the Global Economic and Climate Crises. In A. Naim & V. Devi (Eds.), *Global Applications of the Internet of Things in Digital Marketing* (pp. 1–16). IGI Global. https://doi.org/10.4018/978-1-6684-8166-0.ch001

Khan, S. A. (2023). E-Marketing, E-Commerce, E-Business, and Internet of Things: An overview of terms in the context of small and medium enterprises (SMEs). In A. Naim & V. Devi (Eds.), *Global Applications of the Internet of Things in Digital Marketing* (pp. 332–348). IGI Global. https://doi.org/10.4018/978-1-6684-8166-0.ch017

Khan, S. A., & Magd, H. (2023). New Technology Anxiety and Acceptance of Technology: An Appraisal of MS Teams. In *Advances in Distance Learning in Times of Pandemic*, 105–134. CRC Press, Taylor and Francis. https://doi.org/10.1201/9781003322252-5

Khan, S. A., Magd, H., Al Shamsi, I. R., & Masoom, K. (2022a). Social Entrepreneurship Through Innovations in Agriculture. In H. Magd, D. Singh, R. Syed, & D. Spicer (Eds.), *International Perspectives on Value Creation and Sustainability Through Social Entrepreneurship* (pp. 209–222). IGI Global. https://doi.org/10.4018/978-1-6684-4666-9.ch010

Khan, S. A., Magd, H., Khatri, B., Arora, S., & Sharma, N. (2023). Critical Success Factors of Internet of Things and Digital Marketing. In A. Naim & V. Devi (Eds.), *Global Applications of the Internet of Things in Digital Marketing* (pp. 233–253). IGI Global. https://doi.org/10.4018/978-1-6684-8166-0.ch012

Khan, S. A., Magd, H., & Epoc, F. (2022b). Application of Data Management System in Business to Business Electronic Commerce. In Naim, A., & Malik, P. K. (Ed.), *Competitive Trends and Technologies in Business Management.* (pp. 109–124). Nova Science Publishers.

Khan, S. A., Epoc, F., Gangwar, V. P., Ligori, T. A. A., Ansari, Z. A. (2021). Will Online banking sustain in Bhutan post Covid – 19? A quantitative analysis of the customer e-satisfaction and e-loyalty in the Kingdom of Bhutan. *Transnational Marketing Journal*, *9*(3), 607–624 https://doi.org/10.33182/tmj.v9i3.1288

Khatri, B., Arora, S., Magd, H., & Khan, S. A. K. (2022). Bibliometric Analysis of Social Entrepreneurship. In H. Magd, D. Singh, R. Syed, & D. Spicer (Eds.), *International Perspectives on Value Creation and Sustainability Through Social Entrepreneurship* (pp. 46–60). IGI Global. https://doi.org/10.4018/978-1-6684-4666-9.ch003

Khatri, B., Shrimali, H., Khan, S. A., & Naim, A. (2023). Role of HR Analytics in Ensuring Psychological Wellbeing and Job Security: Learnings from COVID-19. In R. Yadav, M. Sinha, & J. Kureethara (Eds.), *HR Analytics in an Era of Rapid Automation* (pp. 36–53). IGI Global. https://doi.org/10.4018/978-1-6684-8942-0.ch003

Kohli, R., & Melville, N. P. (2019). Digital innovation: A review and synthesis. *Information Systems Journal*, *29*(1), 200–223.

Kurniawan, R., Budiastuti, D., Hamsal, M., & Kosasih, W. (2020). The impact of balanced agile project management on firm performance: The mediating role of market orientation and strategic agility. *Review of International Business and Strategy*, *30*(4), 457–490.

Lee, C. S., Chan, L., & McNabb, D. E. (2017, July). The Role of Strategic Orientation in Business Innovation. In *2017 Portland International Conference on Management of Engineering and Technology (PICMET)* (pp. 1–5). IEEE.

Lee, S. M., & Trimi, S. (2021). Convergence innovation in the digital age and in the COVID-19 pandemic crisis. *Journal of Business Research*, *123*, 14–22.

Lim, K. Y. H., Zheng, P., Chen, C. H., & Huang, L. (2020). A digital twin-enhanced system for engineering product family design and optimization. *Journal of Manufacturing Systems*, *57*, 82–93.

Magd, H. & Khan, S. A. (2022). Strategic Framework for Entrepreneurship Education in Promoting Social Entrepreneurship in GCC Countries During and Post COVID-19. In H. Magd, D. Singh, R. Syed, & D. Spicer (Eds.), *International Perspectives on Value Creation and Sustainability Through Social Entrepreneurship* (pp. 61–75). IGI Global. https://doi.org/10.4018/978-1-6684-4666-9.ch004

Magd, H., Ansari, M., & Negi, S. (2021). The relationship between TQM, knowledge management, and innovation: A framework to achieve organizational excellence in service industry. *Global Business & Management Research*, *13*(3), 283–296.

Magd, H., Jonathan, H., Khan, S. A., & El Geddawy, M. (2023a). Artificial Intelligence—The Driving Force of Industry 4.0. In *A Roadmap for Enabling Industry 4.0 by Artificial Intelligence*, 1–15. Wiley. https://doi.org/10.1002/9781119905141.ch1

Magd, H., Khan, S. A., Khatri, B., Sharma, N., & Arora, S. (2023b). Understanding the Relationship Between IoT and Digital Marketing: A Bibliometric Analysis. In A. Naim & V. Devi (Eds.), *Global Applications of the Internet of Things in Digital Marketing* (pp. 123–140). IGI Global. https://doi.org/10.4018/978-1-6684-8166-0.ch007

Magd, H., Nzomkunda, A., Negi, S., & Ansari, M. (2022). Critical success factors of E-learning implementation in higher education institutions: A proposed framework for success. *Global Business & Management Research*, *14*(2), 20–38.

Moeuf, A., Lamouri, S., Pellerin, R., Tamayo-Giraldo, S., Tobon-Valencia, E., & Eburdy, R. (2020). Identification of critical success factors, risks and opportunities of Industry 4.0 in SMEs. *International Journal of Production Research*, *58*(5), 1384–1400.

Moore, H. L. (2015). Global prosperity and sustainable development goals. *Journal of International Development*, *27*(6), 801–815.

Nassani, A. A., & Aldakhil, A. M. (2023). Tackling organizational innovativeness through strategic orientation: Strategic alignment and moderating role of strategic flexibility. *European Journal of Innovation Management*, *26*(3), 847–861. https://doi.org/10.1108/EJIM-04-2021-0198

Oktari, R. S., Munadi, K., Idroes, R., & Sofyan, H. (2020). Knowledge management practices in disaster management: Systematic review. *International Journal of Disaster Risk Reduction*, *51*, 101881.

Palacios-Marqués, D., Gallego-Nicholls, J. F., & Guijarro-García, M. (2021). A recipe for success: Crowdsourcing, online social networks, and their impact on organizational performance. *Technological Forecasting and Social Change*, *165*, 120566.

Sara, I. M., Saputra, K. A. K., & Utama, I. W. K. J. (2021). The effects of strategic planning, human resource and asset management on economic productivity: A case study in Indonesia. *The Journal of Asian Finance, Economics and Business*, *8*(4), 381–389.

Silvestre, B. S., & Ţîrcă, D. M. (2019). Innovations for sustainable development: Moving toward a sustainable future. *Journal of Cleaner Production*, *208*, 325–332.

Singh, E. H., Khan, S. A., Thoudam, P., & Sharma, P. P. (2019). Factor affecting the choice of cheese in Bhutan: A choice architecture perspective. *International Journal of Engineering and Advanced Technology (IJEAT)*, *8*(5), 1880–1888. (ISSN:2249–8958)

Singh, E. H., Wangda, S., Khan, S., & Khan, S. A. (2020). Exploring the obstacles for start-ups in Bhutan: From a prevented entrepreneur perspective. *International Journal of Innovation, Creativity and Change*, *11*(4), 70–87.

Swanson, E., Kim, S., Lee, S. M., Yang, J. J., & Lee, Y. K. (2020). The effect of leader competencies on knowledge sharing and job performance: Social capital theory. *Journal of Hospitality and Tourism Management*, *42*, 88–96.

Van Veldhoven, Z., & Vanthienen, J. (2022). Digital transformation as an interaction-driven perspective between business, society, and technology. *Electronic Markets*, *32*(2), 629–644.

Wu, S., Cao, L., Xu, D., & Zhao, C. (2023). Historical eco-environmental quality mapping in china with multi-source data fusion. *Applied Sciences*, *13*(14), 8051.

Xue, L. L., Shen, C. C., & Lin, C. N. (2023). Effects of internet technology on the innovation performance of small-scale travel agencies: Organizational learning innovation and competitive advantage as mediators. *Journal of the Knowledge Economy*, *14*(2), 1830–1855.

Zalan, T., & Toufaily, E. (2017). The promise of fintech in emerging markets: Not as disruptive. *Contemporary Economics*, *11*(4), 415–430.

Zareen, S. & Khan, S. A. (2023). Exploring Dependence of Human Resource Management (HRM) on Internet of Things (IoT) and Digital Marketing in the Digital Era. In A. Naim & V. Devi (Eds.), *Global Applications of the Internet of Things in Digital Marketing* (pp. 51–66). IGI Global. https://doi.org/10.4018/978-1-6684-8166-0.ch004

Zhang, Y., Chao, Q., Chen, Y., Zhang, J., Wang, M., Zhang, Y., & Yu, X. (2021). China's carbon neutrality: Leading global climate governance and green transformation. *Chinese Journal of Urban and Environmental Studies*, *9*(3), 2150019.

Zhou, K. Z., Yim, C. K., & Tse, D. K. (2005). The effects of strategic orientations on technology-and market-based breakthrough innovations. *Journal of Marketing*, *69*(2), 42–60.

Zhou, M., Govindan, K., & Xie, X. (2020). How fairness perceptions, embeddedness, and knowledge sharing drive green innovation in sustainable supply chains: An equity theory and network perspective to achieve sustainable development goals. *Journal of Cleaner Production*, *260*, 120950.

6 Role of Human and Artificial Intelligence Biases on Organizational Performance, Efficiency, and Enhanced Productivity

Shoaib Irshad
Namal University, Mianwali, Pakistan

6.1 INTRODUCTION

Both human biases and artificial intelligence (AI) systems have the potential to either improve or harm organizational performance, efficiency, and productivity. However, the interplay between human and machine biases is complex and often overlooked. Davenport and Ronanki (2018) explain that AI promises enhanced decision-making but that it can also amplify problematic biases if caution is not exercised in development and deployment.

Considering the human side, a plethora of research shows how cognitive biases negatively impact organizations. Confirmation bias leads to poor strategic decisions (Kahneman, 2011), in-group favoritism undermines meritocracy (Reskin, 2000), and implicit biases decrease workplace diversity (Bertrand & Mullainathan, 2004). Standardized and audited processes can mitigate individual biases (Bohnet, 2016).

Similarly, AI systems plagued by biased data, algorithms, or unethical applications can entrench discrimination and worsen performance. Zou and Schiebinger (2018) detail flawed word embeddings that encode gender stereotypes, which have been unwittingly amplified through AI text analysis. O'Neil (2016) provides sobering examples of algorithms that punish the poor and reduce opportunity. Thoughtful data practices, testing, and human oversight are necessary to prevent unfair AI (Mehrabi et al., 2021). The intervention of human and AI biases can fuel vicious cycles that are difficult to recognize and can disrupt the flow and processing of information. While AI offers efficiencies, the "black box" nature of many systems obscures embedded biases (Lepri et al., 2018). With careful implementation that accounts for inclusivity and ethics, organizations can avoid AI harm and harness benefits (Morley et al., 2021).

DOI: 10.1201/9781032616810-6

In conclusion, organizational leaders must proactively address both human and AI biases. A nuanced understanding of their interactions is critical for enhancing performance through diversity and measured AI adoption. The use of AI in business and organizations has been growing rapidly in recent years. According to a survey by Deloitte (2018), the number of companies implementing AI grew from 4% in 2016 to 14% in 2018. Key drivers of AI adoption include lowering costs, boosting efficiency, and gaining competitive advantage.

Some of the most common applications of AI in organizations include the following:

- Automation of routine tasks and processes using robotic process automation (RPA), chatbots, etc. This can reduce costs and improve efficiency.
- Enhanced data analytics and decision-making using machine learning (ML) algorithms. AI can analyze large datasets faster than humans and provide insights for better decisions.
- Personalization of customer experiences by leveraging AI capabilities like natural language processing and recommendation engines.
- Fraud and risk detection through pattern recognition and predictive capabilities of AI.

While AI promises various benefits, potential risks and challenges include the following:

- Job losses due to automation of tasks previously done by humans
- Biased or unfair decisions due to flaws in training data or algorithms
- Lack of transparency in how AI systems make decisions
- Cybersecurity vulnerabilities that could be exploited to attack AI systems

Adoption of AI is driven by cost reduction, efficiency, and competitive advantage for organizations. However, the technology also poses risks related to jobs, bias, transparency, and security that need to be proactively managed.

6.1.1 Influence of Human Biases on AI Systems and Organizational Outcomes

AI systems are designed, developed, and deployed by humans. As a result, these systems can inherit human biases present in the data or algorithms used to train them. This can lead to discriminatory and unfair outcomes if not addressed properly. The algorithms can pick the behavioral trends and detect the pattern in the classification, sorting, and filtering of the data. For the very same reason, they are instrumental in uploading unbiased data. Reducing biases in AI requires interdisciplinary efforts as emphasized by Raji and Buolamwini (2019).

Some key ways human biases can seep into AI systems include the following:

- ***Biased data***: The training data may underrepresent certain groups or contain stereotypical associations that the AI absorbs. For example, an AI for recruiting purposes that is trained only based on resumes of previously hired engineers at a tech firm could be biased against women.

- ***Poorly designed algorithms*:** Algorithms themselves may have embedded biases. According to the report by *Harvard Business Review*, ML models, if not properly administered, could place more weight on factors correlated with race or gender rather than direct qualifications.
- ***Unfair performance objectives*:** Maximizing accuracy or efficiency alone without regard for fairness can produce biased outcomes.
- ***Lack of diversity*:** Homogenous teams building AI systems can overlook potential for bias.

These biases can lead to discriminatory outcomes such as the following:

- Biased candidate screening and recruiting by AI tools favoring certain groups.
- Pricing algorithms that charge certain demographics higher prices for the same products/services.
- Flawed AI decision-making in areas like credit lending and criminal justice.

Such biases erode trust in AI; lead to legal, ethical, and PR risks for organizations; and negatively impact brand reputation. Proactive measures like rigorous algorithm testing, cultivating diversity, and setting ethical AI guidelines can help subdue some of these pre-existing biases.

6.1.2 Integration of Human and AI Biases

The dynamic interplay between human biases and artificial intelligence (AI) has become a focal point in both academia and industry. As AI technologies become increasingly integrated into decision-making processes, understanding the potential biases they inherit and propagate is paramount for maintaining fairness, transparency, and effective organizational outcomes (Rastogi et al., 2022).

Overall, human cognitive biases highlight how structure, perspective, and diversity are vital in decision-making. The intersection of these biases with technology can inadvertently find their way into AI systems, affecting both the technology itself and the outcomes it produces. Mitigating these biases requires careful attention to training data representativeness, feedback loops, understanding of the data, and inclusive team composition in AI development. Understanding human cognition provides lessons for engineers to build fairer, more robust AI.

6.2 HUMAN BIASES IN DECISION-MAKING: COGNITIVE PERSPECTIVE

The fundamental biases induced in decision-making can be listed as follows:

6.2.1 Human Biases

- ***Confirmation bias*** refers to seeking and interpreting evidence that confirms existing beliefs (Nickerson, 1998). Confirmation bias caused managers to preferentially seek positive performance reviews for employees they favored, ignoring contradictory feedback.

When developing AI models, human biases in data preparation and data uploading can feed biased patterns to the software that reinforce existing stereotypes and beliefs (Hallihan & Shu, 2013). If the training data is biased, AI algorithms might learn and perpetuate those biases, leading to discriminatory outcomes. For example, if a hiring algorithm is trained on historical data that reflects gender bias, it could inadvertently discriminate against certain genders.

Similarly, confirmation bias can lead executives to filter information in ways that support their preexisting views, resulting in poor strategic decisions that overlook risks or competitive threats. This contributed to underestimating the rise of streaming and filing for bankruptcy.

- ***Affinity bias*** refers to favoring people similar to oneself. Research demonstrates that people subconsciously promote members of their own ethnic group over equally qualified diverse candidates. If diverse perspectives are not considered during the design and development of AI systems, it can lead to biased algorithms that reflect the viewpoints of a homogeneous team. This can result in AI systems that fail to address the needs of diverse viewpoints as demonstrated by Belenguer (2022).

  Cultural biases can influence the data used to train AI models. If the training data predominantly represents a specific culture or group, the AI may struggle to accurately understand or serve other cultural contexts. This can lead to misinterpretation, miscommunication, or inadequate performance in cross-cultural scenarios as the following:
- ***Halo effect*:** Letting overall impression of a person influence evaluations of their traits. Rater biases like the halo effect were shown to negatively impact performance ratings of women and minorities in a meta-analysis by Joshi, Son, and Roh (2015).
- ***Fundamental attribution error*:** Overemphasizing personality factors over situational factors in judging behavior. The fundamental attribution error compares to how AI can fail to consider situational or contextual factors, focusing solely on extracting signals from available data properties.
- ***Availability heuristic*:** Estimating probability based on how easily examples come to mind. AI systems can be vulnerable to the availability heuristic, as they learn from available data. If certain data is more readily available, AI models may prioritize those patterns, potentially leading to biased decisions. This can result in skewed representations of reality, especially if data sources themselves are biased also demonstrated by Schwartz et al. (2022).
- ***Anchoring bias*:** Relying too heavily on initial pieces of information. Anchoring bias in humans parallels how AI systems can be anchored to the specifics of their training data, leading to fixated or biased patterns of judgment, especially in business intelligence as demonstrated by Ni, Arnot, and Gao (2019).

  Anchoring bias causes managers to rely too heavily on initial cost or sales forecasts, even when better data arises. IT projects end up over-budget due to anchoring on original estimates. Companies underinvest due to anchoring on an arbitrary ROI number (Makadok & Gamble, 2019).

- ***Pro-innovation bias*:** Tendency to favor innovation and newness over existing solutions, products, or processes, even when there is little evidence that the innovation is beneficial. Causes of pro-innovation bias include cultural narratives that equate new with better and influence of vested interests that benefit from continuous innovation cycles, and cognitive biases that lead people to focus on novelty over evaluating practical value.
- ***Placebo effect*:** Perceiving improvement from a treatment that is inert. Placebo effects are thought to be triggered by expectations, conditioning, reduced anxiety, and activation of certain areas in the brain related to emotion, reward, and pain modulation.
- ***Survivorship bias*:** Focusing on successes while ignoring failures. Analyzing success stories led entrepreneurs to underestimate real odds and risks. Once deployed, an AI system that makes decisions based on past successes can fall into a feedback loop that reinforces survivorship bias.
- ***Dunning-Kruger effect*:** Overestimating one's own abilities or knowledge. Poor performers consistently overestimated their abilities across many domains. Optimism bias leads people to underestimate risks and challenges. 90% of startups fail likely due to over-optimism about success despite high uncertainty. Over-optimism contributes to cost overruns in large engineering projects like dams or railways.

6.2.2 Organizational Biases

The drive toward data-driven decision-making by using algorithms and AI can yield socially biased outcomes. The increasing role of AI in organizational processes like pricing, lending, hiring, and content moderation invites the potential induction of organizational bias. Organizations are struggling with challenges in detecting bias especially when multiple AI systems interact in complex workflows. Thus, compounding the inequalities in workplace and society (Kordzadeh & Ghasemaghaei, 2022).

- ***Institutional bias*:** Built-in rules, practices, and policies within an organization that intentionally or unintentionally exclude or put certain groups at a disadvantage. This can entrench discriminatory and unfair practices within the organization.
- ***Design bias*:** Workplaces designed primarily for the needs and preferences of the majority group without considering minorities. This can indirectly exclude many groups from full participation and advancement.
- ***Similarity Index bias*:** The unconscious tendency to gravitate toward and promote those who are similar to oneself. This can limit diversity and merit-based advancement within organizations.
- ***Implicit Biases*:** Implicit biases are unconscious stereotypes and attitudes that can heavily influence interactions, decisions, and behaviors. These biases develop over a lifetime through social conditioning, media portrayals, and cultural narratives. At an organizational level, implicit bias also influences workplace culture, feelings of belonging, and access to opportunities.

Microaggressions create environments where underrepresented groups feel isolated and discouraged. Mentorship and sponsorship programs to support minority advancement can help foster inclusion.

Implicit biases in hiring reduce workforce diversity and the range of perspectives, decreasing innovation potential. Experimental studies consistently show equivalent minority candidates receive fewer callbacks. Homogenous teams perform worse on decision-making and problem-solving tasks.

To optimize decision-making, many organizations are pursuing hybrid models that thoughtfully combine the scale of AI automation with human oversight, direction setting, and course correction. Establishing diversity along with governance and accountability structures helps in maximizing organizational productivity and fairness. Organizations must implement structured processes and cultural change initiatives to counteract these effects and enable unbiased hiring, promotion, and an inclusive environment.

6.2.3 AI Biases

Addressing biases in AI is pivotal for optimizing organizational performance and ensuring fairness in decision-making. Varsha (2023) presents systematic literature review on AI biases and highlights the vulnerabilities arising in companies due to the use of biased AI models.

- ***Training data bias*:** When biases in the training data are propagated through the ML model. This can lead to discrimination in areas like hiring, lending, and policing.
- ***Algorithmic biases*:** The architecture, assumptions, or purposes of algorithms can unintentionally introduce bias. For example, facial analysis tools are less accurate for women and darker-skinned faces.
- ***Deployment biases*** Real-world usage and business policies around AI systems can further entrench biases. For example, basing loan decisions solely on AI credit-scoring of applicants.
- ***Automation bias*** The tendency to over-rely on algorithmic decision-making and undervalue human judgment. This can displace oversight and nuanced human analysis.
- ***Embedded bias*** Biased assumptions coded directly into algorithms and can lead to discrimination being scaled through automated decisions.
- ***Measurement biases*** Flawed evaluation metrics that don't account for fairness across groups can overlook harm. For example, driver safety models are based only on aggregate accident data.

Addressing these biases in the context of AI is crucial for creating fair, effective, and ethical AI systems. The cognitive perspective sees biases as byproducts of information processing limitations, reliance on associative memory, selective attention filters, motivational factors, and mental heuristics. These cognitive roots can lead to "debiasing" interventions through reflective processes, structured analysis, and perspective-taking.

Researchers, developers, and policymakers need to actively work to identify, mitigate, and prevent biases in AI systems to ensure that they serve diverse populations, avoid discrimination, and provide accurate and unbiased insights.

6.2.4 The Interplay of Human and AI Biases

Human biases get captured in training data or encoded by programmers, then amplified through AI systems deployed at scale (Suresh & Guttag, 2021). This reinforces and perpetuates existing biases. However, the "black box" nature of many AI models makes it hard to analyze or understand the origin of biases, which may be obscured within multilayer neural networks (Holstein et al., 2019).

Biased feedback loops can develop where AI systems make subtly discriminatory decisions that influence human behaviors and perceptions to become more biased, which further skews AI training data (Hashimoto et al., 2018). Cognitive biases like automation bias lead humans to overly trust AI outputs without verifying for fairness (Wang et al., 2019). Unchecked algorithms can rapidly propagate unfairness.

Intersectional biases are particularly hard to recognize, such as algorithms disadvantaging elderly minority groups (D'Ignazio & Klein, 2020). Testing across demographic combinations is important.

Organizations must implement ongoing auditing at data and algorithmic levels alongside diversity efforts to prevent AI from masking, entrenching, or amplifying unfair biases (Raji et al., 2020). In summary, human and machine biases closely interact, often unintentionally. Thoughtful oversight processes are essential to prevent detrimental impacts from unchecked AI systems.

6.2.5 How Biased AI Systems Can Hurt Organizations

The implicit intrusion of human biases in organizational designs hurts the ethics and civic engagement of corporations and raises concerns regarding the fairness of ML models (Akter et al., 2021). Following are some of the modes that have the capacity to add to the biases in AI and ML models.

Data biases

- Training data that is not representative or suffers from historical biases gets encoded in the models. For example, facial recognition datasets lacked diversity.
- Underrepresented groups in the data lead to worse performance for those groups. Insufficient samples of certain minorities skew the model's decision boundaries.
- Relying on variables like zip codes or social connections that correlate with race or class bakes in discrimination.

Programmer biases

- Explicit or implicit biases of developers lead them to make design choices that disadvantage certain groups.

- Lack of diversity among programmers and testers means biases are not flagged early. Homogenous teams build homogenized systems.
- Shortcuts taken to maximize efficiency rather than fairness infringe on the rights or opportunities for minorities.

Unfair outcomes

- ML optimization functions that focus narrowly on rewards, accuracy, or profit allow biases to emerge.
- Algorithmic systems that benefit the majority or privileged groups are seen as neutral because harms are not visible.
- Without specific countermeasures, noise or randomness in model training can compound existing disparities.

Biased data, developer blindspots, and single-dimension optimization enable AI systems to inherit, amplify, and obscure unfair biases that permeate society. Ongoing scrutiny is required to promote algorithmic fairness.

6.3 AI OPTIMIZATION OF BUSINESS PROCESSES

AI has the potential to optimize various aspects of business processes, leading to increased efficiency, cost savings, and improved decision-making. Aldoseri et al. (2023) presented a structured roadmap outlining the cohesion between the business processes and leading AI technologies. The interlink creates the synergy among the different components of an organization boosting its efficiency and productivity.

Following are some of the key functions that are substantially improved by the renaissance of AI.

1. **Predictive Analytics**: AI-driven predictive analytics can forecast future trends and demands, helping businesses optimize inventory management, supply chain logistics, and resource allocation.
2. **Customer Relationship Management (CRM)**: AI-powered CRM systems use data analytics to personalize customer interactions, leading to more effective marketing campaigns, improved customer retention, and increased sales.
3. **Process Automation**: RPA powered by AI can automate repetitive, rule-based tasks, reducing errors and freeing up employees to focus on higher-value activities.
4. **Supply Chain Optimization**: AI can analyze supply chain data to optimize procurement, production scheduling, and distribution, resulting in reduced costs, minimized waste, and improved delivery times.
5. **Fraud Detection and Risk Management**: AI algorithms can identify unusual patterns and anomalies in financial transactions, helping organizations detect and prevent fraud, as well as manage risk effectively.
6. **Human Resources and Talent Management**: AI-driven HR tools can streamline talent acquisition, assess candidate fit, and predict employee turnover, aiding in the efficient management of human capital.

6.3.1 Challenges of Detecting Algorithms Biases

There are several challenges that make detecting biases difficult when they are embedded within complex AI algorithms. Cheatham et al. (2019) talked about how important it is to take nuanced controls where ML models are in the position of decision-making. Detecting biases in algorithms is a multidisciplinary challenge that involves computer science, ethics, law, and domain-specific expertise.

- The layers in deep neural networks obscure how different variables interact and contribute to outputs. This lacks transparency on how biases influence predictions.
- Biases can manifest indirectly through correlations with other variables not easily interpretable, for example, zip codes correlating to race.
- Intersectional biases against combinations of attributes like race and gender may not be uncovered by evaluating groups separately.
- Edge cases that reflect biases may not be well represented in test and validation data. Real-world data often has skews.
- Biases are often subjective and context dependent. Auditors may not anticipate all possible harms or impacted groups.
- Models that continually update through new data require ongoing bias testing as distributions shift over time. One-time audits are insufficient.
- Once deployed, system complexity combined with user adaptations makes measuring real-world impacts of algorithmic bias challenging.
- Lack of access to proprietary algorithms and data poses barriers to external auditing. Vendors may be evasive about biases.

Overall, identifying potential biases requires a commitment to transparency, explainability, and proactive auditing through techniques like intersectional testing, simulated corruptions, and monitoring of feedback loops. Detecting algorithmic bias remains an inherently challenging problem with no perfect solutions.

6.3.2 Organizational Implications When AI Amplifies or Conceals Human Biases

Here are some potential organizational implications when AI systems amplify or conceal, rather than mitigate, human biases:

- ***Ineffective Decisions*:** Biased AI predictions that overlook certain population segments may result in suboptimal decisions that underserve parts of the market (Danks & London, 2017).
- ***Reputational Damage*:** High-profile cases of algorithmic bias at companies like Amazon, Apple, and Facebook have resulted in public criticism, employee backlash, and loss of user trust (Raji et al., 2020).
- ***Legal Liabilities*:** Biased AI that enables discrimination opens organizations to major legal risks such as lawsuits or regulatory punishments (Lepri et al., 2018).

- ***Loss of Diverse Perspectives*:** Flawed AI systems that reinforce biases could discriminatorily exclude certain groups, depriving organizations of diversity benefits (Hashimoto et al., 2018).
- ***Entrenching Inequity*:** Reliance on algorithms that perpetuate existing societal biases rather than counter them preserves historic inequities (Suresh & Guttag, 2021).
- ***Lack of Transparency*:** The inability to audit black box algorithms for biases reduces accountability and makes organizations vulnerable to unexpected failures (Holstein et al., 2019).

In summary, failure to address biases in AI systems carries major organizational risks including reputational harms, legal liabilities, loss of diversity, poor decisions, and perpetuation of societal inequities. Sustained efforts in algorithmic testing, oversight, and transparency are essential.

6.3.3 Approaches to Break Detrimental Feedback Loops between Humans and AI

Both academia and industry have long been deliberating detrimental effects of systematic and intrinsic biases in algorithms. Fu et al. (2020) discussed five important aspects of algorithmic bias and the challenges in detecting these biases. They also explained the potential sources and reviewed several bias correction methods.

Some approaches that can help break detrimental feedback loops between human and AI biases are as follows:

***External auditing*:** Independent third-party audits assess models for fairness across different demographic groups to identify issues early.

- ***Diverse development teams*:** Having a wide range of perspectives involved in designing, testing, and monitoring AI systems helps surface biases that homogeneous teams could miss.
- ***Simulation of bias*:** Intentionally corrupting data and models with synthetic biases during development stages aids in detecting overlooked unfairness.
- ***Limits on automation*:** Keeping humans involved through oversight, regular review of model decisions, and discretionary override authority on consequential outputs helps halt unfair AI behaviors.
- ***Feedback data sanitization*:** Proactively cleaning real-world data collected for retraining by removing or re-balancing identified biased elements helps prevent amplification.

Overall, taking deliberate steps to monitor for biases involving diverse views, simulating distorted situations, and placing limits on full automation can protect against AI and human biases exacerbating each other.

6.4 VICIOUS CYCLE: REINFORCEMENT OF BIAS

The reinforcement of bias in AI generates lots of ethical and operational inefficacies. Gal et al. (2020) spoke about breaking the vicious cycle of algorithmic management in people analytics tool. They proposed that by introducing alternative technology designs, organizations can draw on sustainable virtue ethics approach.

Here are some examples of how vicious cycles can develop between human bias and AI bias:

- Biased data or algorithms produce discriminatory AI outputs that go unnoticed and get relied upon. This further skews real-world data collected for retraining.
- Pre-existing prejudice causes people to under-scrutinize unfair outputs as long as they align with expectations, entrenching the bias.
- Homogenous teams of developers fail to notice unfair model behavior on minority groups, leading to deployment of flawed systems.
- Humans overly trust AI judgments due to automation bias, causing them to make preemptive biased decisions that eventually feed back into training data (Madras et al., 2018).
- Biased AI recruitment or promotion systems discriminatorily exclude certain groups, feeding historical inequities back into an organization's practices and data.
- Once discriminatory algorithms are deployed at scale, their unfair outputs impact many human decisions and behaviors, complicating debiasing.

Careful oversight is required to detect and break these vicious cycles before AI systems and real-world environments become mutually and increasingly biased.

6.4.1 Aligning or Maligning of AI in Retrospective of Organizational Performance

AI is being rapidly adopted for high-impact applications like hiring, lending, pricing, content moderation, facial recognition, and more. This expands the potential scale of any biases.

Early real-world examples have demonstrated how seemingly neutral algorithms can discriminate due to built-in biases. Ferrer et al. (2021) talked about digital discrimination and how it marginalizes certain factions of society in a diverse range of fields. For instance, facial analysis AI falsely identified some darker-skinned individuals as gorillas. Lack of diversity among AI developers and training data limitations have surfaced as key reasons why bias creeps into AI models and goes unnoticed.

There are increasing calls for algorithmic transparency and accountability as AI systems make or aid consequential decisions that affect human lives and opportunities. But the "black box" nature of many AIs prevents scrutiny. Organizations are struggling with challenges in detecting bias, especially when multiple AI systems interact in complex workflows. Unconscious bias in human-AI collaboration also appears widespread.

High-profile cases of biased algorithms at companies like Amazon, Facebook, and Goldman Sachs have heightened awareness and pressures for ethical AI practices. New regulations like the EU's proposed AI Act are beginning to impose legal obligations for reducing bias and discrimination risks from AI systems. AI that is transparent, fair, and accountable from the start is amenable to responsible scaling without harmful societal impacts (Arnold et al., 2019). Overall, all signs point to AI bias becoming a bigger priority and problem if left unaddressed. Organizations across sectors will need to invest in bias mitigation to fulfill ethical duties and reduce legal, reputation, and performance risks.

- AI systems can inherit and amplify biases in several ways. Biased data used for training can skew algorithms. Programmer biases get built into code and models. Lack of diversity among AI developers is another source (Silberg & Manyika, 2019).
- Real-world examples show how seemingly neutral AI can produce discriminatory outcomes. Facial recognition has higher error rates for women and people of color. Resume screening algorithms disadvantage non-traditional candidates. Targeted ads can exclude or exploit marginalized groups.
- Overreliance on automated decisions without human oversight removes nuance and context. AI predictions should augment rather than replace human judgment. Lack of transparency around AI systems hides unfair impacts.
- Biased AI can significantly harm organizations that deploy such systems by making discriminatory decisions on hiring, performance reviews, qualification screening, target marketing, and more. This leads to reputational damage, lack of diversity, and legal liabilities.
- Techniques like testing with diverse data, algorithmic auditing, external ethical reviews, and getting broader inputs into system design help reduce biases. But mitigating AI bias requires ongoing vigilance, monitoring, and a commitment to fairness.
- Understanding how bias manifests in AI and incorporating ethics into its development is crucial for preventing exclusions, improving opportunities, and allowing organizations to fully benefit from AI capabilities.

6.4.2 Understanding the AI Model Thresholds

Model thresholds are often used in binary classification problems, where the model needs to decide between two classes, such as "positive" or "negative." The threshold represents the probability above which a sample is classified as the positive class, and below which it's classified as the negative class. The choice of model architecture and hyperparameters can impact bias (Blondheim, 2022). Some models may inherently favor certain groups, and tuning hyperparameters can magnify or mitigate this bias.

Some of the methods that can be used to adjust model thresholds and error costs when implementing AI systems are,

- Varying classification thresholds – For binary classifiers, adjust the decision threshold between classes to allow more false positives or negatives based on impact.
- Cost-sensitive training – Set different penalty weights for false positives versus negatives when training models to influence minimizing more costly errors.
- Performance metrics – Optimize models for metrics like balanced accuracy that account for relative class importance versus raw accuracy.

- Probability calibration – For probabilistic models, calibrate predicted probabilities to match true class proportions.
- Up sampling – Increase samples of minority classes in training data to improve detection.
- Class weights – Set higher weights for errors on critical classes during model training.
- Partial predictions – Abstain from high-risk predictions when model confidence is low.

The goal is to tune the models to align with real-world costs and harms associated with different error types. This is context dependent – false positives may be costlier in fraud detection, while false negatives are riskier for medical diagnosis. The technical adjustments should reflect organizational priorities and impact. Adjusting model thresholds and error costs is a crucial aspect of implementing AI systems, especially in applications where precision, recall, or other performance metrics are critical.

6.4.3 Choosing the Right Threshold

Selecting the appropriate threshold depends on the specific business problem and its associated costs and benefits. Here are a few considerations:

- ***ROC Curve Analysis*:** Receiver Operating Characteristic (ROC) curve analysis can help you visualize the trade-off between true positive rate (recall) and false positive rate as the threshold varies. You can choose the threshold that balances these trade-offs according to your business goals.
- ***Cost-Benefit Analysis*:** Conduct a cost-benefit analysis to determine the impact of false positives and false negatives. Consider the financial, operational, and reputational costs associated with each type of error. Adjust the threshold to minimize the overall cost.
- ***Domain Knowledge*:** Your domain expertise is invaluable here. Sometimes, certain misclassifications are more tolerable than others. Adjust the threshold accordingly based on your industry knowledge.

6.4.4 Adjusting Error Costs

The information ecosystem evolved by integrating AI and ML in human-operated organizations shows some significant limitations while addressing and aiming for the resolution of these models (Walton, 2018). Error costs refer to the consequences of misclassifications, and they can vary significantly between different applications. Here's how you can adjust error costs:

***Assign Costs to Errors*:** Quantify the costs associated with false positives and false negatives. These costs may include financial losses, customer dissatisfaction, or legal liabilities.

***Cost-Sensitive Learning*:** Some ML algorithms offer cost-sensitive learning, allowing you to assign different misclassification costs to different classes or instances. This can be particularly useful in imbalanced datasets.

***Iterative Optimization*:** Fine-tune error costs iteratively. Start with initial cost assignments, evaluate the model's performance, and adjust the costs based on the observed errors.

6.5 MITIGATION STRATEGIES FOR HUMAN AND AI BIASES

Diversity and bias training programs raise awareness of implicit biases and provide strategies to counteract them. Meta-analyses find these can effectively improve attitudes and behaviors, provided they are sustained over time rather than one-off sessions. Standardized hiring, evaluation, and promotion processes limit individual discretion and subjective judgment that allow bias to creep in. Structured interviews, blind resume screening, and performance rubrics help assess merit more objectively (Castilla, 2015).

Data collection, auditing, and feedback mechanisms are vital to identify areas where bias manifests and correct them. For example, orchestras instituted blind auditions to conceal gender and increased female hires by 30–55% as a result. Incentives and accountability for improving diversity, such as tying manager compensation to equitable hiring or making diversity goals public, are used increasingly. They demonstrate organizational commitment and prioritization (Dobbin & Kalev, 2016).

Emphasizing organizational values of impartiality, meritocracy, and inclusiveness establishes norms that counteract biased behaviors. However, values must align with actual policies and practices to be effective (Carnes et al., 2012). Underrepresentation in data diversity occurs when certain groups or perspectives are inadequately represented in training datasets. The most effective approach combines technical bias mitigation techniques with thoughtful organizational policies and review processes to address bias risks in AI systems on multiple fronts. This upholds ethical obligations and helps build public trust.

Evidence-based interventions can minimize the detrimental impacts of deeply ingrained human biases within organizations and lead to more objective, performance-focused workplaces. A multipronged approach is highly effective in this regard. Some of the dangers that can arise from overreliance on automated decisions made by AI systems without human oversight can be listed as follows:

Lack of nuance and context

- AI models rely on simplified correlations and patterns. They cannot fully account for complex societal factors behind data or consider edge cases.
- This can lead to unfair outcomes when predicting risk scores, determining qualifications, or profiling individuals. Humans better understand individual contexts.

Unchecked biases

- Without ongoing human auditing and correction, biases encapsulated in training data or algorithms get blindly propagated through automated decisions.
- Historic inequities around race, gender, class, etc., will persist if systems are left to make biased decisions at scale (Mitchell et al., 2021).

Bias amplification

- Automated systems can perpetuate and exacerbate existing biases present in the data used to train them. Without oversight, these biases can go unchecked and lead to discriminatory outcomes.

No room for discretion

- Automated systems work based on rigid rules rather than holistic weighing of circumstances and reasonable judgment.
- This removes discretion needed in many real-world cases with ambiguity, unintended consequences, or ethical dilemmas.

Lack of transparency

- The black box nature of complex AI models obscures how and why certain decisions are made.
- This absence of transparency prevents accountability and the ability to diagnose harmful failures.

AI automation has advantages in scaling decisions, overreliance without ongoing human oversight, auditing, and discretion carries many risks that organizations must proactively address.

6.5.1 Bias Reduction Approaches

Here are some promising techniques researchers and practitioners propose to reduce harmful biases in AI systems:

- ***Diverse and unbiased training data*:** Gebru et al. (2021) proposed ensuring data represents population diversity and is tested for any bias toward any particular faction of the society.
- ***Testing with intersectional samples*:** Evaluate models across combinations of gender, race, and age groups to uncover biases. For example, evaluating image classifiers separately on dark-skin minorities (Buolamwini, 2017).
- ***Adding contextual factors*:** Supplementing models with societal context variables like income levels help avoid superficial correlations that disadvantage minorities.
- ***Human-in-the-loop approaches*:** Getting human feedback on initial model decisions to correct errors and biases before automated deployment.
- ***Algorithmic auditing*:** Techniques to diagnose model fairness across groups by proactively detecting biases (Adebayo et al., 2020).

Overall, reducing biases requires ongoing vigilance, not just one-time fixes. But thoughtful data practices, testing methods, and human oversight provide ways to promote algorithmic fairness.

6.5.2 Leveraging AI to Improve Organizational Performance

Here are some ways organizations can thoughtfully leverage AI to improve performance, while being mindful of potential downsides:

- ***Automating routine tasks*:** AI can take over high-volume, repetitive tasks like paperwork, data entry, and customer service inquiries, freeing up human employees for higher-value work (Wilson et al., 2017).
- ***Optimizing business processes*:** Algorithms excel at finding efficiencies in complex workflows like supply chains, manufacturing, and inventory management, leading to cost and time savings (Tambe et al., 2019).
- ***Enhanced data-driven decision-making*:** AI predictive analytics provide nuanced insights from large datasets that complement human intuition and experience. AI-driven predictive maintenance can help organizations reduce downtime and extend the lifespan of machinery and equipment by identifying maintenance needs before failures occur.
- ***Personalization at scale*:** ML models allow customization of products, content, and recommendations for each individual user, leading to sales boosts (Reshma & MeenaKumari, 2021).

However, to avoid detrimental impacts, AI should not be overextended beyond suitable use cases. Integrating human oversight and auditing for ethics are vital best practices.

6.5.3 AI and Standardization of Tasks

Here are some details on how AI can automate routine, repetitive tasks to free up human employees to focus on more productive work:

- Chatbots and virtual assistants handle large volumes of simple customer service queries, freeing up humans to manage more complex issues (Logesh et al., 2021).
- RPA uses AI to mimic clerical tasks like data entry, form filling, and report generation faster and at lower cost.
- ML techniques can generate customized sales proposals, invoices, contracts automatically with greater efficiency (Surden & Williams, 2020).
- Language processing AI summarizes lengthy documents into concise highlights allowing knowledge workers to extract key insights faster.
- Computer vision AI performs quality assurance on manufacturing lines, image categorization, and other visual inspections more quickly and consistently (Park et al., 2021).

Automating these mundane tasks through AI allows human employees to shift their efforts toward creative problem-solving, advanced analysis, relationship building, and other activities that create greater organizational value. AI has the potential to automate routine tasks, liberating human workers to focus on more complex and creative endeavors. This can significantly boost productivity and efficiency within organizations.

6.5.4 Practical Implications and Case Studies of Organizations using AI to Become More Efficient and Profitable

Certainly, there are several organizations that have successfully leveraged AI to become more efficient and profitable (Sharma & Biros, 2021). Here are some case studies with references and a bibliography to support them:

1. **Amazon**: Amazon uses AI extensively in its supply chain and logistics operations. AI-powered algorithms predict demand, optimize inventory, and manage delivery routes, contributing to significant cost savings and efficient operations.
2. **Netflix**: Netflix employs AI-driven recommendation algorithms to personalize content recommendations for its subscribers. This has led to increased user engagement, reduced churn rates, and higher profitability.
3. **IBM Watson Health**: IBM Watson Health utilizes AI to analyze vast amounts of medical data for healthcare providers. This AI-powered analysis helps in disease diagnosis, treatment recommendations, and drug discovery, ultimately improving patient outcomes and healthcare efficiency.
4. **General Electric (GE)**: AI monitoring of wind turbines detects operational anomalies early and guides preventative maintenance. This has reduced turbine downtime by 10–20% (Davenport & Ronanki, 2018).
5. **Starbucks**: Starbucks employs AI in its mobile app and in-store operations to optimize order processing and personalize marketing offers. This has led to increased sales, better customer experiences, and improved profitability.
6. **Ford Motor Company**: Ford uses AI in its manufacturing processes to improve quality control and optimize production efficiency. AI-driven robotics and analytics have reduced defects and enhanced production speed.
7. **JPMorgan Chase**: JPMorgan Chase utilizes AI for fraud detection and risk management in its financial operations. This has helped the bank prevent fraudulent activities, reduce losses, and enhance profitability.
8. **Walmart**: Walmart employs AI for demand forecasting, inventory management, and supply chain optimization. AI has allowed the company to reduce costs, optimize stocking levels, and improve overall efficiency. Also adopted AI for optimizing staff schedules and placement to align workers with expected store traffic patterns. The system boosted productivity by over $2 billion in its first year.

6.6 SUMMARY

Organizations rely on human effort and automation to increase efficiency and productivity. The biggest challenge is to ensure the transparency of the processes. Human and digital biases violate the basic governance mechanism and invite discrimination and prejudice. Human biases, if unchecked, shape the culture of the organization and collectively they (human and organizational biases) can seep deep into ML and AI. Though there are mechanisms to reduce the biases induced in the ML and algorithmic models, a majority of the corporations do not have the technical or human skills to modify the design models or adjust the error costs of thresholds set for these models. Though there is no perfect solution for this; however, this challenge presents the opportunity to evolve further AI and ML toward the impartiality necessary to increase the efficiency of the organizations in line with their ethical standards and mission.

REFERENCES

Adebayo, J., Gilmer, J., Muelly, M., Goodfellow, I., Hardt, M., & Kim, B. (2020). Sanity checks for saliency maps. *Advances in Neural Information Processing Systems*, *33*, 9525–9536.

Akter, S., McCarthy, G., Sajib, S., Michael, K., Dwivedi, Y. K., D'Ambra, J., & Shen, K. N. (2021). Algorithmic bias in data-driven innovation in the age of AI. *International Journal of Information Management*, *60*, 102387.

Aldoseri, A., Al-Khalifa, K., & Hamouda, A. (2023). A Roadmap for Integrating Automation with Process Optimization for AI-powered Digital Transformation.

Arnold, M., Bellamy, R. K., Hind, M., Houde, S., Mehta, S., Mojsilović, A., … Varshney, K. R. (2019). Fact Sheets: Increasing trust in AI services through supplier's declarations of conformity. *IBM Journal of Research and Development*, *63*(4/5), 6.

Belenguer, L. (2022). AI bias: Exploring discriminatory algorithmic decision-making models and the application of possible machine-centric solutions adapted from the pharmaceutical industry. *AI and Ethics*, *2*(4), 771–787.

Bertrand, M., & Mullainathan, S. (2004). Are Emily and Greg more employable than Lakisha and Jamal? A field experiment on labor market discrimination. *The American Economic Review*, *94*(4), 991–1013.

Blondheim Jr, D. (2022). Improving manufacturing applications of machine learning by understanding defect classification and the critical error threshold. *International Journal of Metalcasting*, *16*(2), 502–520.

Bohnet, I. (2016). *What works: Gender equality by design*. Harvard University Press.

Buolamwini, J. A. (2017). *Gender shades: Intersectional phenotypic and demographic evaluation of face datasets and gender classifiers* (Doctoral dissertation, Massachusetts Institute of Technology).

Carnes, M., Devine, P. G., Isaac, C., Manwell, L. B., Ford, C. E., Byars-Winston, A., … Sheridan, J. (2012). Promoting institutional change through bias literacy. *Journal of Diversity in Higher Education*, *5*(2), 63.

Castilla, E. J. (2015). Accounting for the gap: A firm study manipulating organizational accountability and transparency in pay decisions. *Organization Science*, *26*(2), 311–333.

Cheatham, B., Javanmardian, K., & Samandari, H. (2019). Confronting the risks of artificial intelligence. *McKinsey Quarterly*, *2*(38), 1–9.

D'Ignazio, C., & Klein, L. F. (2020). *Data feminism*. MIT Press.

Danks, D., & London, A. J. (2017, August). Algorithmic bias in autonomous systems. In *Ijcai* (Vol. 17, No. 2017, pp. 4691–4697).

Davenport, T., & Ronanki, R. (2018). Artificial intelligence for the real world. *Harvard Business Review*, *96*(1), 108–116.

Deloitte, (2018). *State of AI in the Enterprise*, 2nd Edition.

Dobbin, F., & Kalev, A. (2016). Why diversity programs fail. *Harvard Business Review*, *94*(7), 14.

Ferrer, X., van Nuenen, T., Such, J. M., Coté, M., & Criado, N. (2021). Bias and discrimination in AI: A cross-disciplinary perspective. *IEEE Technology and Society Magazine, 40*(2), 72–80.

Fu, R., Huang, Y., & Singh, P. V. (2020). Ai and algorithmic bias: Source, detection, mitigation and implications. *Detection, Mitigation and Implications (July 26, 2020).*

Gal, U., Jensen, T. B., & Stein, M. K. (2020). Breaking the vicious cycle of algorithmic management: A virtue ethics approach to people analytics. *Information and Organization, 30*(2), 100301.

Gebru, T., Morgenstern, J., Vecchione, B., Vaughan, J. W., Wallach, H., III, & Crawford, K. (2021). Datasheets for datasets. *Communications of the ACM, 64*(12), 86–92.

Hashimoto, T., Srivastava, M., Namkoong, H., & Liang, P. (2018, July). Fairness without demographics in repeated loss minimization. In *International Conference on Machine Learning* (pp. 1929–1938). PMLR.

Hallihan, G. M., & Shu, L. H. (2013). Considering confirmation bias in design and design research. *Journal of Integrated Design and Process Science, 17*(4), 19–35.

Holstein, K., Wortman Vaughan, J., Daumé III, H., Dudik, M., & Wallach, H. (2019, May). Improving fairness in machine learning systems: What do industry practitioners need? In *Proceedings of the 2019 CHI conference on human factors in computing systems* (pp. 1–16).

Joshi, A., Son, J., & Roh, H. (2015). When can women close the gap? A meta-analytic test of sex differences in performance and rewards. *Academy of Management Journal, 58*(5), 1516–1545.

Kahneman, D. (2011). *Thinking, fast and slow*. Macmillan.

Kordzadeh, N., & Ghasemaghaei, M. (2022). Algorithmic bias: Review, synthesis, and future research directions. *European Journal of Information Systems, 31*(3), 388–409.

Lepri, B., Oliver, N., Letouzé, E., Pentland, A., & Vinck, P. (2018). Fair, transparent, and accountable algorithmic decision-making processes. *Philosophy & Technology, 31*(4), 611–627.

Logesh, R., Subramaniyaswamy, V., Malathi, D., Siva Kumaar, E., Vijayakumar, V., Li, X., & Wang, H. (2021). Intelligent chatbot design framework for campus guidance and services. *Computers & Electrical Engineering, 93*, 107234.

Madras, D., Creager, E., Pitassi, T., & Zemel, R. (2018). Learning adversarially fair and transferable representations. In *International Conference on Machine Learning* (pp. 3384–3393). PMLR.

Makadok, R., & Gamble, J. (2019). Wall Street vs. Main Street: How firms optimize capital allocation decisions. *Academy of Management Perspectives, 33*(2), 225–242.

Mehrabi, N., Morstatter, F., Saxena, N., Lerman, K., & Galstyan, A. (2021). A survey on bias and fairness in machine learning. *ACM Computing Surveys (CSUR), 54*(6), 1–35.

Mitchell, S., Potash, E., Barocas, S., D'Amour, A., & Lum, K. (2021). Algorithmic fairness: Choices, assumptions, and definitions. *Annual Review of Statistics and Its Application, 8*, 141–163.

Morley, J., Machado, C. C., Burr, C., Cowls, J., Joshi, I., Taddeo, M., & Floridi, L. (2021). The ethics of AI in health care: A mapping review. *Social Science & Medicine, 270*, 113172.

Ni, F., Arnott, D., & Gao, S. (2019). The anchoring effect in business intelligence supported decision-making. *Journal of Decision Systems, 28*(2), 67–81.

Nickerson, R. S. (1998). Confirmation bias: A ubiquitous phenomenon in many guises. *Review of General Psychology, 2*(2), 175–220.

O'Neil, C. (2016). *Weapons of math destruction: How big data increases inequality and threatens democracy*. Broadway Books.

Park, K. H., Um, T., Lee, J. H., & Park, S. W. (2021). Automated visual inspection of stamped workpieces using convolutional neural networks. *Applied Sciences, 11*(19), 9132.

Raji, I. D., & Buolamwini, J. (2019, January). Actionable auditing: Investigating the impact of publicly naming biased performance results of commercial AI products. In *Proceedings of the 2019 AAAI/ACM Conference on AI, Ethics, and Society* (pp. 429–435).

Raji, I. D., Smart, A., White, R. N., Mitchell, M., Gebru, T., Hutchinson, B., … Barnes, P. (2020, January). Closing the AI accountability gap: Defining an end-to-end framework for internal algorithmic auditing. In *Proceedings of the 2020 conference on fairness, accountability, and transparency* (pp. 33–44).

Rastogi, C., Zhang, Y., Wei, D., Varshney, K. R., Dhurandhar, A., & Tomsett, R. (2022). Deciding fast and slow: The role of cognitive biases in ai-assisted decision-making. *Proceedings of the ACM on Human-Computer Interaction*, *6*(CSCW1), 1–22.

Reshma, K. K., & MeenaKumari, M. K. (2021). The role of artificial intelligence in marketing. *Materials Today: Proceedings*, *37*, 627–632.

Reskin, B. F. (2000). The proximate causes of employment discrimination. *Contemporary Sociology*, *29*(2), 319–328.

Sharma, M., & Biros, D. (2021). AI and its implications for organisations. In *Information Technology in Organisations and Societies: Multidisciplinary Perspectives from AI to Technostress* (pp. 1–24). Emerald Publishing Limited.

Schwartz, R., Vassilev, A., Greene, K., Perine, L., Burt, A., & Hall, P. (2022). Towards a standard for identifying and managing bias in artificial intelligence. NIST special publication, 1270(10.6028).

Silberg, J., & Manyika, J. (2019). Notes from the AI frontier: Tackling bias in AI (and in humans). *McKinsey Global Institute*, *1*(6), 1–31.

Surden, H., & Williams, M. A. (2020). Technological opacity, predictability, and self-driving cars. *Cardozo Law Review*, *38*, 121.

Suresh, H., & Guttag, J. (2021). A framework for understanding sources of harm throughout the machine learning life cycle. In *Equity and access in algorithms, mechanisms, and optimization* (pp. 1–9).

Tambe, P., Cappelli, P., & Yakubovich, V. (2019). Artificial intelligence in human resources management: Challenges and a path forward. *California Management Review*, *61*(4), 15–42.

Varsha, P. S. (2023). How can we manage biases in artificial intelligence systems–A systematic literature review. *International Journal of Information Management Data Insights*, *3*(1), 100165.

Walton, P. (2018). Artificial intelligence and the limitations of information. *Information*, *9*(12), 332.

Wang, D., Yang, Q., Abdul, A., & Lim, B. Y. (2019, May). Designing theory-driven user-centric explainable AI. In *Proceedings of the 2019 CHI conference on human factors in computing systems* (pp. 1–15).

Wilson, H. J., Daugherty, P. R., & Morini-Bianzino, N. (2017). The jobs that artificial intelligence will create. *MIT Sloan Management Review*, *58*(4). https://sloanreview.mit.edu/article/will-ai-create-as-many-jobs-as-it-eliminates/

Zou, J., & Schiebinger, L. (2018). AI can be sexist and racist—it's time to make it fair. *Nature*, *559*(7714), 324–326.

7 Role of Human Capital in the Supply Chain Management

Muhammad Zulkifl Hasan
University of Central Punjab, Lahore, Pakistan

Muhammad Zunnurain Hussain
Bahria University Lahore Campus, Lahore, Pakistan

Sonia Umair
Al Zahra College for Women, Muscat, Oman

Umair Waqas
University of Buraimi, Al Buraimi, Oman

7.1 INTRODUCTION

In today's fast-paced and competitive business environment, supply chain efficiency has become increasingly vital. If a company is aiming for greatness, disregarding this issue might possibly hinder its attempts to get there, which would be a shame. It is becoming more and more obvious that successful logistics and operations are not the only factors that contribute to the success of a firm. Instead, the capability and experience of the workers are essential components of a productive supply chain management system (Khurosani et al., 2021).

It is becoming more understood how essential human capital is to the management of supply chains. For the foreseeable future, one of the primary areas of focus for research and government is expected to be skilled workers and the role they play in the supply chain. The corporate world is gradually coming to grips with the idea that the success of the supply chain is dependent on the people involved just as much as it is on the processes that are already in place and the technology that is available (Pham et al., 2019).

In the context of the management of supply chains, human capital does not only refer to a group of knowledgeable individuals; rather, it emphasizes making the most of the abilities possessed by all individuals (Alshurideh et al., 2022). There is a possibility that a variety of repercussions will be brought upon the skilled workers in the supply chain. The labor market for skilled workers is particularly sensitive to even the most minute shifts in any of these interrelated elements. It is not enough to merely have the right people working for you; you also need to give them the necessary tasks

DOI: 10.1201/9781032616810-7

inside the system to which they are accountable. When it comes to the supply chain, human resources are given a great deal of attention and discussion. We will talk about the significance of talent and the process of cultivating it, as well as the consequences of the various interwoven ties that make up the supply chain. The management of the complex linkages that exist between a company's skilled staff and its supply chain is the single most important factor in determining the company's long-term performance (Agyabeng-Mensah et al., 2020).

7.2 HUMAN CAPITAL AND SUPPLY CHAIN PERFORMANCE

7.2.1 Defining Human Capital in the Supply Chain Context

The concept of human capital, within the context of corporations and other organizational settings, encompasses the collective combination of an employee's competencies, expertise, and personal attributes. These cover not only an individual's formal education but also their practical experience, skill set, and idiosyncratic traits. When an organization engages in the process of recruiting and selecting new individuals, it can be perceived as an endeavor to augment its reservoir of human capital. The aforementioned asset is not confined to an individual but rather undergoes a process of evolution over time and across an organization, culminating in the formation of its collective knowledge, creativity, and efficiency (Huo, 2016).

The significance of human capital in the field of supply chain management has experienced substantial growth in recent times. Due to the intricate nature of the supply chain and the multitude of dynamic elements it encompasses, a diverse set of talents is required to ensure its seamless operation. The domain of general knowledge encompasses a comprehensive understanding of management and operations concepts, whereas specialized knowledge pertains to specific facets within the supply chain. The presence of skilled labor ensures the coordination and integration of many components within the supply chain, including procurement policies, inventory management, manufacturing processes, and customer service, in order to attain maximum operational efficiency and effectiveness (Santa et al., 2022).

The primary responsibility of a supply chain manager is to effectively manage and synchronize the various interconnected activities that constitute the supply chain. Consequently, individuals are required to possess a diverse array of both broad and specific knowledge in order to successfully fulfill the numerous tasks assigned to them. The comprehensive procedure encompasses many stages, beginning with the identification of prospective suppliers, followed by conducting thorough quality assessments on the items and concluding with addressing consumer comments. In essence, the role of a supply chain manager involves leveraging human resources to effectively coordinate the many interdependencies within the supply chain, with the ultimate objective of ensuring the efficient and timely delivery of items to clients while optimizing costs (Alshurideh et al., 2022).

7.2.2 The Impact of Human Capital on Supply Chain Performance

The importance of human resources in the management of the supply chain is of utmost relevance and should not be underestimated. The fundamental factors that

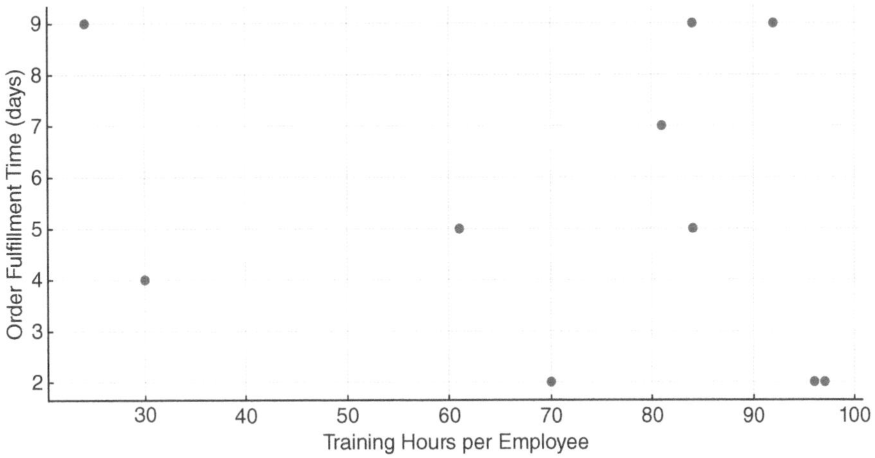

FIGURE 7.1 Training hours per employee vs. order fulfillment time (Siddique, 2023h).

contribute to operational efficiency, quality control, and flexibility are contingent upon the capital of human resources inside a business. The correlation between effective human resource management and improved performance metrics is not accidental but rather causative. Neglecting this connection would indicate a misunderstanding of the essential dynamics that contribute to a successful supply chain (Khurosani et al., 2021).

The correlation between staff training hours and order processing times does not appear to be linear, as seen by the graph (Figure 7.1). Companies with average training hours may show some clustering in terms of order fulfillment timeframes. Additional information and analysis are needed to identify any observable patterns.

1. The results show that the order processing time does not go down or up in a straight line in relation to the number of hours spent in training. It seems to reason that the more time is spent training, the quicker orders will be processed. But there seems to be no underlying cause-and-effect relationship between the two.
2. The data points seem to cluster around certain total training hour thresholds. Order processing times appear to be very similar for companies with similar average training hours. This pooling of data indicates that there are various training thresholds or levels at which order processing speeds rise.
3. However, the correlation between training time and order processing times is not well justified by the graph. Stronger conclusions and distinct trends cannot be attained without further data and research.
4. To have a better grasp on the nature of this correlation, we need to consider the involvement of other factors that may affect order processing timeframes. Internal variables like order volume and complexity, as well as external factors like seasonality and fluctuations in customer demand, can affect how long it takes to complete an order.

5. Several variables, including diminishing returns on training investment, might influence the non-linear relationship between training hours and order processing times. It's possible that the productivity benefits from more training time will be large at first but that the rate of improvement may eventually plateau.
6. Data analysis might utilize cutting-edge statistical techniques, such as regression analysis or machine learning models, to unearth hidden patterns or tendencies. Non-linear connections, interaction effects with other variables, and outliers are some examples of the types of data irregularities that might be uncovered using these methods.
7. It would be helpful to draw some conclusions from an examination of the correlation between training and order processing times. By adjusting training programs, standardizing order-taking methods, and other data-informed activities, businesses may increase efficiency and customer satisfaction.

There is more complexity than meets the eye in the correlation between training hours and order processing times that is shown by the statistics. A more sophisticated understanding and actionable findings require further investigation and examination of other components.

This scatterplot suggests a weak correlation between employee retention rates and inventory turnover rates (Figure 7.2). This finding implies that organizations with longer tenures of employees have a leg up when it comes to inventory control.

7.2.3 Operational Productivity and Human Capital

If investments are made in human capital, there is the potential for there to be a considerable rise in the operational productivity of supply chain activities. This gain might be of significant magnitude. Employees who possess the required levels of

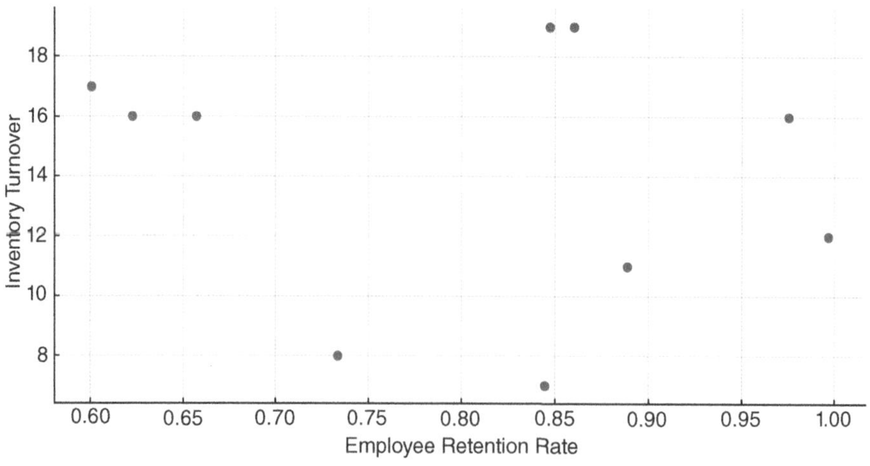

FIGURE 7.2 Employee retention rate vs. inventory turnover (Siddique, 2023h).

education as well as the requisite amount of experience are the backbone of every thriving business. Because of their high level of competence as well as the many years of experience they have gained, they are able to evaluate intricate processes, discover the sources of waste, and make proposals for improvements that would result in reduced costs. This is made possible by the fact that they have gathered an extensive amount of knowledge. Consequently, greater levels of operational productivity will inevitably follow from investments in the education, mentorship, and growth of people. When it comes to determining whether or not a supply chain has been appropriately optimized, the caliber of the human capital that is used is, in the great majority of cases, the component that proves to be decisive (Khurosani et al., 2021).

7.2.4 The Role of Skilled Personnel in Quality Assurance

In the quality assurance industry, it is essential to have staff who are qualified. These individuals are the ones who, in the end, decide the level of quality that a product or service possesses. Because of their expertise and years of experience, they are able to monitor, assess, and guarantee that the goods satisfy requirements at every stage of the production process. In today's environment, when consumers have access to what seems like an infinite number of alternatives and are more discerning than they ever have been, it is very essential to maintain the highest possible level of quality. Skilled workers who are provided with the relevant information are in the best position to guarantee that the final goods or services meet or exceed the expectations of the consumers. This, in turn, minimizes the number of customers who are dissatisfied with the company's products or services (Figure 7.3).

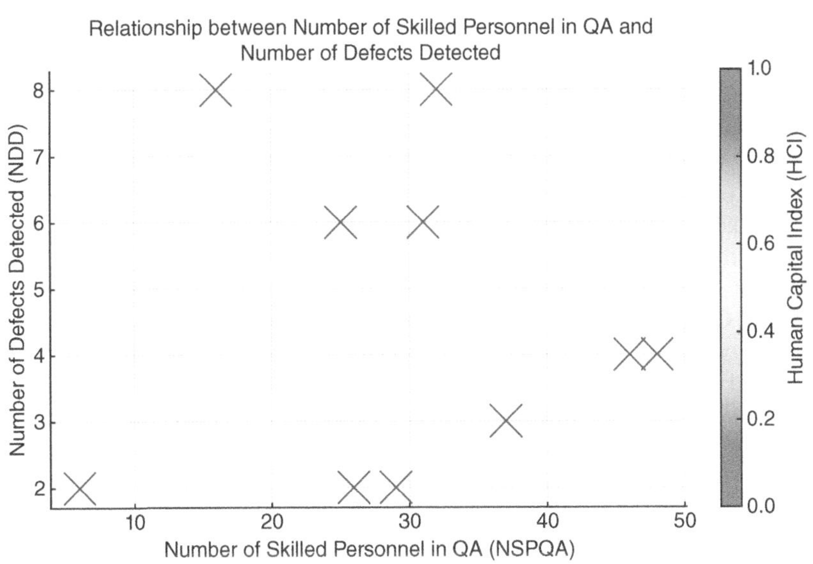

FIGURE 7.3 Number of skilled personnel (Siddique, 2023f).

- The x-axis represents the Number of Skilled Personnel in Quality Assurance (NSPQA).
- The y-axis represents the Number of Defects Detected (NDD).
- The size of each point is proportional to the Human Capital Index (HCI).

7.2.5 Collaboration, Coordination, and Human Capital

The development of collaboration and coordination among the many stakeholders is essential to the functioning of the supply chain, which is why its success is dependent on these two factors. The presence of a large pool of knowledgeable people also contributes to the success of the partnership. When an efficient team is formed, there is a marked improvement in the flow of communication, there is a marked acceleration in the sharing of information, and the process of problem-solving becomes more instinctual. The supply chain functions as a cohesive and effective system, with each component functioning in harmony with one another and being aided by skilled human resources. This allows for the chain to function as efficiently as possible. The practice of cooperating eventually becomes second nature, which fosters an environment that is amenable to the rapid discovery and resolution of problems (Moharana et al., 2010).

A comprehensive understanding of the principles of collaboration, coordination, and the value of human capital might potentially provide benefits for supply chain management and other organizational contexts. The following are a few essential concepts:

Resource-Based View (RBV): The RBV posits that the primary determinant of an organization's success lies in its unique combination of resources and capabilities. The significance of human capital as a valuable resource within the realm of supply chain management cannot be overstated. Maximizing the utility of this asset may be achieved via the promotion of collaboration and coordination among relevant parties (Colbert, 2004).

Transaction Cost Theory: The concept of transaction costs, originally formulated by Ronald Coase and further developed by Oliver Williamson, elucidates the significance of minimizing expenses associated with company transactions. Through the facilitation of more efficient information exchange and the reduction of intermediary involvement, cooperation and coordination have the potential to contribute to the reduction of transaction costs (Ketokivi et al., 2020).

Social Capital Theory: The concept of "social capital" refers to the cumulative worth of an individual's or organization's social relationships and networks that develop and evolve over a period of time. Robust social capital is often seen as the fundamental basis for successful collaboration and coordination, as it facilitates the development of trust and cooperation among diverse stakeholders (DiClemente et al., 2002).

Resource Dependency Theory: The Resource Dependence Theory posits that enterprises are reliant on external assistance in order to effectively operate. When an organization ensures a continuous and reliable supply of resources, it becomes less vulnerable to external disruptions through effective collaboration and coordination within the supply chain (Kim et al., 2020).

Agency Theory: The discipline of agency theory examines the challenges associated with the delegation of authority from one individual (the principal) to another (the agent). The implementation of mechanisms that provide efficient coordination and collaboration among players in the supply chain serves to align their aims and mitigate the challenges arising from agency problems (Mitnick, 2015).

Knowledge Management: The proficient management of information is a critical factor in facilitating efficient supply chain collaboration and coordination. Organizations have the potential to enhance their knowledge acquisition, dissemination, and utilization by employing knowledge management theories and methodologies (Tiwari, 2022).

Organizational Learning Theory: Based on the theoretical framework of organizational learning, it is posited that the facilitation of knowledge advancement inside an organization is achieved through the dual mechanisms of collaboration and coordination. The dynamics of supply chain ecosystems necessitate enterprises to cultivate a culture of learning and adaptability in order to effectively address the issues they encounter (Ali et al., 2023).

Game Theory: The application of game theory enables a more comprehensive understanding and modeling of the strategic interactions among supply chain stakeholders. The analysis of group decision-making processes and the enhancement of collaborative effectiveness are valuable outcomes that may be derived from this approach (AlOmari, 2023).

Human Capital Theory: Educating, training, and developing the workforce are all illustrative instances of investments that are considered valuable according to the Human Capital Theory. The proficiency and knowledge possessed by employees are of paramount importance in facilitating the coordination and collaboration within supply chain operations (Nie et al., 2023).

Team Effectiveness Models: Models aimed at maximizing team performance acknowledge the need for establishing high-performing teams to effectively address intricate challenges. Tuckman's stages of group development and Belbin's team roles are two prominent models within the field of team effectiveness, offering valuable insights into enhancing collaborative dynamics (Verwijs et al., 2023).

Organizational Culture: The organizational culture has the potential to either enhance or impede the organization's capacity to collaborate efficiently. The likelihood of collaboration thriving is higher in an atmosphere characterized by a robust focus on open conversation, mutual trust, and a shared objective (Assoratgoon & Kantabutra, 2023).

Systems Thinking: The supply chain is a complex system characterized by interdependent components. By adopting a holistic perspective, one may enhance their understanding of how elements such as collaboration and synchronization impact the entirety of the supply chain (Gammelgaard, 2023).

The comprehension and improvement of supply chain management and organizational performance may be furthered by the use of the ideas and theories expounded about in this discourse. The significance of these factors in achieving effective and prosperous supply chain operations, as well as their interconnectedness, is underscored.

The graph depicts the Collaboration, Coordination, and Human Capital scores of 10 prominent supply chain firms. J.B. Hunt and C.H. Robinson thrive in collaboration, while UPS, Maersk, and Kuehne + Nagel shine in coordination (Figure 7.4). In terms of human capital, DHL and Maersk top the pack. Conversely, Panalpina regularly scores worse across all three criteria, indicating possible areas for development. This hypothetical data gives a picture of each company's strengths and shortcomings in five important areas, enabling in strategic decision-making and benchmarking.

7.2.6 Navigating Change with Creativity and Adaptability

A number of factors, including shifts in markets, developments in technology, and alterations in client preferences, contribute to the ever-evolving nature of the modern corporate environment. Because of its inherent adaptability and creative potential, human capital is a company's only chance in the unpredictable commercial world of today. Employees who are able to think critically and adapt quickly will be extremely important in helping the firm weather these changes and maintain its competitive edge. Companies that are inclined toward creativity, flexibility, and responsiveness may be able to avoid being purely reactive in their planning and instead anticipate, prevent, and seize opportunities and obstacles as they arise with the support of these traits.

The chart compares the results of 10 consisting of firms on "Navigating Change with Creativity and Adaptability." Xi LLC. scores highly in the category of innovation,

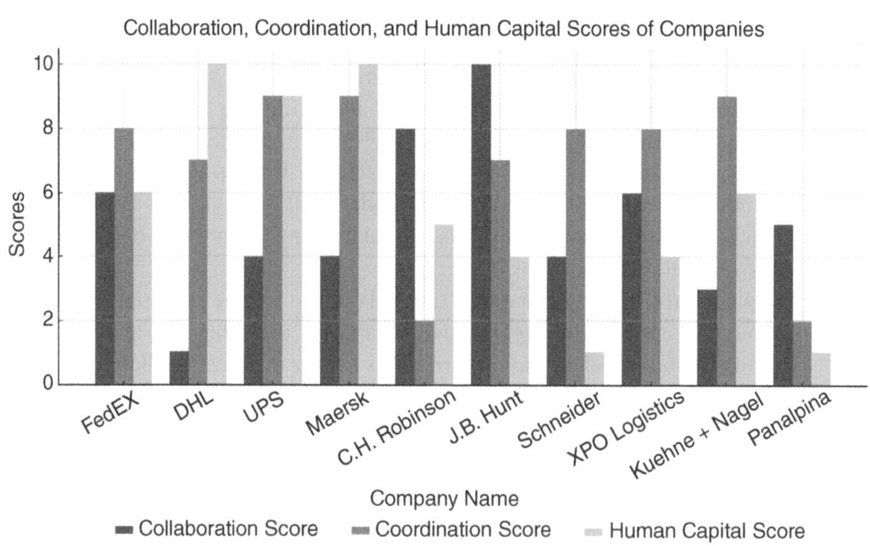

FIGURE 7.4 Collaboration, coordination, and human capital (Siddique, 2023c).

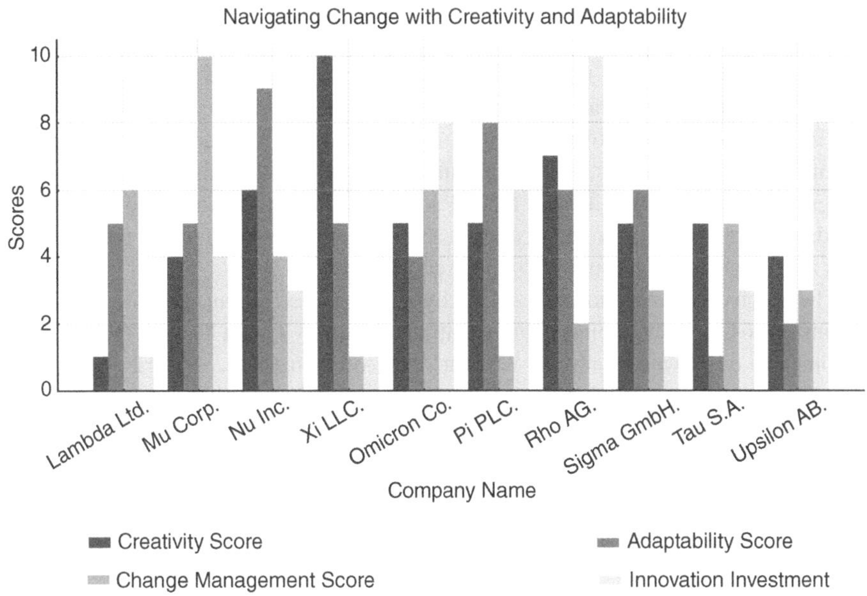

FIGURE 7.5 Creativity and adaptability (Siddique, 2023b).

whereas Nu Inc. does well in the category of flexibility. Rho AG. is the leader in innovation investment, whereas Mu Corp. excels in change management. Tau S.A. and Upsilon AB., on the other hand, have some room for improvement. This infographic highlights the several ways in which businesses may prioritize innovation, flexibility, and creativity in order to successfully deal with change (Figure 7.5).

7.3 DEVELOPING HUMAN CAPITAL IN SUPPLY CHAIN MANAGEMENT

The role of human capital is crucial within the dynamic realm of supply chain management (SCM). The role of human resources is crucial in the implementation of an effective supply chain strategy. This involves several tasks such as the establishment of standardized procedures, the exploration of innovative techniques, and the enhancement of communication channels among all relevant stakeholders. The growth of human capital in this particular domain necessitates continuous education, the acquisition of novel skills, and the cultivation of a proactive mindset. As supply chains continue to evolve through worldwide operations and the integration of advanced technology, the development of competencies and skills becomes imperative. In order to guarantee that the workforce is adequately equipped to address the continuously developing needs of the business landscape, it is imperative to provide resources toward staff development programs, leadership training initiatives, and specialized courses aimed at bridging knowledge gaps (Agyabeng-Mensah et al., 2020).

The process is fostering an environment in which individuals from diverse backgrounds feel at ease expressing their viewpoints, while aggressively seeking novel ideas. By prioritizing collaboration and fostering a team-oriented environment, organizations may effectively leverage the collective knowledge and expertise of their whole workforce, leading to enhanced decision-making processes and the development of comprehensive plans. The development and maintenance of a highly skilled and unified workforce is of utmost importance in facilitating the long-term success of supply chain operations, particularly in light of the growing significance of global issues such as sustainability and ethical sourcing. The proficiency and cooperative mindset of a company's workforce are essential factors in determining the effectiveness and longevity of its supply chain (Acquah et al., 2021).

The graph compares and analyzes the hypothetical performance of 10 well-known organizations in "Developing Human Capital in Supply Chain Management." When it comes to employee retention, Walmart and Amazon shine, whereas Toyota is a leader in training and development. Both Apple and Nike have highly engaged workforces, with Nike also having effective recruitment strategies. In contrast, Samsung and McDonald's fall short in certain key areas, highlighting development opportunities. This infographic highlights potential differences in approach and emphasis between these main actors when it comes to developing and supervising human capital in supply chain operations (Figure 7.6).

7.3.1 Recruitment and Selection

In today's competitive business environment, it is essential to have the right personnel in place to manage and enhance supply chain operations. The supply chain is crucial to every business; therefore, it's important that those working in it are not

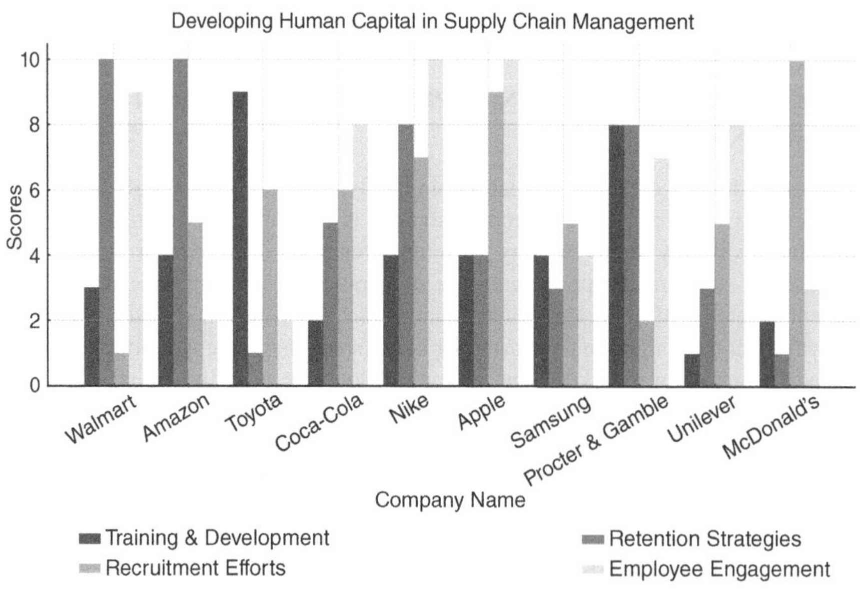

FIGURE 7.6 Developing human capital in supply chain (Siddique, 2023d).

just technically proficient but also subject matter experts in their respective fields. When employing new employees, businesses should prioritize finding those with the knowledge and experience to keep their supply chains running smoothly. In order to keep the supply chain running smoothly, it is essential to find personnel that have both subject expertise and problem-solving skills.

Moreover, despite the apparent requirement for technical acumen, the importance of soft skills cannot be understated in the context of supply chain management. Skillsets like demand forecasting and inventory management are crucial to the technical functioning of the supply chain. Communication and cooperation are the "hard" abilities that provide a responsive supply chain to fluctuating market conditions and allow teams to work together effectively. Therefore, businesses need to strike a balance between the value of technical skills and interpersonal skills throughout the recruiting process. This comprehensive plan ensures that workers have access to all of the tools they need to drive supply chain excellence. Some theories:

Resource-Based View (RBV) Theory: The RBV theory, also known as the resource-based perspective, is a theoretical framework that focuses on the internal resources and capabilities of a firm as the primary drivers of competitive advantage. As per this particular theoretical framework, an organization's competitive advantage may be attributed to its distinct array of resources and capabilities. The RBV theory emphasizes the need to recruit and select supply chain personnel who possess specialized knowledge and skills that may contribute to the overall value creation of the organization (Syed et al., 2022).

Person-Job Fit Theory: Based on the Person-Job Fit Theory, employees tend to exhibit higher levels of productivity and job satisfaction when there is a strong alignment between their own talents and interests and the tasks they are assigned. This approach emphasizes the need to align applicant skills and interests with the demands of supply chain roles (Alqhaiwi et al., 2023).

Person-Organization Fit Theory: Based on the Person-Organization Fit Theory, the likelihood of workers achieving success is higher when there is congruence between their personal views, objectives, and cultural preferences and those of their employer. When engaging in the recruitment process for supply chain roles, it is of utmost importance to identify individuals whose personal values and professional principles align harmoniously with those upheld by the organization (Ruiz-Palomino et al., 2023).

Competency-Based Hiring: Competency-based hiring is an approach that involves the identification and subsequent utilization of fundamental skills and knowledge required for supply chain roles as the basis for applicant selection. Skills like problem-solving, critical thinking, and adaptability are very advantageous in this specific domain (Zamkova et al., 2023).

Talent Pipelining: The process of establishing a talent pipeline involves actively and strategically identifying and nurturing individuals for supply chain roles in anticipation of future job openings. This approach ensures a consistent availability of competent personnel to assume crucial positions in the case of a vacancy (Mitchell, 2023).

Behavioral Interviewing: To predict an applicant's future performance, behavioral interviewing assesses their prior experiences in handling comparable scenarios. The evaluation of candidates' ability to effectively address intricate challenges and exercise sound judgment within the practical realm of supply chain management may be accomplished by employing behavioral interviews (Vann Yaroson et al., 2023).

Assessment Centers: Workplace simulations serve as the foundation for assessment centers, which are a structured method of conducting evaluations. The utilization of assessments in the supply chain industry enables the evaluation of prospective personnel by observing their behaviors and reactions within a controlled environment (Saïah et al., 2023).

Diversity and Inclusion Strategies: Organizations are increasingly broadening their recruitment endeavors to encompass a deliberate emphasis on diversity and inclusion tactics in conjunction with technical proficiencies. The inclusion of individuals with diverse backgrounds and experiences within supply chain teams yields significant advantages. Two potential strategies for enhancing diversity in the workplace are actively engaging with underrepresented groups and implementing more inclusive hiring practices (Oberlack et al., 2023).

Succession Planning: Succession planning is a strategic approach aimed at guaranteeing the future availability of competent persons to assume critical roles within the supply chain. Succession planning ensures the assurance of a pool of competent internal candidates (Larino et al., 2023).

This bar chart compares and contrasts 10 well-known businesses in terms of "Recruitment and Selection in HR in the supply chain." FedEx's outreach is remarkable, while Boeing's rigorous hiring practices are outstanding. Diversity and inclusion efforts at both Philips and Siemens are exemplary. In terms of how long it takes to fill a position, certain companies, like FedEx, Nestle, Philips, and Siemens, appear to be more efficient than others, like Boeing and Ford. This infographic gives us a fictitious peek into how these significant organizations conduct their hiring processes, highlighting their strengths and places for improvement (Figure 7.7).

7.3.2 Training and Development

In order for managers in the supply chain to keep up with the increasing need for skilled laborers, they are required to participate in various types of continuous education and professional development. It is of the utmost importance to give employees opportunities to raise their level of technical skill and education in the most recent best practices and new trends in their respective sectors. Training programs have to be adapted to the requirements of individual professions, but they should also offer opportunities for cross-functional learning in order to enhance communication and cooperation throughout the whole supply chain (Adamson et al., 2023).

This chart compares and analyzes the hypothetical performance of 10 leading corporations in the area of "Training and Development in HR in the supply chain." While Walmart and Boeing succeed at their respective onboarding and induction

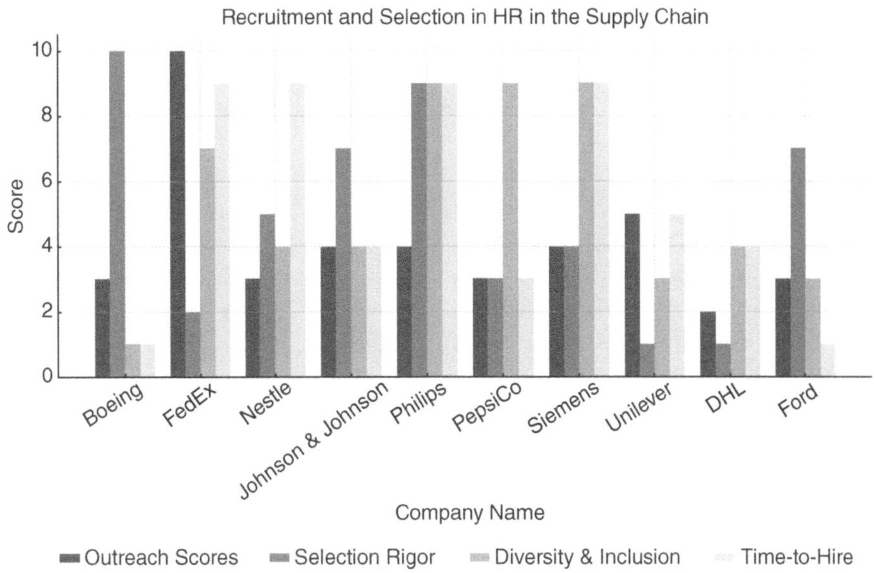

FIGURE 7.7 Selection in HR (Siddique, 2023c).

procedures, companies like Nestle and Amazon excel in technical training. Soft skills education is a priority for Apple, Coca-Cola, and Nike, while Walmart and Toyota place a premium on continuous training and education. The strengths and opportunities in the supply chain can be better understood through this graphic representation of the prospective commitments of these main participants (Figure 7.8).

7.3.3 Performance Management and Recognition

When an organization's goals align with those of the supply chain as a whole, everyone in the chain, from individuals to organizations, stands to profit. These methods, when effectively implemented, ensure that all parts of the supply chain work together toward a unified aim, with the goals of each individual being in sync with the overarching goal. When all of the people involved in the supply chain have a shared understanding of their role in the process and how it contributes to the whole, it not only enables operations to be improved, but it also contributes to the maximization of productivity (Ralston et al., 2023).

It is critical that one understands the significance of motivation in terms of achieving success in any effort. The implementation of award programs, frequent performance evaluations, and the provision of constructive criticism are three strategies that can help businesses boost employee engagement and productivity. When employees believe that their ideas and efforts are valued by their employers, they are more likely to put their whole effort into their work. Because of this, the efficiency of the supply chain is substantially increased as a consequence of the creation of an atmosphere in which people feel motivated to offer their absolute best effort on a consistent basis. It has been demonstrated that one of the most effective ways to

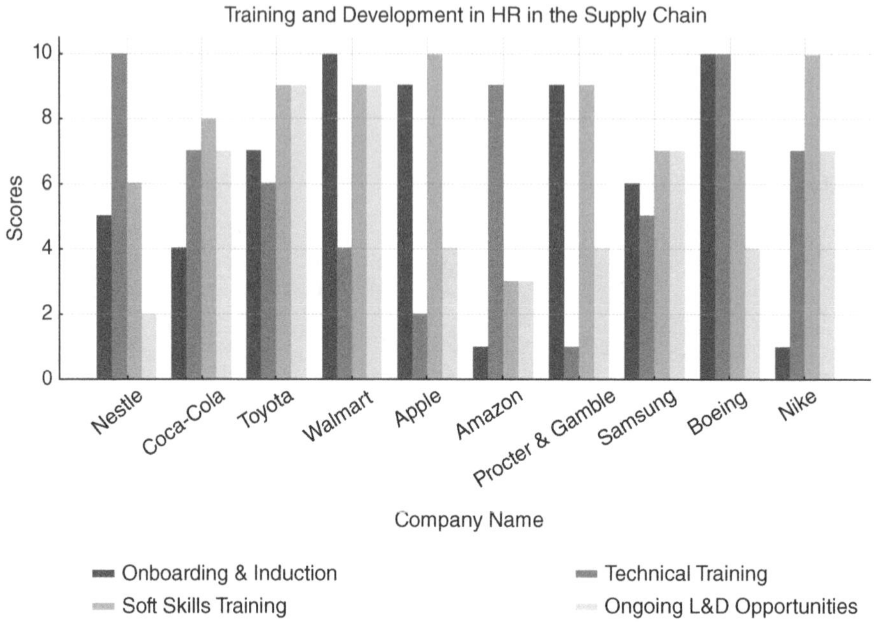

FIGURE 7.8 Training and development in HR.

encourage high levels of productivity that benefits everyone involved is to create an environment in the workplace that is characterized by an attitude of thankfulness and praise (Ralston et al., 2023) (Figure 7.9).

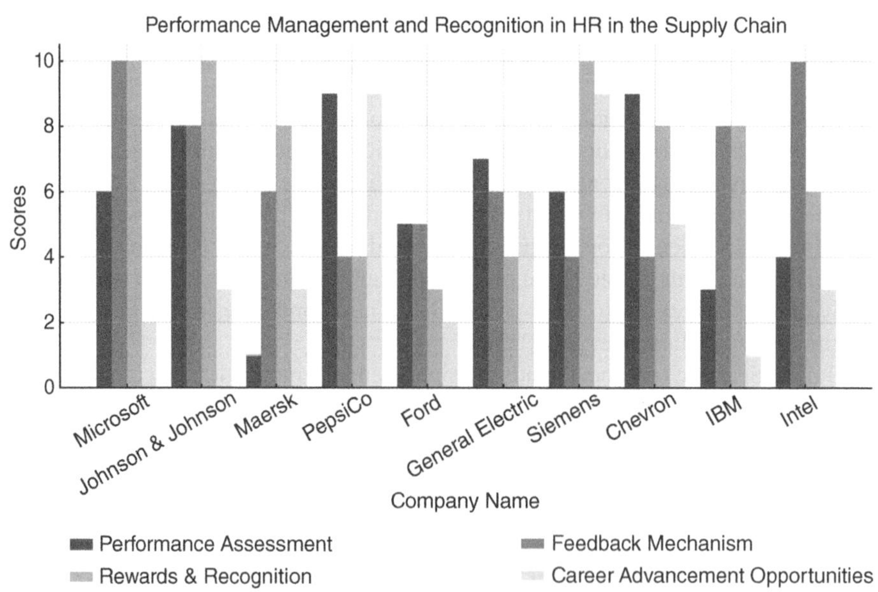

FIGURE 7.9 Performance management in HR in supply chain.

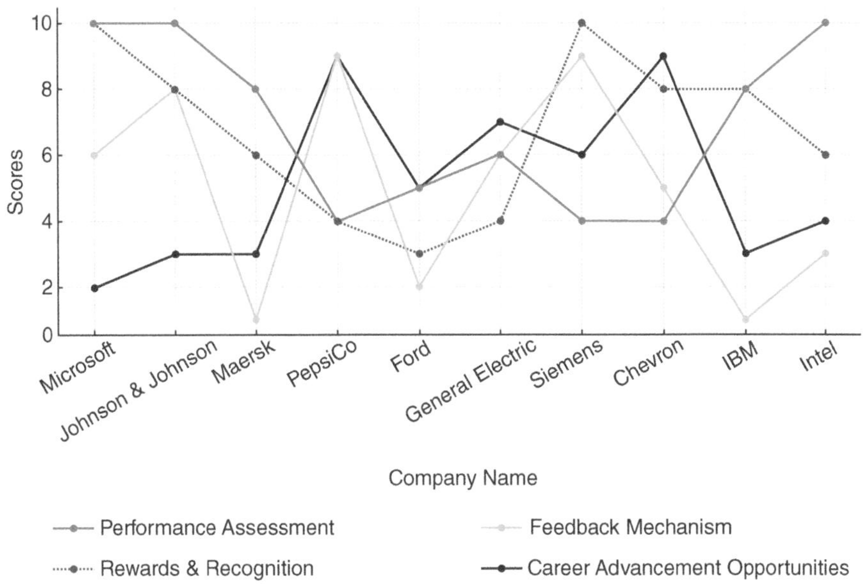

FIGURE 7.10 Performance management and recognition trends in HR in supply chain.

These graphs result of 10 prominent businesses in the field of "Performance Management and Recognition in HR in the supply chain." While Maersk and PepsiCo place a premium on efficient feedback channels, Microsoft and J&J are industry leaders when it comes to performance evaluation. Companies like Ford, GE, and Siemens have particularly innovative approaches to employee recognition and awards. Both Chevron and IBM place a premium on rewarding hard work with promotions and raises (Figure 7.10). This infographic provides insight into the supply chain management objectives and strengths of some of the world's largest companies.

7.4 LEVERAGING HUMAN CAPITAL IN SUPPLY CHAIN MANAGEMENT

7.4.1 Collaboration and Knowledge Sharing

The establishment of an environment that is conducive to the promotion of collaboration and the facilitation of information transmission in a seamless manner throughout the supply chain can result in improved performance for an organization. When companies use this strategy, they are able to harness the pooled knowledge, experiences, and insights of their staff more effectively, which ultimately results in improved decision-making processes and greater operational efficiency. Workers are able to actively participate in the operations of an organization when they are provided with platforms and tools that make it easier for them to communicate effectively, share data, and work together to find solutions to problems. People are able to share their direct observations, advocate for the most effective approaches, and disseminate the information they have gained from their positions. This makes it easier for employees inside the company to share information that might be helpful to one another, which is beneficial to the firm as a whole (AlQershi, N., 2021).

In addition, when a firm makes it a priority to encourage open lines of communication and collaborative problem-solving, it unintentionally creates an atmosphere that fosters lifelong learning and provides opportunities for professional development. The existence of knowledge that is easily available enables workers at all stages of the supply chain to constantly improve their skills, get new perspectives from their coworkers, and maintain familiarity with the most recent industry standards and most effective methods. This kind of environment not only helps to foster the growth of individuals who are highly skilled and informed, but it also helps to establish an atmosphere that promotes ongoing innovation and enhancement. Maintaining organizational flexibility, resilience, and long-term performance over the long run requires the incorporation of continual education and development into the network of the supply chain (Figure 7.11).

The line chart provides a visual representation for the purpose of comparing 10 prominent organizations with regard to their level of collaboration and knowledge sharing in the human resources aspect of the supply chain. Adidas and Shell have exceptional proficiency in cultivating inter-departmental collaboration, while Netflix and Airbus exhibit notable expertise in establishing repositories of information.

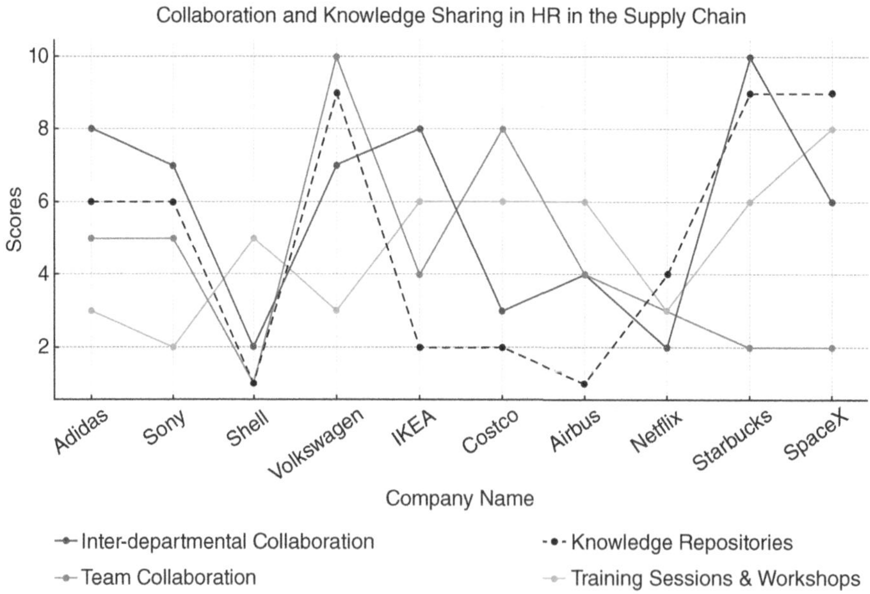

FIGURE 7.11 Collaboration and knowledge in HR (Siddique, 2023a).

The strength of team cooperation is particularly evident in the case of Shell and SpaceX, whereas Sony, Costco, and Starbucks are distinguished by their notable dedication to conducting training sessions and workshops. The graphic highlights the diverse goals across these organizations, indicating different strategies for promoting cooperation and knowledge sharing in their human resources operations within the framework of the supply chain (Siddique, 2023g).

7.4.2 Empowering Decision-Making

Because they are intricate networks for the distribution of products, supply chains demand a high degree of reactivity and agility from their operators in order to function at the highest possible level. It is possible to greatly improve these two essential aspects by providing workers with the authority to make decisions within the scope of their responsibilities. Instead of depending on directions from higher-ups, individuals who are given the authority to make decisions based on accurate and current information are able to quickly address issues or inefficiencies in the workplace. Their capacity to make well-informed judgments is improved as a result of the deployment of cutting-edge decision support technology, which in turn helps to cultivate an environment that is conducive to rapid decision-making and is free of unneeded bureaucratic obstructions (Pessot et al., 2023).

In addition, the empowerment of those involved not only makes it easier to streamline the decision-making processes but also serves a number of other functions. Employees are more likely to acquire a higher sense of commitment toward the outcomes of their decisions when they are given the liberty to oversee their separate domains and when there is an atmosphere of trust inside the workplace. This cultivates a proactive approach in which employees are continually alerted to exploring opportunities to optimize processes and swiftly resolving any disruptions or impediments observed within the supply chain. Specifically, the supply chain may be disrupted when an obstacle is encountered. Individuals take on an active role in affecting the efficiency and success of the supply chain, as opposed to just following the directives that are given to them by higher authorities in the organization. Individuals being given greater control over their work via increased ownership and adopting a more proactive approach typically result in the development of innovative solutions and a more resilient supply chain in general (Pessot et al., 2023).

The human resources department of "Celestial Solutions" is given wide latitude to make personnel choices across the supply chain. On the other hand, "Pulsar Inc." displays the least independence. When we compare companies based on their supply chain efficiency, "Nova Industries," "Stellar Corp.," and "Orion Enterprises" come out on top. The "Orion Enterprises" and "Lunar Group" teams, on the other hand, are less effective. The high levels of satisfaction reported by "Celestial Solutions" employees show that giving HR more leeway might boost morale. On the other hand, "Pulsar Inc." has the unhappiest workers (Figures 7.12–7.14).

Essentially, the revised graphics show possible connections between HR freedom, supply chain effectiveness, and employee happiness at various organizations, highlighting the significance of enabling HR choices along the supply chain.

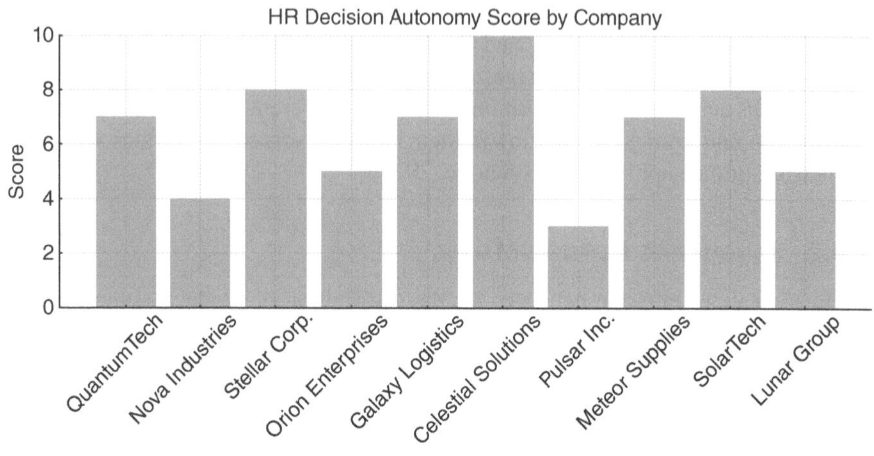

FIGURE 7.12 HR decision autonomy score.

FIGURE 7.13 Supply chain efficiency score.

7.4.3 Leadership and Talent Management

The optimization of human capital is of the utmost significance in fields such as the supply chain, where the successful coordination of numerous parts is essential to the business' overall performance. Effective leadership is at the center of this, playing an essential role in ensuring that the available human resources are utilized to their fullest potential. Strong leaders are responsible for planting the seeds of

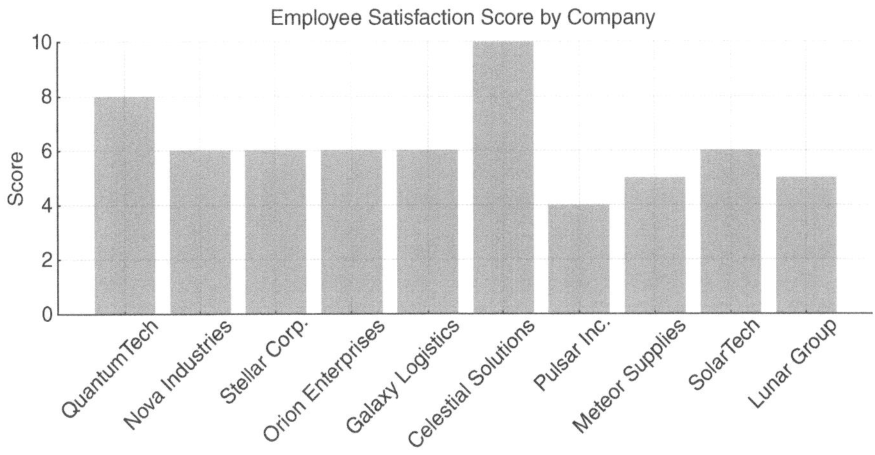

FIGURE 7.14 Employee satisfaction score.

trust, cooperation, and innovation in their teams, in addition to creating operational efficiency. It is possible for leaders to embed these guiding principles in their teams in order to consistently elicit excellent performance from their personnel. When employees come to work in an environment where there is an emphasis placed on trust, collaboration is encouraged, and innovation is rewarded, they are more likely to feel inspired and involved in the work that they do (Prabhu et al., 2023).

In addition to this, the long-term success of any firm is dependent on the organization's ability to cultivate future leaders and preserve a consistent supply of skilled personnel. In this regard, effective tactics for talent management are absolutely essential. Planning for a company's succession may help an organization get ready for the future by locating and cultivating potential leaders who will be able to take over when the time is right. In addition, organizations may be able to improve employee retention and motivation by fostering a sense of appreciation among workers and outlining a route to promotion for employees by offering career advancement possibilities. This may be accomplished by providing employees with opportunities to grow in their professions. These steps not only help in retaining top individuals, but they also ensure a crew that is ready to take on new challenges and contribute to the supply chain's continuous success. This is an important factor in ensuring continued success (Figures 7.15–7.17).

Organizations such as "Helix Enterprises" and "MoleculeTech" demonstrate commendable proficiency in several aspects of human resources management. Conversely, "Neutron Industries" has notable competencies in some domains, however, would benefit from reassessing and adjusting its tactics in other areas. Data-driven insights provide firms with the ability to refine their human resources strategies in order to achieve optimal outcomes throughout the supply chain.

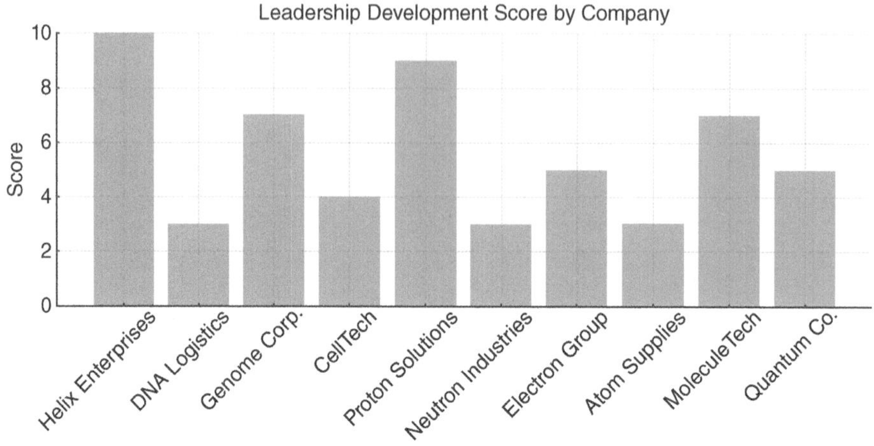

FIGURE 7.15 Leadership and talent management in HR (Siddique, 2023e).

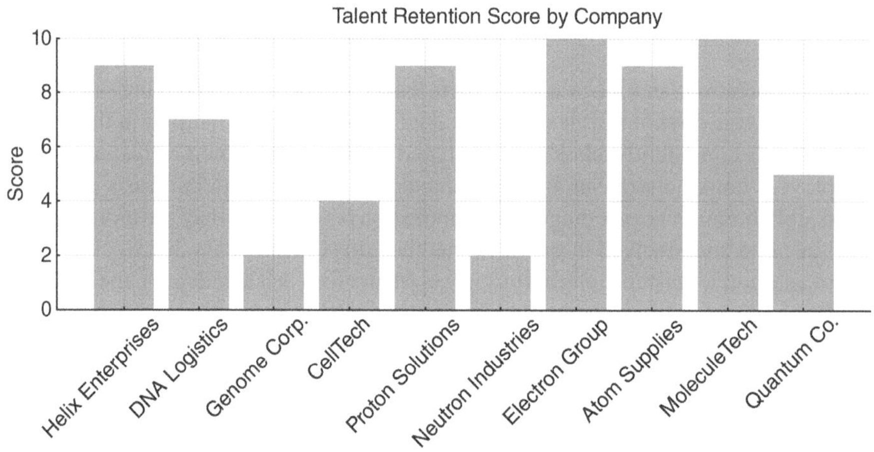

FIGURE 7.16 Talent score (Siddique, 2023e).

7.5 CONCLUSION

The administration of supply chains is a complicated process that places significant demands on the knowledge and talents of human resources. A highly experienced workforce that possesses a full understanding of the nuances and intricacies of the whole ecosystem is required for the proper coordination of logistics, procurement, and distribution. This is because the entire ecosystem is intricate. In the context of businesses, recognizing the crucial significance of competent individuals goes beyond merely recognizing their talents; rather, it constitutes an acknowledgment of the direct relationship between talented employees and the efficient operation of the supply chain. This is an important step in the process of recognizing the crucial

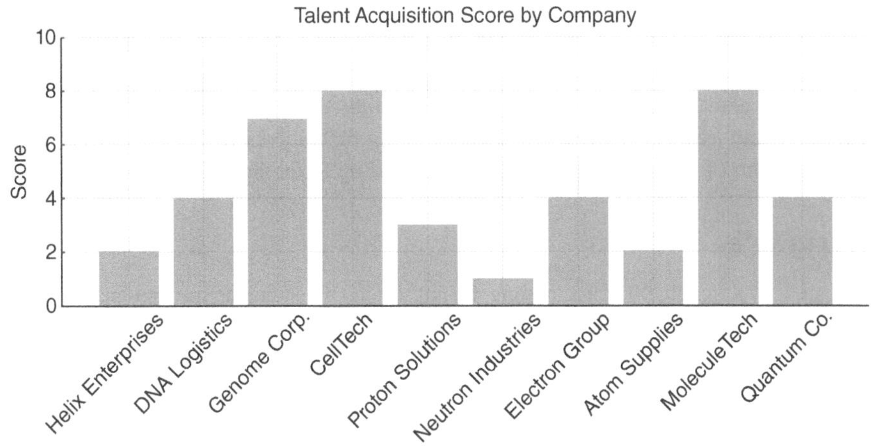

FIGURE 7.17 Talent acquisition score (Siddique, 2023e).

significance of competent individuals. It is not a secondary or tertiary responsibility to invest in the professional development of these employees; rather, it is a purposeful and calculated action that may greatly affect the efficiency, flexibility, and overall achievement of a company's operations.

The obligation rests not only in the recruitment of qualified workers but also in the continuous improvement of their abilities in order for them to be able to adjust to the ever-changing nature of supply chain management. This requires not just the provision of training programs but also the creation of an environment that is supportive and supports the development of skills, the invention of new ideas, and the ability to respond to changing conditions. This strategy ensures that employees are not just reacting to shifts in the market but also anticipating shifts in the market and modifying their tactics to account for these developments. It is possible that the existence of a team that is highly competent and creative will have a big influence on the success of a company, distinguishing it from rivals that struggle to adapt to changes in the market.

The involvement of human resources in the management of supply chains goes beyond the operational components of these chains. Their efforts to stimulate innovation respond to challenges and ensure sustainability are the sources of the priceless assets they currently hold. It is possible for businesses to improve their supply chain operations by encouraging a culture of collaboration, providing staff with the authority to take initiative, and putting effective leadership techniques into place. In the extremely competitive field of supply chain management, the inclusion of human connection, collective expertise, and the sharing of experiences among employees may considerably steer a firm in the direction of unparalleled accomplishments.

The fundamental focus of future endeavors should be around implementing comprehensive programs that aim to foster the ongoing growth and empowerment of supply chain employees. In conjunction with routine training sessions, it is imperative to cultivate an environment that fosters proactivity in problem-solving and a willingness to adapt to dynamic market circumstances. Acknowledging the significance of skilled individuals and allocating resources toward their professional growth enables organizations to cultivate teams capable of not just adjusting to shifts in the market but also anticipating and perhaps shaping them. The use of a systematic approach may assist enterprises in differentiating themselves within the highly competitive realm of supply chain management, leading to unparalleled achievements.

REFERENCES

Acquah, I. S. K., et al., 2021. Examining the link among green human resource management practices, green supply chain management practices and performance. *Benchmarking an International Journal*, p. 23. https://doi.org/10.1108/BIJ-05-2020-0205

Adamson, H. K., et al., 2023. An exploration of communication skills development for student diagnostic radiographers using simulation-based training with a standardised patient: UK-based focus-group study. *Journal of Medical Imaging and Radiation Sciences*, p. 7. https://doi.org/10.1016/j.jmir.2023.06.004

Agyabeng-Mensah, Y. et al., 2020. Examining the influence of internal green supply chain practices, green human resource management and supply chain environmental cooperation on firm performance. *Supply Chain Management: An International Journal*, p. 14. https://doi.org/10.1108/SCM-11-2019-0405

Ali, A., Jiang, X., & Ali, A. (2023). Enhancing corporate sustainable development: Organizational learning, social ties, and environmental strategies. *Business Strategy and the Environment*, 32(4), pp. 1232–1247.

AlOmari, A. M. H., 2023. Game theory in entrepreneurship: A review of the literature. *Journal of Business and Socio-economic Development*, p. 12. https://doi.org/10.1108/JBSED-01-2023-0005

AlQershi, N., 2021. Strategic thinking, strategic planning, strategic innovation and the performance of SMEs: The mediating role of human capital, [Online] https://doi.org/10.5267/j.msl.2020.9.042 Available at: http://growingscience.com/beta/msl/4391-strategic-thinking-strategic-planning-strategic-innovation-and-the-performance-of-smes-the-mediating-role-of-human-capital.html [Accessed September 2023].

Alqhaiwi, Z. O. et al., 2023. Linking person-job fit and intrinsic motivation to salespeople's service innovative behavior. *Journal of Services Marketing*, p. 15. https://doi.org/10.1108/JSM-04-2023-0154

Alshurideh, M. T., Al Kurdi, B., Alzoubi, H. M., Ghazal, T. M., Said, R. A., AlHamad, A. Q., et al. (2022). Fuzzy assisted human resource management for supply chain management issues. *Annals of Operations Research*, 2, pp. 1–19.

Assoratgoon, W. and Sooksan Kantabutra, 2023. Toward a sustainability organizational culture model. *Journal of Cleaner Production*, p. 12. https://doi.org/10.1016/j.jclepro.2023.136666

Colbert, B. A., 2004. The complex resource-based view: Implications for theory and practice in strategic human resource management. *Academy of Management Review*, p. 13. https://doi.org/10.2307/20159047

DiClemente, R. J. et al., 2002. *Emerging theories in Health Promotion practice and research: Strategies for improving public health*. p. 28.

Gammelgaard, B., 2023. Editorial: Systems approaches are still providing new avenues for research as the foundation of logistics and supply chain management. *International Journal of Logistics Management*, p. 4. https://doi.org/10.1108/IJLM-01-2023-601

Huo, B. E. A., 2016. The impact of human capital on supply chain integration and competitive performance. *International Journal of Production Economics*, p. 11. https://doi.org/10.1016/j.ijpe.2016.05.009

Ketokivi, M. et al., 2020. Transaction cost economics as a theory of supply chain efficiency. *Production and Operations Management*, p. 21. https://doi.org/10.1111/poms.13148

Khurosani, A. et al., 2021. The role of green human resource management in creating green supply chain culture in a service industry. *Uncertain Supply Chain Management*, p. 12. https://doi.org/10.5267/j.uscm.2021.7.003

Kim, S. T. et al., 2020. Logistics integration in the supply chain: A resource dependence theory perspective. *International Journal of Quality Innovation*, p. 6. https://doi.org/10.1186/s40887-020-00039-w

Larino, J. P. et al., 2023. Of succession stories and scenarios: The challenges and options facing first-generation family-Owned Restaurants. *American Journal of Multidisciplinary Research and Innovation*, p. 24. https://doi.org/10.54536/ajmri.v2i2.1201

Mitchell, D., 2023. Strategies for the implementation of supply chain internships: A case of meaningful boundary spanning relationships. *University of Arkansas, Fayetteville*, p. 11. https://scholarworks.uark.edu/scmtuht/20

Mitnick, B. M., 2015. Agency Theory. *Wiley Encyclopedia of Management*, p. 6. https://doi.org/10.1002/9781118785317.weom020097

Moharana, H. et al., 2010. Coordination, collaboration and integration for Supply Chain Management. *Interscience Management Review*, p. 4. https://doi.org/10.47893/IMR.2010.1044

Nie, P.-Y. et al., 2023. Innovation and competition with human capital input. *Managerial and Decision Economics*, p. 6. https://doi.org/10.1002/mde.3782

Oberlack, C. et al., 2023. With and beyond sustainability certification: Exploring inclusive business and solidarity economy strategies in Peru and Switzerland. *World Development*, p. 13. https://doi.org/10.1016/j.worlddev.2023.106187

Pessot, E. et al., 2023. Empowering supply chains with Industry 4.0 technologies to face megatrends. *Journal of Business Logistics*, p. 13. https://doi.org/10.1111/jbl.12360

Pham, N. T. et al., 2019. Green human resource management: A comprehensive review and future research agenda. *International Journal of Manpower*, p. 28. https://doi.org/10.1108/IJM-07-2019-0350

Prabhu, H. M. et al., 2023. CEO transformational leadership, supply chain agility and firm performance: A TISM modeling among SMEs. *Global Journal of Flexible Systems Management*, p. 9. https://doi.org/10.1007/s40171-022-00323-y

Ralston, P. M. et al., 2023. The building blocks of a supply chain management theory: Using factor market rivalry for supply chain theorizing. *Journal of Business Logistics*, p. 18. https://doi.org/10.1111/jbl.12320

Ruiz-Palomino, P. et al., 2023. How temporary/permanent employment status and mindfulness redraw employee organizational citizenship responses to person-organization fit. *Revista de psicología del trabajo y de las organizaciones*, p. 13. https://doi.org/10.5093/jwop2023a3

Saïah, F. et al., 2023. Process modularity, supply chain responsiveness, and moderators: The Médecins Sans Frontières response to the Covid-19 pandemic. *Production and Operations Management*, p. 21. https://doi.org/10.1111/poms.13696

Santa, R. et al., 2022. An investigation of the impact of human capital and supply chain competitive drivers on firm performance in a developing country. *PloS one*, p. 12. https://doi.org/10.1371/journal.pone.0274592

Siddique, M. R., 2023a. Collaboration and Knowledge dataset. [Online] Available at: https://www.kaggle.com/datasets/razasiddique/collaboration-and-knowledge-dataset [Accessed 2023].

Siddique, M. R., 2023b. Creativity and Adaptability. [Online] Available at: https://www.kaggle.com/datasets/razasiddique/creativity-and-adaptability [Accessed 2023].

Siddique, M. R., 2023c. HR in supply chain. [Online] Available at: https://www.kaggle.com/datasets/razasiddique/hr-in-supply-chain [Accessed 2023].

Siddique, M. R., 2023d. Human Capital in Supply Chain Dataset. [Online] Available at: https://www.kaggle.com/datasets/razasiddique/human-capital-in-supply-chain-dataset [Accessed 2023].

Siddique, M. R., 2023e. Leadership and Talent Management in HR. [Online] Available at: https://www.kaggle.com/datasets/razasiddique/leadership-and-talent-management-in-hr [Accessed 2023].

Siddique, M. R., 2023f. Quality Assurance in Supply Chain. [Online] Available at: https://www.kaggle.com/datasets/razasiddique/quality-assurance-in-supply-chain [Accessed 2023].

Siddique, M. R., 2023g. Selection in HR in supply chain. [Online] Available at: https://www.kaggle.com/datasets/razasiddique/selection-in-hr-in-supply-chain [Accessed 2023].

Siddique, M. R., 2023h. Supply Chain Performance. [Online] Available at: https://www.kaggle.com/datasets/razasiddique/supply-chain-performance [Accessed 2023].

Syed, I. K. et al., 2022. Supply chain agility and organization performance: A resource based view. [Online] Available at: https://ieomsociety.org/proceedings/2023houston/144.pdf [Accessed 2023].

Tiwari, S. P., 2022. Knowledge management strategies and emerging technologies – an overview of the underpinning concepts. [Online] https://doi.org/10.31235/osf.io/uzq5c Available at: http://arxiv.org/abs/2205.01100 [Accessed 2023].

Vann Yaroson, E. et al., 2023. The role of power-based behaviours on pharmaceutical supply chain resilience. *Supply Chain Management: An International Journal*, p. 16. https://doi.org/10.1108/SCM-08-2021-0369

Verwijs, C. et al., 2023. A theory of Scrum team effectiveness. *ACM Transactions on Software Engineering and Methodology*, p. 50. https://doi.org/10.1145/3571849

Zamkova, N. et al., 2023. Training of future logistics and supply chain managers: A competency approach. *Financial and Credit Activity Problems of Theory and Practice*, p. 13. https://doi.org/10.55643/fcaptp.1.48.2023.3946

8 Understanding Artificial Intelligence in Supply Chain and Innovation Performance

Muhammad Zulkifl Hasan
University of Central Punjab, Lahore, Pakistan

Muhammad Zunnurain Hussain
Bahria University Lahore Campus, Lahore, Pakistan

Sonia Umair
Al Zahra College for Women, Muscat, Oman

Umair Waqas
University of Buraimi, Al Buraimi, Oman

8.1 INTRODUCTION

The exponential growth of artificial intelligence (AI) across diverse industries signifies a notable transformation in the execution of operations and procedures. Supply chain management is an industry that is now going through a significant period of transformation. The integration of AI-driven solutions is currently being employed to simplify supply chains, which have traditionally been perceived as intricate and multidimensional procedures. These technological solutions not only provide improved operational efficiency but also introduce creative approaches to address supply chain difficulties. AI is increasingly transforming the landscape of supply chain management by enhancing many aspects, such as demand prediction and logistics route optimization. This advancement enables supply chain operations to become more proactive, agile, and flexible (Baryannis et al., 2019).

The ability of AI to effectively analyze and interpret extensive datasets and then generate well-informed conclusions significantly enhances the overall decision-making procedures within supply chain management. These breakthroughs are facilitating the emergence of a novel era of innovation, characterized by not just the enhancement of existing processes but also their complete reconceptualization. One example of the application of AI is the utilization of predictive analytics, which enables organizations to get valuable insights into forthcoming market demands.

DOI: 10.1201/9781032616810-8

This, in turn, facilitates the adjustment of production schedules in alignment with these anticipated demands. In addition, machine learning algorithms have the potential to contribute to inventory optimization by minimizing carrying costs and ensuring the timely availability of items. In addition, the utilization of AI-powered chatbots and automated customer support systems has the potential to offer clients timely notifications regarding their purchases, therefore augmenting the overall quality of the customer experience. The integration of AI has led to the emergence of several breakthroughs that offer concrete advantages for enterprises, including cost reduction, enhanced operational effectiveness, and heightened consumer contentment (Sharma et al., 2022).

The implementation of AI in the domain of supply chain management presents some obstacles. Many companies face challenges during the first phases of incorporating AI into their business processes, mostly attributed to the need for substantial investments in technology and training. Another problem lies in the imperative task of guaranteeing the accuracy and relevance of the vast volumes of data that are necessary for the optimal functioning of AI (Dubey et al., 2021). The preservation of data privacy and security is of utmost importance, particularly in the context of utilizing cloud-based AI technologies. Moreover, it is important for enterprises to recognize that the adoption of new technologies entails a learning process, and it is crucial for them to guarantee that their employees receive sufficient training and are well-prepared to collaborate effectively with AI-driven systems (Dlamini et al., 2020). Essentially, although the advantages of integrating AI are significant, it is crucial for organizations to be aware of the accompanying difficulties and be ready to properly tackle them (Sharma et al., 2022).

8.2 AI APPLICATIONS IN THE SUPPLY CHAIN

8.2.1 Demand Forecasting and Planning

The accurate prediction and strategic preparation of demand are key elements within the supply chain. The introduction of AI has led to notable enhancements in the precision and effectiveness of these procedures (Ahmed et al., 2020; Kelly et al., 2019). AI algorithms can analyze extensive quantities of historical data, encompassing previous sales records, market trends, and even patterns in client behavior. Through the analysis and interpretation of this data, AI has the capability to make precise and reliable predictions regarding demand (Gao et al., 2021).

The following are some advantages associated with the utilization of (AI) in the context of demand forecasting and planning:

- Optimized inventory management involves the utilization of AI to assist enterprises in effectively maintaining an appropriate level of stock. This enables them to mitigate the risk of excessive inventory, which can result in cash being tied up or insufficient inventory, which can result in missed sales opportunities.
- The occurrence of stockouts can result in the loss of potential sales and have a detrimental impact on a brand's reputation. AI has the capability to

forecast the depletion of a product and afterward initiate the necessary procedures for replenishment.
- The provision of items at the desired time by businesses results in heightened levels of satisfaction among consumers, hence fostering a higher likelihood of repeat purchases and favorable referrals.

The blue line shows the product's historical demand over the last 50 days. AI's projection of future demand is shown as a red line. The observed pattern shows that the predicted demand nearly reflects the historical demand, with a minor change (Figure 8.1). This is common for projections, as they don't always coincide with real demand. Variations might emerge because of things like incorrect assumptions made in the model, unexpected occurrences, or other unforeseen impacts (Chopra, A., 2019).

The product's daily inventory is depicted by the green bars. Inventory levels appear to be in flux. Demand, supply chain restrictions, and replenishment tactics may all play a role in this fluctuation. Supply chain management relies heavily on accurate stock counts to both satisfy customer demand and prevent the high costs associated with overstocking (Figure 8.2).

Through the use of these visualizations, firms can get valuable information pertaining to their demand patterns, forecast accuracy, and inventory management. Subsequent modifications might be implemented to the supply chain strategy in response.

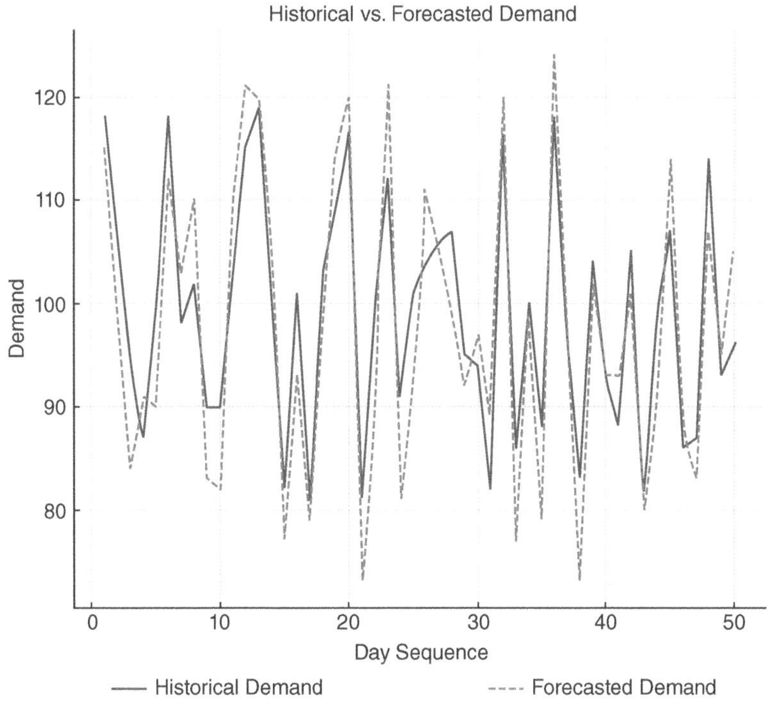

FIGURE 8.1 Historical vs. forecasted demand using AI (lastman0800, 2023c).

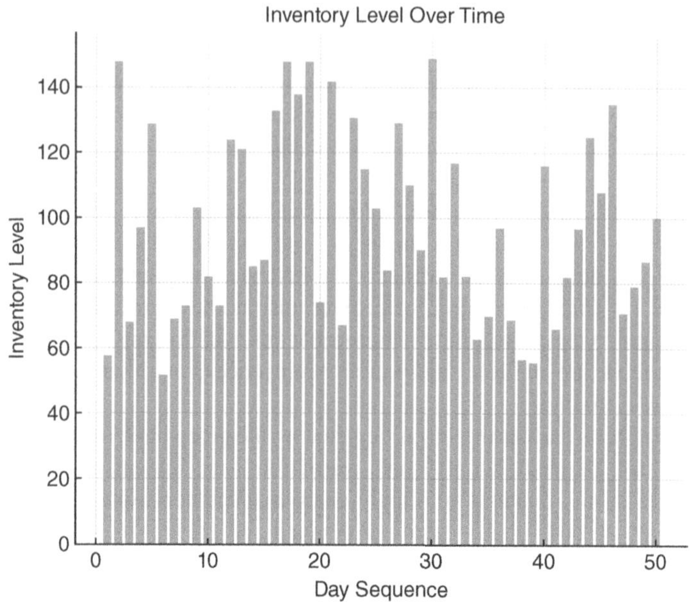

FIGURE 8.2 Inventory Level (lastman0800, 2023c).

8.2.2 Warehouse and Inventory Management

Warehouse and inventory management is essential for the timely and smooth shipping of goods stored there. An integral part of these processes is played by AI, which may be used to examine how often items are retrieved to optimize warehouse layout. As a result, AI may provide suggestions on how the warehouse should be set up such that frequently used items are easier to find. Using AI, stock may be tracked in real time. This method ensures that supplies will always be restocked on schedule and that surplus stock will never build up. Order fulfillment is one area where AI has been proven to greatly improve operational efficiency (Gregory, 2023). By using AI capabilities, it becomes possible to accurately anticipate the likelihood of certain goods being purchased together. Furthermore, AI provides valuable insights to optimize the selection of picking routes for warehouse personnel, hence streamlining the whole order fulfillment process (Pournader et al., 2021).

The "Day of the Week," from Monday through Sunday, is now plotted along the x-axis. As was previously seen, the actual sales for each product are rather close to the expected sales (Figures 8.3 and 8.4).

The horizontal axis is indicative of the "Day of the Week." The observed patterns in inventory levels exhibit a similar tendency, as seen by the preceding graphs, wherein there is a notable decline in inventory across all goods during the week (Figures 8.5 and 8.6).

8.2.3 Logistics and Transportation Optimization

Transportation and logistics serve as the fundamental components of the supply chain. AI has the potential to catalyze transformative advancements within this industry. The utilization of AI in route planning has been optimized to enhance the efficiency of delivery truck operations. By analyzing various factors such as traffic conditions, historical data, and weather forecasts, AI algorithms provide recommendations for the most optimal routes (Abduljabbar et al., 2019). Load balancing involves the usage of AI to provide recommendations on the most effective method of distributing loads within vehicles, with the objectives of maintaining stability and maximizing space efficiency. For delivery destinations and several relevant criteria, AI has the capability to generate optimized delivery plans, so assuring punctual and timely deliveries (Dubey et al., 2021).

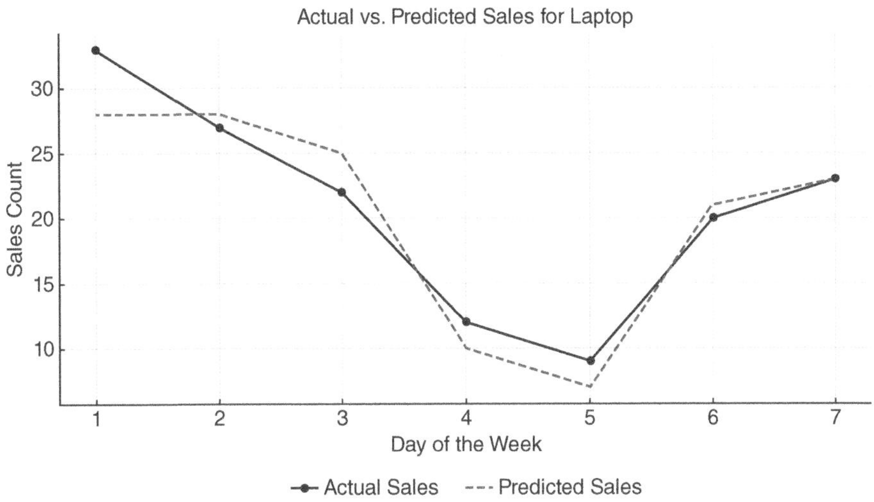

FIGURE 8.3 Sales prediction of laptop using AI (lastman0800, 2023e).

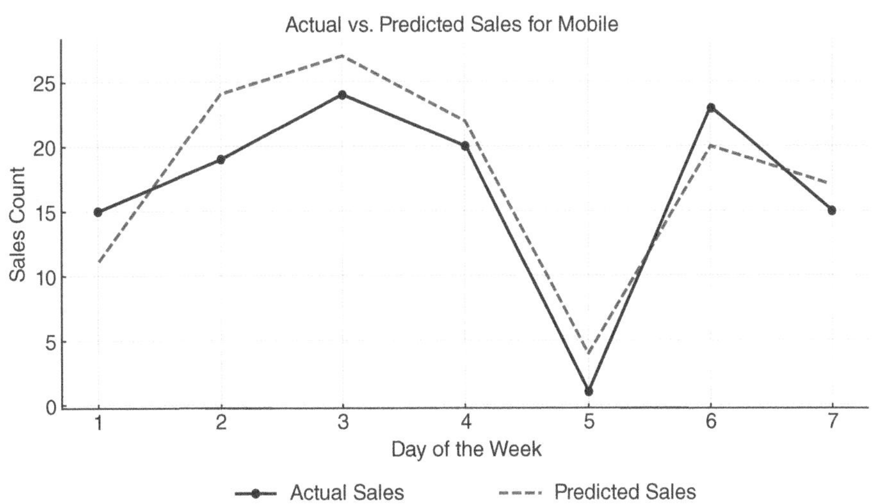

FIGURE 8.4 Sales Prediction of mobile using AI (lastman0800, 2023e).

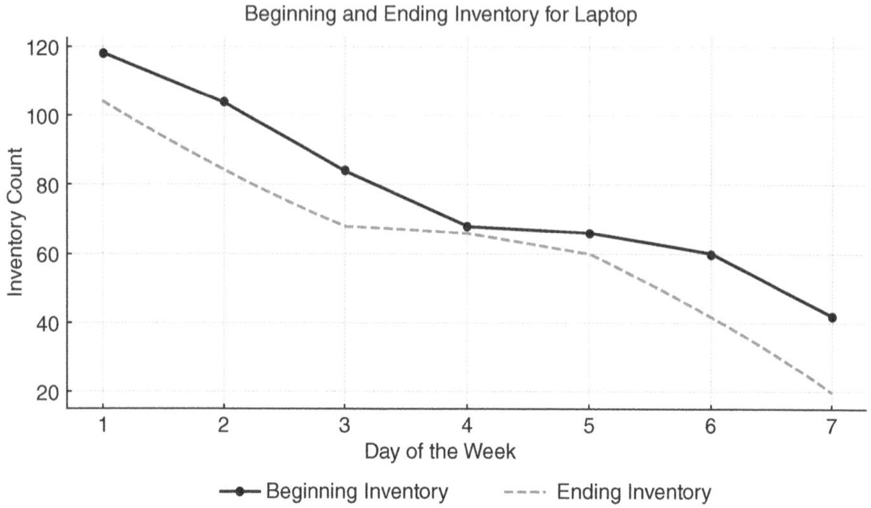

FIGURE 8.5 Inventory prediction of laptop (lastman0800, 2023e).

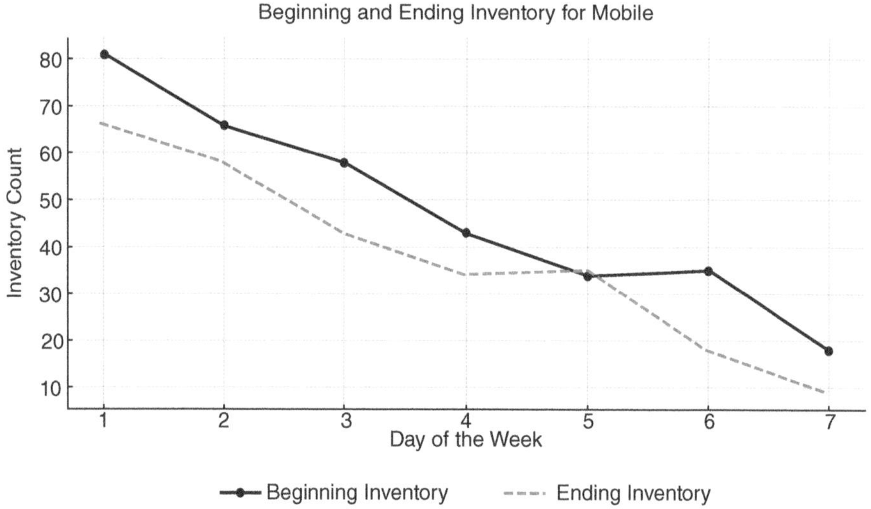

FIGURE 8.6 Inventory prediction of mobile (lastman0800, 2023e).

Both actual and AI-predicted shipping costs, black and white, are charted here over a period of a month. The accuracy of the forecast may be evaluated by contrasting it with the observed data (Figure 8.7).

The cost differences between the AI forecast and reality are represented by the black bars. If the value is positive, the AI-predicted cost was lower than the actual cost, and if it is negative, the AI-predicted cost was greater than the actual cost (Figure 8.8).

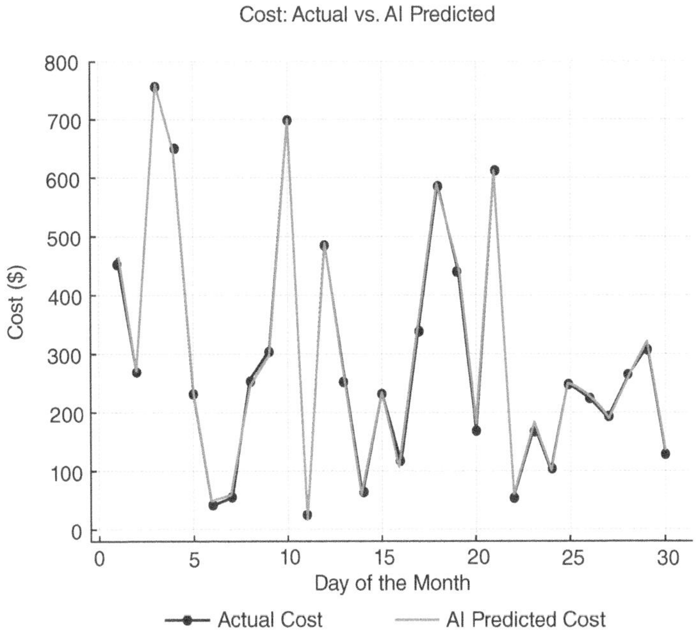

FIGURE 8.7 Cost vs AI Predicted 1 (lastman0800, 2023a).

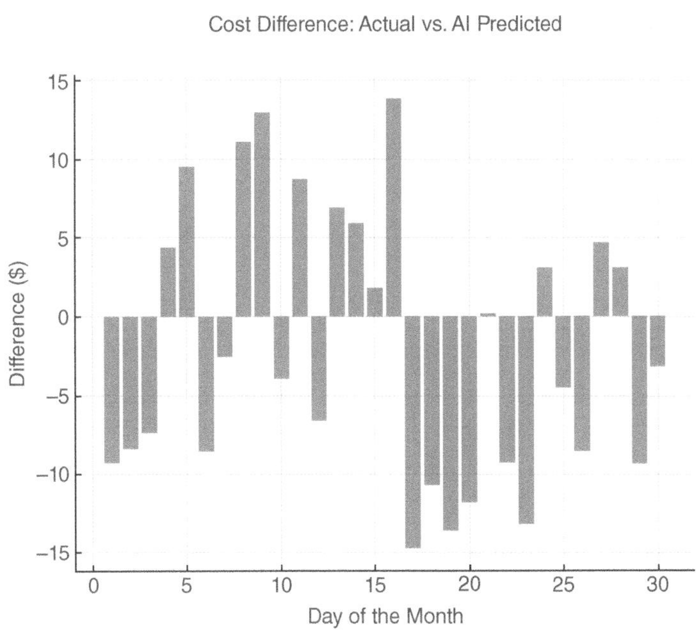

FIGURE 8.8 Cost vs AI Predicted 2 (lastman0800, 2023a).

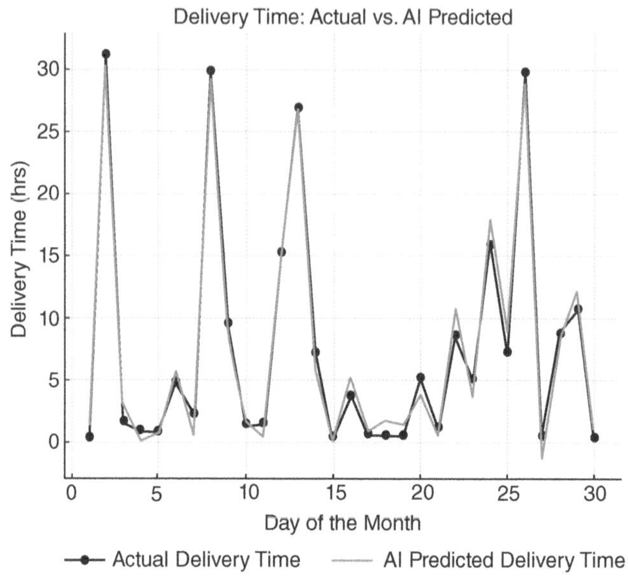

FIGURE 8.9 Delivery time with AI vs without AI 1 (lastman0800, 2023a).

Delivery times predicted by AI are shown in black, while actual delivery timings are shown in blue. Once again, you may evaluate the precision of the AI forecasts (Figure 8.9).

The black bars represent the deviations, in hours, between the actual delivery times and the times predicted by AI. If the value is positive, the AI prediction was quicker than the actual delivery time; if it is negative, the AI prediction was less accurate (Figure 8.10).

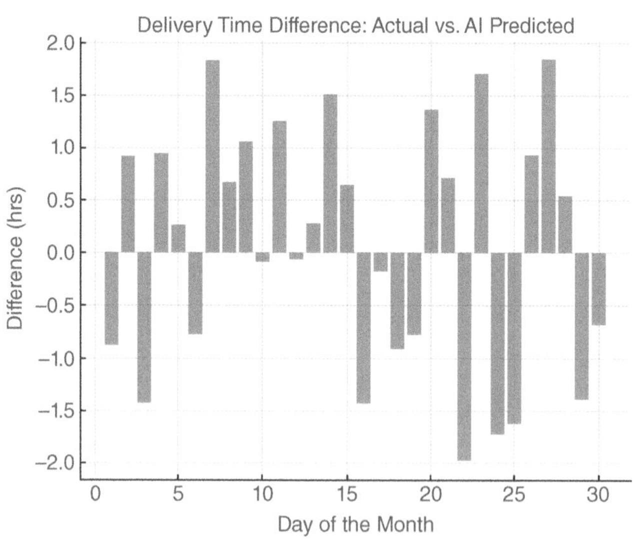

FIGURE 8.10 Delivery time with AI vs without AI 2 (lastman0800, 2023a).

Logistics and transportation managers may use these visuals to evaluate the efficacy of their AI models and make data-driven choices.

8.2.4 Supplier and Vendor Management

AI has revolutionized supply chain management by speeding up and simplifying the procedure of choosing suppliers. Previously, selecting vendors necessitated tedious and error-prone manual examinations of massive volumes of data. Businesses can quickly evaluate crucial variables like cost, quality, and turnaround time without sacrificing accuracy by using AI to filter through massive amounts of data. Using such objective criteria to select the most efficient suppliers has been shown to increase productivity and performance (Modgil et al., 2022).

The success of your supply chain depends on your ability to select dependable suppliers and monitor their performance. The usage of AI-driven technologies allows for continuous monitoring and analysis of critical suppliers' key performance indicators (KPIs). AI algorithms are ideally suited for tracking delivery delays, evaluating product quality, and gauging consumer response because of their speed, accuracy, and capacity to analyze massive volumes of data. In addition to providing valuable insight into supplier performance, this facilitates rapid response to quality issues, which is crucial for maintaining a competitive edge (Dubey et al., 2021).

AI's risk-factor recognition capabilities are particularly useful for vendor and supplier management. Supply chain interruptions, signals of financial trouble, and regulatory compliance infractions may all be detected in real time with the use of AI algorithms monitoring supplier data. These discoveries serve as a type of early warning, making possible preventative steps that would not be possible for firms without them. By applying AI in this preventative manner, supply networks can be kept in excellent shape and issues can be fixed before they become catastrophic (Toorajipour et al., 2021).

The quantity of orders placed with each vendor is displayed via a bar chart for easy comparison. It has been determined that Supplier D receives the most orders, whereas Supplier E receives the fewest (Figure 8.11).

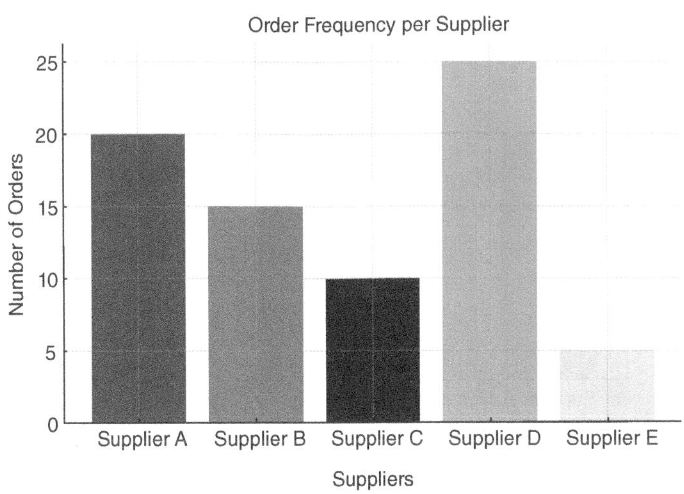

FIGURE 8.11 Order frequency (lastman0800, 2023d).

The average number of days it takes to deliver an order is displayed as a bar chart, while the reliability of orders is shown as a line chart. The average delivery time from Supplier D is the quickest, and they consistently provide accurate orders (Figure 8.12).

This graph displays the productivity boost brought about by AI. The most significant AI-driven efficiency boost, 15%, has been realized by Supplier C (Figure 8.13).

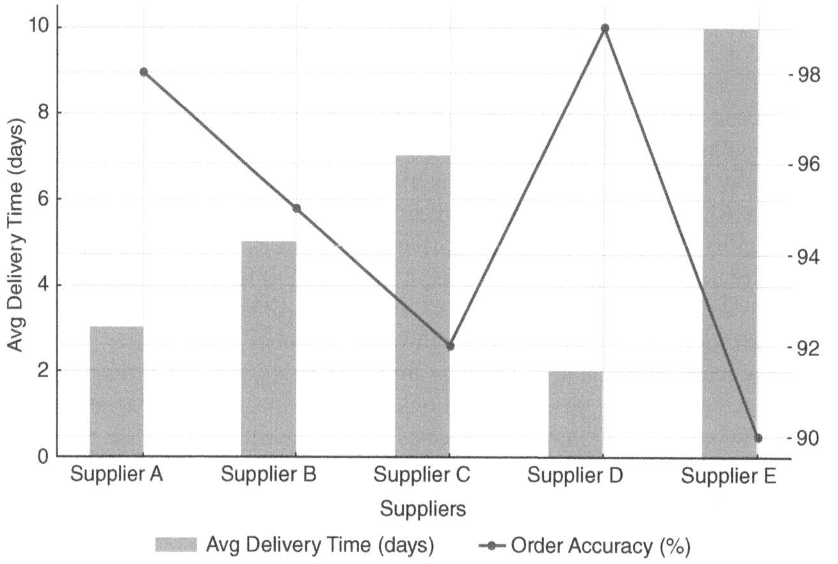

FIGURE 8.12 Average delivery time and order accuracy per supplier (lastman0800, 2023).

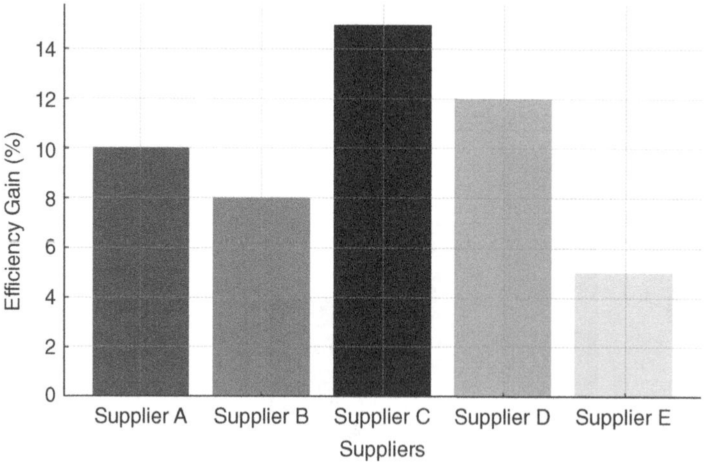

FIGURE 8.13 AI efficiency gain per supplier (lastman0800, 2023d).

8.3 BENEFITS OF AI IN INNOVATION PERFORMANCE

8.3.1 Enhanced Decision-Making

AI has revolutionized big data processing and interpretation. Traditional methods make it difficult to analyze large datasets, whereas AI systems excel. They efficiently sift massive data volumes and find patterns and trends that human analysts miss. AI's speed and accuracy in data processing make it essential in many sectors. Supply chain management benefits from AI insights (Kelly et al., 2019). Supply chain experts must make difficult decisions based on a deep understanding of several factors. AI insights will help these specialists make faster, more accurate decisions. This will speed up decision-making, reduce mistake, and enhance innovation (Dlamini et al., 2020). AI can improve supply chains' agility, reactivity, and efficiency to better serve consumers and deliver goods and services faster. AI's enhanced skills may boost supply chain strategic decision-making beyond data analysis (Dong et al., 2020). AI for scenario analysis can help professionals simulate various situations and assess their outcomes. In an ever-changing market, this knowledge is invaluable for firms planning for the unexpected. AI may reveal supply chain gaps and risks, aiding risk assessment. AI can also optimize resource use and supply chain performance by evaluating several parameters in an optimization model. All these qualities show how AI is changing logistics and business strategy (Ghahremani Nahr et al., 2021).

8.3.2 Process Optimization

AI automation has been transformative for businesses in the quest to save time and money. With the help of AI, businesses can eliminate a lot of repetitive tasks. As a result, work is done more quickly and accurately with less room for error. As a result, employees are free to focus on their own ideas and the business at hand. It's possible that adopting this perspective will completely change businesses by giving them a competitive edge and fresh insights. Automation powered by AI has many other advantages as well. By automating routine procedures, businesses may potentially decrease time to market and react to market changes and consumer demands more quickly (Nayak et al., 2018). Flexibility matters more in today's global economy (Nayarisseri, A. et al., 2021). Automation frequently promotes ongoing development in a company. People and communities take the initiative to improve, creating a cycle of growth and development. This culture's dominance may determine an organization's long-term success (Figure 8.14).

8.3.3 Predictive Analytics

Trend prediction, market research, and spotting customer preferences, all powered by AI, are all under the purview of predictive analytics. By planning ahead, businesses may ensure they can meet the evolving needs of their clientele. Using predictive analytics, companies can do much more than simply identify sales opportunities; they can even invent completely new products and services. Their foresight has put them in a favorable position to succeed in today's fiercely competitive business climate (Toorajipour et al., 2021).

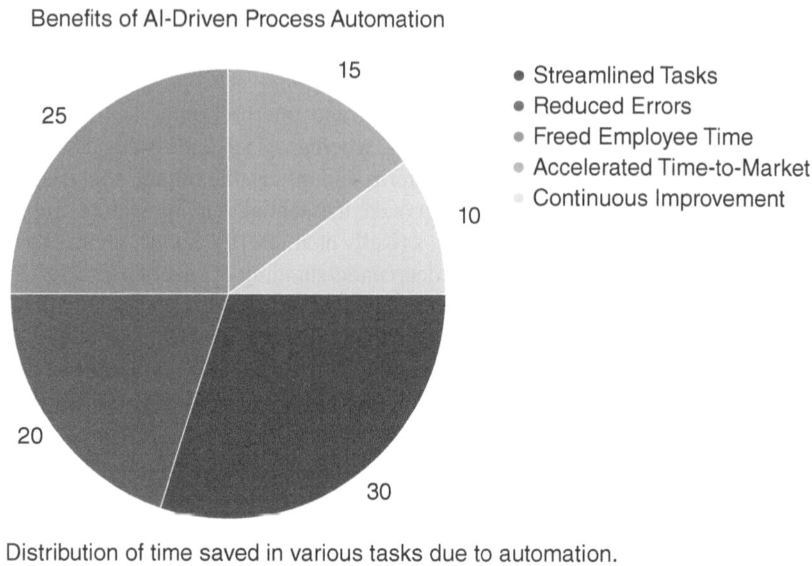

FIGURE 8.14 Benefits of AI (lastman0800, 2023b).

8.4 CHALLENGES AND CONSIDERATIONS

8.4.1 Data Quality and Availability

The quality and usefulness of AI algorithms are highly dependent on the data they are given to work with. Businesses can maximize the value of these algorithms and the insights they provide by ensuring the data they use is accurate, consistent, and complete. Without these features, the AI's findings might be deceiving or incorrect. The quality of the data, its accessibility, and availability are crucial. This is a challenge for many organizations, particularly when attempting to put together their supply chain using many data sources. Though essential for a full picture and understanding, challenges in achieving this integration may restrict the efficacy of AI-driven initiatives (Çıtak, 2023; Nishant et al., 2020).

8.4.2 Change Management and Workforce Adaptation

Changes to the company's culture and staff are necessary for successful AI implementation. The proliferation of AI systems has led to the development of new ways of thinking about problems and solving them. Because of this, employees will need to take action, and the business as a whole will need to prepare. Businesses must take measures to ensure a seamless transition to using AI in their supply chain. Change management programs should get substantial funding to aid employees in making the necessary adjustments. Workers require access to training and development programs that provide them with the information and skills they'll need to interact effectively with AI systems. It is crucial to address workers' fears and apprehensions about using AI technology in order to develop an atmosphere that is more responsive and collaborative (Nishant et al., 2020).

8.5 ETHICAL AND LEGAL CONSIDERATIONS

There are many moral and legal issues that must be resolved as AI develops and becomes more complicated. It is becoming increasingly necessary for organizations to address issues related to data privacy, algorithmic bias, and a lack of transparency in AI-driven operations. These problems are as much the result of the evolution of this technology as they are of the rules and morals of modern society. Businesses should emphasize creating procedures that adhere to existing laws and ethical norms to guarantee the legality and morality of AI systems (Olan, F. et al., 2022). This is essential if they want to keep the support of the law and their constituents. Integrating AI into today's technologically advanced society hinges on its ethical growth and implementation (Safdar et al., 2020).

8.6 FUTURE OUTLOOK AND CONCLUSION

AI technology has changed supply chain optimization, allowing businesses to reap substantial productivity improvements. Automating complicated decision-making processes like routing and scheduling may help enhance demand forecasts and inventory optimization by using sophisticated algorithms. Machine learning models' real-time analysis of vast volumes of data allows for quicker reactions to shift in the market and disruptions in the supply chain. Supply chains may be able to save money, get products to customers faster, and be more efficient with the assistance of more automation and data-driven decision-making.

However, there is pushback to implementing AI in supply chain management. The success of the rollout depends on careful planning and effective implementation. To effectively manage and comprehend AI outputs, businesses need to solve challenges with data quality, system integration, and personnel upskilling. When AI systems are entrusted with making judgments that can have a significant impact on people's financial well-being or even their lives, questions of ethics and transparency arise. Putting off addressing these issues might reduce productivity and irritate key stakeholders.

Complete AI integration into supply chain processes requires both ongoing learning and adaptability. AI models gain in efficiency and effectiveness when they are exposed to more data and trained for increasingly niche applications. It is essential for businesses to invest in the ongoing education and improvement of both their human employees and their AI systems. To achieve this goal, one has to routinely track key performance metrics, be flexible enough to make minor modifications, and keep abreast of developments in AI. Firms may maintain dependable and effective supply chain operations by continuously implementing new innovations in response to shifting market circumstances and new threats.

Embracing AI-driven innovation in supply chain management is a great way to set yourself apart in today's competitive business environment. Successful adopters of AI technologies in the supply chain will have a considerable competitive advantage.

They'll have an easier time enhancing customer service, adapting to new market realities, and streamlining internal processes. Long-term success in today's competitive market may depend on a company's ability to employ AI to develop a more agile and efficient supply chain.

REFERENCES

Abduljabbar, R. et al., 2019. Applications of Artificial Intelligence in transport: An overview. *Sustainability*, p. 16. https://doi.org/10.3390/su11010189

Ahmed, Z. et al., (2020. Artificial intelligence with multi-functional machine learning platform development for better healthcare and precision medicine. *The Journal of Biological Databases and Curation*, p. 18. https://doi.org/10.1093/database/baaa010

Baryannis, G. et al., 2019. Supply chain risk management and artificial intelligence: State of the art and future research directions. *International Journal of Production Research*, p. 15. https://doi.org/10.1080/00207543.2018.1530476

Chopra, A., 2019. AI in Supply & Procurement in 2019 Amity International Conference on Artificial Intelligence (AICAI), p. 7. https://doi.org/10.1109/AICAI.2019.8701357

Çıtak, E., 2023. AI has many obstacles in its way. [Online] Available at: https://dataconomy.com/2023/07/06/challenges-in-artificial-intelligence/ [Accessed 2023].

Dlamini, Z. et al., 2020. Artificial intelligence (AI) and big data in cancer and precision oncology. *Computational and Structural Biotechnology Journal*, p. 18. https://doi.org/10.1016/j.csbj.2020.08.019

Dong, Y. et al., 2020. The impact of R&D intensity on the innovation performance of artificial intelligence enterprises-based on the moderating effect of patent portfolio. *Sustainability*, p. 20. https://doi.org/10.3390/su13010328

Dubey, R. et al., 2021. Facilitating artificial intelligence powered supply chain analytics through alliance management during the pandemic crises in the B2B context. *Industrial Marketing Management*, p. 40. https://doi.org/10.1016/j.indmarman.2021.05.003

Gao, P. et al., 2021. An introduction to key technology in artificial intelligence and big data driven e-learning and e-education. *Mobile Networks and Applications*, p. 12. https://doi.org/10.1007/s11036-021-01777-7

Ghahremani Nahr, J. et al., 2021. Green supply chain based on artificial intelligence of things (AIoT). *International Journal of Innovation in Management, Economics and Social Sciences*, p. 12. https://doi.org/10.52547/ijimes.1.2.56

Gregory, M., 2023. The role of artificial intelligence and machine learning in warehouse management. [Online] Available at: https://www.linkedin.com/pulse/role-artificial-intelligence-machine-learning-michael-gregory [Accessed 2023].

Kelly, C. J. et al., 2019. Key challenges for delivering clinical impact with artificial intelligence. *BMC Medicine*, p. 12. https://doi.org/10.1186/s12916-019-1426-2

lastman0800, 2023a. AI-Optimized Logistics. [Online] Available at: https://www.kaggle.com/datasets/lastman0800/ai-optimized-logistics [Accessed 2023].

lastman0800, 2023b. AI-driven process. [Online] Available at: https://www.kaggle.com/datasets/lastman0800/ai-driven-process [Accessed 2023].

lastman0800, 2023c. kaggle lastman0800/datasets. [Online] Available at: https://www.kaggle.com/datasets/lastman0800/demand-forecasting-and-planning-using-ai [Accessed 2023].

lastman0800, 2023d. Supplier and Vendor Management using AI. [Online] Available at: https://www.kaggle.com/datasets/lastman0800/supplier-and-vendor-management-using-ai [Accessed 2023].

lastman0800, 2023e. Warehouse Management using AI. [Online] Available at: https://www.kaggle.com/datasets/lastman0800/warehouse-manageent-using-ai [Accessed 2023].

Modgil, S. et al., 2022. Artificial intelligence for supply chain resilience: Learning from Covid-19. *International Journal of Logistics Management*, p. 22. https://doi.org/10.1108/IJLM-02-2021-0094

Nayak, M. et al., 2018. Artificial intelligence driven process optimization for cleaner production of biomass with co-valorization of wastewater and flue gas in an algal biorefinery. *Journal of Cleaner Production*, p. 18. https://doi.org/10.1016/j.jclepro.2018.08.048

Nayarisseri, A. et al., 2021. Artificial Intelligence, big data and Machine Learning approaches in precision medicine & drug discovery. *Current Drug Targets*, p. 12. https://doi.org/10.2174/18735592MTEzsMDMnz

Nishant, R. et al., 2020. Artificial intelligence for sustainability: Challenges, opportunities, and a research agenda. *International Journal of Information Management*, p. 15. https://doi.org/10.1016/j.ijinfomgt.2020.102104

Olan, F. et al., 2022. The role of Artificial Intelligence networks in sustainable supply chain finance for food and drink industry. *International Journal of Production Research*, p. 22. https://doi.org/10.1080/00207543.2021.1915510

Pournader, M. et al., 2021. Artificial intelligence applications in supply chain management. *International Journal of Production Economics*, p. 13. https://doi.org/10.1016/j.ijpe.2021.108250

Safdar, N. M. et al., 2020. Ethical considerations in artificial intelligence. *European Journal of Radiology*, p. 20. https://doi.org/10.1016/j.ejrad.2019.108768

Sharma, R. et al., 2022. The role of artificial intelligence in supply chain management: Mapping the territory. *International Journal of Production Research*, p. 20. https://doi.org/10.1080/00207543.2022.2029611

Toorajipour, R. et al., 2021. Artificial intelligence in supply chain management: A systematic literature review. *Journal of Business Research*, p. 15. https://doi.org/10.1016/j.jbusres.2020.09.009

9 Ecological Strategic Orientation and Sustainable Development

Zaheer Ahmed Khan
Mazoon College Muscat, Muscat, Oman

Ijaz Nawaz
National College of Business and Management Lahore, Lahore, Pakistan

Hyder Kamran
University of Buraimi, Al Buraimi, Oman

9.1 INTRODUCTION

The ecological strategic orientation (ESO) describes how organizations approach their strategic decision-making and perform operations in a sustainable and environmentally responsible manner as a part of sustainable strategy (Baumgartner, 2014). To minimize negative impacts on the environment and promote sustainable practices, it involves integrating ecological considerations into various aspects of business activities. To grow long-term, organizations must consider ecological concerns and sustainability. With their approach to management, sustainability cognizant organizations strive to maintain a balance between environmental stewardship and profitability (Epstein & Buhovac, 2014). Philosophically, ESO and sustainable development stem from eco-ethical and intergenerational justice principles. ESO as a strategy urges to make decisions based on a long-term ecological perspective, acknowledging the relationship between human actions and the environment (Norgaard, 1994). There is intrinsic value in all living beings and a recognition of their right to exist. A similar concept is sustainable development, which aims to meet present needs without compromising the ability of future generations to meet theirs (United Nations, 1987). To ensure the well-being of all life types, both approaches emphasize the importance of protecting Earth's ecosystems and maintaining ecological balance.

The strategic approach of ESO consists of four key components (Criado-Gomis *et al.*, 2017). First, identifying potential environmental risks and impacts associated with business operations by conducting comprehensive environmental assessments. This process involves assessing product lifecycles, waste generation, and disposal

 DOI: 10.1201/9781032616810-9

options. Second, ecologically oriented organizations aim to use environmentally friendly products and services throughout their value chains. Therefore, procurement policies should rest on green procurement principles and encourage suppliers to reduce their environmental footprints and use energy-efficient production processes. Third, product development and sustainable innovation lead to eco-friendly product development by investing in research and development (Öberg, Huge-Brodin & Björklund, 2012; Lion, Donovan & Bedggood, 2013). This approach not only mitigates environmental harm but also creates opportunities for competitive advantage and market differentiation. Fourth, eco-strategy orientation requires employees and stakeholders to promote environmental awareness and responsibility. The role of stakeholders is vital in embracing ecological and sustainability approaches (Sharma & Henriques, 2005). Energy used for heating and cooling, cooking food, providing artificial lighting, as well as powering the working places, industries, agriculture, and transportation depends on the right energy forms and amounts being available in the right places (WCED, 1987).Therefore, companies need to effectively communicate sustainability implementation procedures and educate stakeholders.

With an ecological strategic approach, organizations can build a positive brand image, lower regulatory compliance risks, and reduce resource scarcity risks (Le et al., 2022). Furthermore, they play a significant role in building a more resilient world to achieve global environmental sustainability. As the term implies, sustainable development means meeting present-day needs without compromising the ability to meet those same needs for future generations. The balance of efforts on social, economic, and environmental factors is therefore essential to achieving long-term prosperity. Sustainable development seeks innovative and equitable environmental supporting solutions to address sustainability issues such as poverty, inequality, pollution, and climate change. Ecosystems and social equity rely on resourced efficiency with a harmonious balance between economic growth, social progress, and environmental preservation. Sustainable development offers societies with opportunities to achieve economic prosperity while conserving natural resources and ensuring social justice (Evans & Vladimirova, 2017). Therefore, major premises of sustainable development policy are to promote renewable and clean energy sources, adopt sustainable agriculture practices, conserve biodiversity, and promote inclusive and participatory governance. Implementation also considers global challenges and tackles them through international collaboration and cooperation (Oelze *et al.*, 2016). Therefore, contemplating the macro concerns of sustainability is necessary to change mindsets, policies, and practices to create a more sustainable, equitable, and resilient world.

9.2 STRATEGIC PLANNING, ESO, AND SUSTAINABLE DEVELOPMENT

The concept of an ecological orientation revolves around understanding and valuing nature because all living things are interdependent, and earth's resources are finite. Ecological orientation ensures that humans do not irreversibly damage the environment with their activities. The balance between economic, social, and environmental factors is essential for long-term prosperity and well-being (Guerry *et al.*, 2015). Sustainable development involves integrating economic growth, social progress, and environmental protection. Sustainable development and ecological orientation are closely related and complementary, and in a synergistic approach, both reinforce each other (Hansmann, Mieg, & Frischknecht, 2012). Gann *et al.* (2019) suggest the following initiatives to integrate eco principles into policies, practices, and individual behavior to ensure a prosperous and balanced future.

- Responsible resource management
- Mitigating environmental impact
- Enhancing resilience
- Public awareness and participation

Strategic planning covers a wide range of topics including sustainable entrepreneurship, corporate social responsibility, and environmental management. The sustainability strategy serves as a manifesto for an organization by balancing profits with environmental and societal concerns. Organizers' vision and mission describe how they will grow by conserving resources and being socially responsible. Sustainability manifesto guides corporate culture, supply chain practices, and product development. To ensure long-term viability and ethical practices, the company informs customers, employees, investors, and regulators. Therefore, a business makes a clear statement about its intention to thrive while positively influencing the environment and society at large (Crane, Matten, & Spence, 2014).

Researchers believe that sustainable development move will drive the next industrial revolution (Carvalho *et al.*, 2018). The concept "Green to Gold" describes how smart companies innovate, create value, and gain competitive advantage through environmental strategy (Esty & Winston, 2009). This concept presents a framework for achieving ecological sustainability while maximizing profitability for businesses. In addition to providing practical insights, this concept integrates sustainability into various aspects of business, including strategic planning, to emphasize the importance of natural resources and ecosystems.

Hence, the nexus of ecological orientation and sustainable development approach guides for integrating sustainability into core business functions including strategic planning and innovation. According to Hawken (1993), in economic terms, sustainable development and ecological orientation go together and yield ecological commerce as an act of sustainability. Use of ecological principles urges for a business model that addresses the shortcomings of traditional models and includes sustainable and ecological thinking as key components of strategic decision-making.

9.3 THE PRINCIPLES OF ESO

A long-term perspective emphasizes the protection of natural resources and sustainability. Sustainable business models take a long-term view of social and environmental issues. Since businesses operate within complex networks of interdependent relationships, system thinking is vital for a long-term perspective. This encourages organizations to consider the ecological context by focusing on the interconnectedness of various components (Molderez & Ceulemans, 2018). ESO recommends a precautionary to prevent irreversible environmental damage. The preservation of ecosystems and biodiversity is a high priority for ESO. Healthy ecosystems contribute to the preservation of biodiversity and provide essential services to the environment. To accomplish this, ESO collaborates with local communities, environmental groups, and Indigenous groups (Wals, 2019). There is a commitment to sustainability throughout the entire value chain, including the sources, the production, and the distribution of products (Wolf, 2014).

9.4 ENVIRONMENTAL PILLAR OF SUSTAINABLE DEVELOPMENT

Fundamentally, the literature narrates three pillars of sustainable development as economic, social, and environmental (Dhahri & Omri, 2018; Purvis *et al.*, 2019; Ranjbari *et al.*, 2021). The economic pillar relates to social and environmental factors. Humanity's future depends on socially and environmentally equitable economic growth. To achieve sustainable development, the economic pillar aims at four key outcomes: (1) inclusive and growing economies, (2) efficient economies, (3) innovative economies, and (4) collaborative world orders. Providing economic prosperity for all while protecting the planet's natural resources and ecosystems is the economic pillar of sustainable development (Huttmanová & Valentiny, 2019).

The social dimension of sustainability refers to improving the well-being, equity, and quality of life of women and children, as well as facilitating their improvement in the future. A major focus of this pillar is the advancement of social justice, the promotion of fair labor practices, and the reduction of inequality. It focuses on issues such as poverty, discrimination, and socioeconomic exclusion, as well as nurturing and building inclusive communities (Murphy, 2014).

Changing climates, as well as environmental and social changes, require humanity to develop more sustainably by preserving and safeguarding the natural environment, ecosystems, and resources that support all life on earth. It consists of practices and policies designed to minimize environmental degradation, conserve biodiversity, mitigate climate change, and ensure the responsible use of natural resources such as water, land, and air. To ensure a healthy planet for future generations, the environmental pillar emphasizes the need to address ecological challenges such as pollution, habitat loss, deforestation, and resource depletion (Mikulčić *et al.*, 2017). UNO's explanation of Sustainable Development Goals (SDGs) in 2018 emphasized the importance of institutional pillars (Littlewood & Holt, 2018). Institutions, policies, and structures support sustainability.

Legislation, regulations, and mechanisms must aid in protecting human rights, promote economic stewardship, and ensure equitable resource allocation (Guha & Chakrabarti, 2019). Therefore, bringing diverse stakeholders into decision-making processes, ensuring transparency and accountability in government systems, and enforcing the rule of law are important for sustainable development.

9.5 ECONOMIC AND SOCIAL DIMENSIONS OF ESO

From a strategic ecological perspective, environmental sustainability integrates into strategic decision-making processes. Business operations must consider their environmental impact not only to mitigate negative effects but also to promote positive ones. A sustainable business model integrates ecological concerns with overarching business objectives. A complex interaction exists between ESOs and economic aspects (Partidario & Gomes, 2013). Investing in eco-friendly practices helps reduce operating costs and improves operational efficiency (Kapiki, 2012). Utilizing energy-efficient technologies and minimizing waste can significantly reduce operating costs. Environmentally conscious businesses have the potential to attract environmentally conscious consumers, improve their brand's reputation, and carve out their own niche within the market. A strategic ecological approach can pose financial challenges when investing in eco-friendly processes and technologies. Despite the substantial upfront cost, the long-term benefits can often be greater than the cost (Katsikeas, Leonidou, & Zeriti, 2016).

From a societal perspective, ESO has the potential to significantly impact a variety of stakeholders. Organizations must also demonstrate an unwavering commitment to environmental sustainability to contribute to society and future generations. They address environmental problems such as climate change, pollution, and resource depletion (Harris & Roach, 2013). The dynamics of ESO may vary depending on the context, regional factors, and industry of an organization. There is a possibility to use sustainable practices in partnership with government agencies and non-profit organizations for use and development of renewable energy resources (Bartle & Leuenberger, 2014).

9.6 INTERPLAY BETWEEN ESO, INNOVATIVENESS, AND CAPABILITY OF ORGANIZATION

Adapting to the rapidly changing business environment depends on balancing ESO with innovation and organizational capability. An organization's ability to navigate sustainability and environmental responsibility challenges depends on its ESO, innovativeness, and organizational capability. An ESO should incorporate environmental concerns into strategic decision-making. The organization's goals and activities incorporate several sustainability principles, including reducing pollution, minimizing resource consumption, and promoting social responsibility. ESOs guide organizations' development and innovation (Albort-Morant *et al.*, 2020).

An eco-strategic organization emphasizes sustainability as well as innovation and capability building. Innovative business models (Ritala, Golnam, & Wegmann, 2014) develop and implement innovative ideas, products, processes, or business models. Sustainable environments require environmentally friendly practices, technologies, and products. Developing creative solutions for ecological improvement requires an initiative-taking mindset. The collective skills, resources, and competencies of an organization determine its success. Technologies, managerial capabilities, financial resources, and organizational culture all contribute to an organization's effectiveness. Developing skills, infrastructure, and systems is essential to ensuring environmental responsibility. A company's strategy guides strategic decisions and identifies a vision for environmental stewardship (Kuo, Fang, & LePage, 2022). The implementation of ecological strategies relies heavily on innovation. A company's environmental management system enables it to develop and implement environmentally friendly products, technologies, and practices. The creation of sustainable business models requires innovative methods for reducing environmental impact, increasing resource efficiency, and enhancing resource efficiency (Altuntas & Karaboga, 2019).

Successful organizations can execute ecological strategic initiatives and implement innovative approaches. The implementation of sustainability strategies requires a combination of skills, resources, and systems. Green technology adoption and development require technological capabilities, financial resources, and environmental resvponsibility cultures. There is a dynamic relationship between an organization's ability, innovation, and ecological orientation. Environmental strategic orientation innovation fuels companies to seek sustainable solutions to environmental problems. Through innovation, organizations can develop new competencies, adapt to changing conditions, and adapt to changing conditions (Zhu & Sarkis, 2016). Therefore, effective implementation of innovative ideas is important to ensure long-term sustainability. In organizations with an integrated approach, environmental strategies, innovativeness, and organizational capabilities play a greater role.

9.7 SYNERGIES AND INTERDEPENDENCIES

Eco-strategic orientation is strategic alignment with ecological and environmental factors having implications of operations of an organization (Makhloufi *et al.*, 2023). Organizations incorporate sustainable practices and principles into their core strategies and operations. ESO achieves sustainability goals by integrating synergies and interdependencies. Renewable energy sources can reduce carbon footprint, such as solar and wind power. It is possible to address both climate change risks and environmental concerns (eco-orientation) by investing in renewable energy infrastructure. A combination of carbon reduction initiatives such as energy efficiency and renewable energy can strengthen eco-friendliness and strategic alignment. A sustainable supply chain is also an integral part of an eco-friendly strategy. Promotion of eco-friendly sourcing practices is an essential part of a company's supply chain management strategy. When selecting suppliers, eco-friendly companies consider their environmental concerns. Sustainable procurement can improve ecological orientation, reduce resource scarcity risks, and ensure regulatory compliance (Hansmann, Mieg, & Frischknecht, 2012).

To manage waste in an ecologically friendly manner, we must embrace a circular economy model. Resource efficiency and waste reduction are goals of ecological frameworks like the circular economy. Contrary to traditional linear economies, the circular economy strives for closed-loop systems instead of take-make-dispose. This model emphasizes long-term durability, repairability, and recycling of products and materials. It is a widespread practice to refurbish, remanufacture, or recycle products that have reached the end of their useful life. This approach reduces natural resource depletion, pollution, and overall environmental impact. Furthermore, it promotes sustainable consumption and production practices, thereby fostering a more sustainable and environmentally friendly economy, since it minimizes waste while maximizing resource efficiency (Kirchherr, Reike, & Hekkert, 2017; Murray, Skene, & Haynes, 2017). Increasing the life cycle of products, recycling, and reducing waste can contribute to a more sustainable and resource-efficient economy. Therefore, this ESO can also result in cost savings, loyalty increases, and improved brand reputations.

An effective ecological strategy orientation involves collaboration and engagement with stakeholders, such as customers, employees, communities, and regulators. Engaging stakeholders in decision-making processes enhances company's ecological reputation. An eco-strategic orientation involves addressing environmental risks in collaboration with NGOs, such as through partnerships (Rhee, Park, and Petersen, 2021). Poor resource allocation, regulatory changes, and ineffective implementation can damage a company's reputation. Therefore, organizations should manage these risks proactively to ensure the long-term sustainability of eco-performance and strategic objectives. Incorporating ecological factors into decision-making processes results in synergies and interdependencies. Aligning sustainability goals with strategic objectives can help an organization manage risks, optimize resource utilization, and gain a competitive advantage.

9.8 ORGANIZATIONAL LEVEL STRATEGIES

It is possible to view sustainability as either an external factor that enhances profitability and competitiveness or a burden that imposes additional constraints on the traditional "single" bottom line (Pinelli & Maiolini, 2017). To enhance their strategic position, companies use sustainability-as-a-means strategies (SAMS) to leverage social-environmental synergies. A combination of ESO strategies and organizational-level strategies will assist in achieving sustainability objectives (Baumgartner & Ebner, 2010; Hsu et al., 2016).

- *Implementing an Environmental Management System (EMS)*: The environmental management system provides a framework through which organizations manage their environmental responsibilities effectively. Environmental management includes setting environmental objectives, implementing policies and procedures, conducting regular audits, and continuously improving performance (Bravi *et al.*, 2020). EMS makes ecological strategic implementation easy by identifying, assessing, and managing environmental impacts. (Sammalisto & Brorson, 2008). It is possible to develop a quality management system according to the ISO 14001 standard from the International Organization for Standardization (ISO).
- *Environmentally friendly product development*: Adopting eco-design principles helps in producing environmentally friendly products and services. A product's life cycle should include the option to recycle or dispose of it when it has reached the end of its useful life as well as the use of sustainable materials and reducing energy consumption. Organizations need to engage stakeholders, such as employees, customers, local communities, and NGOs, to understand stakeholders' expectations and concerns about sustainability. Engaging stakeholders in decision-making processes can build trust and enhance ESO. Throughout the lifecycle of a product or service, organizations can assess how it impacts the environment. Organizations can reduce environmental impacts and enhance sustainability by identifying hotspots and areas for improvement (Gautam, 2020). Sustainable supply chain management involves taking environmental considerations into account. Working with suppliers helps promote

circular economies, reduce transportation emissions, optimize packaging, and ensure responsible sourcing.

- *Reporting the practices*: Transparent reporting practices enhance an organization's ESO. Disclosure of environmental performance, goals, and progress demonstrates accountability and builds credibility (Burritt & Schaltegger, 2010). Organizations must adopt tools and approaches that facilitate environmentally conscious decision-making and long-term sustainability. It is easy to implement this strategy with the help of the following tools.
- *Life Cycle Assessment (LCA)*: LCA assesses the environmental impacts associated with all stages of a product's life cycle, from raw material extraction to disposal (Kasai, 1999). Life cycle assessment helps in improving environmental performance in different areas of business operations.
- *Carbon Footprint Analysis*: Organizations use carbon footprint to measure the total amount of greenhouse gases such as carbon dioxide- and methane-generated actions. Using this tool, the company identifies emission sources and mitigation strategies for addressing effects of climate change (Wandana *et al.*, 2021).
- *Sustainable Supply Chain Management*: From sourcing to production to distribution, a sustainability approach integrates sustainability principles throughout the supply chain. The goal is to promote environmental responsibility, improve efficiency, and minimize environmental impacts (Nayal *et al.*, 2022).
- *Natural Capital Accounting*: This tool enables users to quantify and value the diverse benefits ecosystems provide, including clean water, air, and biodiversity. The conservation and sustainable use of ecosystems and natural resources for their economic value needs an informed decision-making (Ruijs *et al.*, 2019). These strategies make it easier for an organization to incorporate ESO and sustainability into its operations.

9.9 EXAMPLES OF ORGANIZATIONS EMBRACING ESO

9.9.1 Interface Inc. – Mission Zero

Founded in 1980, Interface Inc. is one of the world's largest manufacturers of modular carpet tiles. With its Mission Zero initiative, the company is committed to ecological sustainability, aiming to eliminate any negative environmental impact by 2020. Interface achieved significant reductions in waste, water consumption, and carbon emissions through innovative approaches such as recycling, renewable energy adoption, and sustainable sourcing. Innovative modular carpet tiles manufacturer Interface Inc. is a leader in the industry for sustainable business practices. Mission Zero, the company's initiative to eradicate any negative environmental impact by 2020, exemplifies the company's commitment to ecological sustainability. Interface Inc.'s Mission Zero relies on the following key pillars (United Nations, 2020).

- *Elimination of Waste*: Manufacturing processes at the company focus on eliminating waste. The company uses the optimal number of raw materials by minimizing waste generation. They generate less waste and recycle it. Where possible, they also reuse the leftovers of raw material.

- *Sustainable Energy*: Interface Inc. is committed to using only renewable energy sources in the future. The company continuously strives to increase the use of clean energy throughout its operations, including solar and wind projects.
- *Carbon Neutrality*: A third party has granted *Carbon Neutral Enterprise* certification to Interface Inc. This marks an important milestone toward our goal of becoming a carbon-negative enterprise by 2040 by neutralizing our carbon footprint across our entire business.

9.9.2 Unilever – Sustainable Living Plan

Unilever's Sustainable Living Plan integrates ESO into its business strategy. In addition to reducing greenhouse gas emissions and promoting sustainable sourcing, the plan aims to improve smallholder farmers' livelihoods. ESO efforts by Unilever provided positive environmental outcomes in addition to enhancing the brand's reputation and driving growth for the company.

Enhanced livelihoods, improved health, and well-being, and reduced environmental impact are the three pillars of the plan. Unilever's Sustainable Living Plan illustrates its profit motives with purpose through corporate social innovation. Under its ambitious sustainability agenda, the company is tackling issues such as climate change, plastic pollution, and inequality. The factories generate most of their electricity from renewable sources and decarbonize their energy usage.

Unilever offers a program called Unilever Living Plan (ULP) that helps people live healthier lives. Unilever aims to contribute to a sustainable society in three ways: (1) improved health, (2) livelihoods, and (3) welfare to reduce the negative effects on the environment. Unilever offers health programs, including healthy school programs, world handwashing day, dental health month, healthy village programs, healthy market programs, maternal and child health programs, and care for health programs in the local community. Moreover, Unilever is committed to maintaining public health through its products to achieve health financing indicators (Unilever, 2021).

9.9.3 Patagonia – Environmental Activism

Outdoor apparel company Patagonia participates actively in environmental activism (Patagonia, 2021). The company advocates for environmental safety and encourages customers to repair and reuse products, and the company supports environmental causes. If clothes are in good condition, Patagonia repairs and reuses them before recycling. As one of the most environmentally conscious brands, they use recycled fabrics and implement programs to reduce their ecological impact, so they can make another product with it or figure out another way to recycle. Recycling, repair, reselling, and waste are all essential elements of this company's eco-consciousness. Patagonia prioritizes recycling as part of its sustainability efforts. There is a critical role for recycling in reducing waste pollution. The company dedicates its efforts to playing a role in driving positive social and environmental change in the sports community. The Patagonia Company recycles its products and uses recycled materials for its products. The company invests in reforestation and regenerative agriculture projects to improve soil health and social development.

9.10 CONCLUSION

An ESO is essential to achieving sustainable development in today's rapidly changing world. An integral part of this approach is the recognition of the interrelationship between human activities and the natural environment. Integrating ecological considerations into our decision-making processes can contribute to economic growth, social well-being, and environmental protection. Through responsible and ethical practices, this approach integrates business and the environment for long-term success. This strategy with an ecological orientation fosters growth, resilience, and positive impacts – both for the environment and society. An initiative-taking approach to environmental challenges can help the organization contribute to a sustainable and equitable future. The application of ecological considerations to organizational strategy results in long-term viability because it reduces environmental impact, saves money, provides a competitive advantage, and complies with regulations. There are a variety of reasons why sustainability is essential to any business strategy, including its positive effects on the environment, the company, and society. Sustainable practices have some compelling reasons to be embraced. Furthermore, adopting environmentally friendly practices can lead a company to innovate and think creatively. This will enable the company to innovate and gain a competitive advantage. It is possible to benefit an organization from a sustainable business strategy including financial gains as well as long-term viability, reputation, and positive social and environmental effects. Besides demonstrating ethical behavior and demonstrating responsibility, sustainable business practices provide smart business opportunities. It is possible to achieve environmental and societal benefits, as well as create a positive impact on the bottom line of a company. Not integrating sustainability strategies into business may appear cost-saving but has significant consequences in the long run. It can lead to many negative consequences for the environment, society, and businesses, such as inefficient operations, and reputational risks.

AUTHOR'S DECLARATION

The scholarly work presented in this book chapter is our original research. The information provided is authentic, and its creation was free of plagiarism and unethical practices. By citing sources and providing references, we acknowledge the sources.

REFERENCES

Albort-Morant, G., Leal-Millán, A., & Fernández-Menéndez, J. (2020). The interplay between strategic orientation, innovation capability, and sustainable performance. *Business Strategy and the Environment*, *29*(3), 1103–1119.

Altuntas, C., & Karaboga, T. (2019). The interplay between sustainable innovation, dynamic capability, and organizational performance. *Journal of Business Research*, *104*, 374–382.

Bartle, J. R., & Leuenberger, D. Z. (2014). *Sustainable development for public administration*. Routledge.
Baumgartner, R. J. (2014). Managing corporate sustainability and CSR: A conceptual framework combining values, strategies and instruments contributing to sustainable development. *Corporate Social Responsibility and Environmental Management*, *21*(5), 258–271.
Baumgartner, R. J., & Ebner, D. (2010). Corporate sustainability strategies: Sustainability profiles and maturity levels. *Sustainable Development*, *18*(2), 76–89.
Bravi, L., Santos, G., Pagano, A., & Murmura, F. (2020). Environmental management system according to ISO 14001: 2015 as a driver to sustainable development. *Corporate Social Responsibility and Environmental Management*, *27*(6), 2599–2614.
Burritt, R. L., & Schaltegger, S. (2010). Sustainability accounting and reporting: Fad or trend? *Accounting, Auditing & Accountability Journal*, *23*(7), 829–846.
Carvalho, N., Chaim, O., Cazarini, E., & Gerolamo, M. (2018). Manufacturing in the fourth industrial revolution: A positive prospect in sustainable manufacturing. *Procedia Manufacturing*, *21*, 671–678.
Crane, A., Matten, D., & Spence, L. (Eds.). (2014). *Corporate social responsibility: Readings and cases in a global context*. Routledge.
Criado-Gomis, A., Cervera-Taulet, A., & Iniesta-Bonillo, M. A. (2017). Sustainable entrepreneurial orientation: A business strategic approach for sustainable development. *Sustainability*, *9*(9), 1667.
Dhahri, S., & Omri, A. (2018). Entrepreneurship contribution to the three pillars of sustainable development: What does the evidence really say? *World Development*, *106*, 64–77.
Epstein, M. J., & Buhovac, A. R. (2014). *Making sustainability work: Best practices in managing and measuring corporate social, environmental, and economic impacts*. Berrett-Koehler Publishers.
Esty, D. C., & Winston, A. (2009). *Green to gold: How smart companies use environmental strategy to innovate, create value, and build competitive advantage*. John Wiley & Sons.
Evans, J., & Vladimirova, D. (2017). Eco-innovation for sustainable development: Evidence from energy and water industry. *Technological Forecasting and Social Change*, *118*, 128–137.
Gann, G. D., McDonald, T., Walder, B., Aronson, J., Nelson, C. R., Jonson, J., … Dixon, K. (2019). International principles and standards for the practice of ecological restoration. *Restoration Ecology*, *27*(S1), S1–S46.
Gautam, V. (2020). Examining environmentally friendly behaviors of tourists towards sustainable development. *Journal of Environmental Management*, *276*, 111292.
Guerry, A. D., Polasky, S., Lubchenco, J., Chaplin-Kramer, R., Daily, G. C., Griffin, R., … Vira, B. (2015). Natural capital and ecosystem services informing decisions: From promise to practice. *Proceedings of the National Academy of Sciences*, *112*(24), 7348–7355.
Guha, J., & Chakrabarti, B. (2019). Achieving the Sustainable Development Goals (SDGs) through decentralisation and the role of local governments: A systematic review. *Commonwealth Journal of Local Governance*, *22*, 1–21.
Hansmann, R., Mieg, H. A., & Frischknecht, P. (2012). Principal sustainability components: empirical analysis of synergies between the three pillars of sustainability. *International Journal of Sustainable Development & World Ecology*, *19*(5), 451–459.
Harris, J. M., & Roach, B. (2013). *Environmental and natural resource economics: A contemporary approach*. ME Sharpe.
Hawken, P. (1993). The ecology of commerce*: A declaration of sustainability*. (No Title).
Hsu, C. C., Tan, K. C., & Mohamad Zailani, S. H. (2016). Strategic orientations, sustainable supply chain initiatives, and reverse logistics: Empirical evidence from an emerging market. *International Journal of Operations & Production Management*, *36*(1), 86–110.
Huttmanová, E., & Valentiny, T. (2019). Assessment of the economic pillar and environmental pillar of sustainable development in the European Union. *European Journal of Sustainable Development*, *8*(2), 289–289.

Kapiki, S. (2012). Implementing sustainable practices in Greek eco-friendly hotels. *Journal of Environmental protection and Ecology*, *13*, 1117–1123.

Kasai, J. (1999). Life cycle assessment, evaluation method for sustainable development. *JSAE Review*, *20*(3), 387–394.

Katsikeas, C. S., Leonidou, C. N., & Zeriti, A. (2016). Eco-friendly product development strategy: antecedents, outcomes, and contingent effects. *Journal of the Academy of Marketing Science*, *44*, 660–684.

Kirchherr, J., Reike, D., & Hekkert, M. (2017). Conceptualizing the circular economy: An analysis of 114 definitions. *Resources, Conservation, and Recycling*, *127*, 221–232.

Kuo, F. I., Fang, W. T., & LePage, B. A. (2022). Proactive environmental strategies in the hotel industry: Eco-innovation, green competitive advantage, and green core competence. *Journal of Sustainable Tourism*, *30*(6), 1240–1261.

Le, T. T., Vo, X. V., & Venkatesh, V. G. (2022). Role of green innovation and supply chain management in driving sustainable corporate performance. *Journal of Cleaner Production*, *374*, 133875.

Lion, H., Donovan, J. D., & Bedggood, R. E. (2013). Environmental impact assessments from a business perspective: Extending knowledge and guiding business practice. *Journal of Business Ethics*, *117*, 789–805.

Littlewood, D., & Holt, D. (2018). How social enterprises can contribute to the Sustainable Development Goals (SDGs)–A conceptual framework. In *Entrepreneurship and the sustainable development goals* (Vol. 8, pp. 33–46). Emerald Publishing Limited.

Makhloufi, L., Zhou, J., & Siddik, A. B. (2023). Why green absorptive capacity and managerial environmental concerns matter for corporate environmental entrepreneurship? *Environmental Science and Pollution Research*, *30*(46), 102295–102312. https://doi.org/10.1007/s11356-023-29583-6.

Mikulčić, H., Duić, N., & Dewil, R. (2017). Environmental management as a pillar for sustainable development. *Journal of Environmental Management*, *203*, 867–871.

Molderez, I., & Ceulemans, K. (2018). The power of art to foster systems thinking, one of the key competencies of education for sustainable development. *Journal of Cleaner Production*, *186*, 758–770.

Murphy, K. (2014). The social pillar of sustainable development A literature review and framework for policy analysis. *The ITB Journal*, *15*(1), 4.

Murray, A., Skene, K., & Haynes, K. (2017). The circular economy: An interdisciplinary exploration of the concept and application in a global context. *Journal of Business Ethics*, *140*, 369–380.

Nayal, K., Raut, R. D., Yadav, V. S., Priyadarshinee, P., & Narkhede, B. E. (2022). The impact of sustainable development strategy on sustainable supply chain firm performance in the digital transformation era. *Business Strategy and the Environment*, *31*(3), 845–859.

Norgaard, R. B. (1994). *Development betrayed: The end of progress and a coevolutionary revisioning of the future*. Routledge.

Öberg, C., Huge-Brodin, M., & Björklund, M. (2012). Applying a network level in environmental impact assessments. *Journal of Business Research*, *65*(2), 247–255.

Oelze, N., Hoejmose, S. U., Habisch, A., & Millington, A. (2016). Sustainable development in supply chain management: The role of organizational learning for policy implementation. *Business Strategy and the Environment*, *25*(4), 241–260.

Partidario, M. R., & Gomes, R. C. (2013). Ecosystem services inclusive strategic environmental assessment. *Environmental Impact Assessment Review*, *40*, 36–46.

Patagonia. (2021). Environmental & Social Initiatives. https://www.patagonia.com/static/on/demandware.static/-/Library-Sites-PatagoniaShared/default/dwca48f93c/PDF-US/2020EnvironmentalandSocialInitiatives.pdf

Pinelli, M., & Maiolini, R. (2017). Strategies for sustainable development: Organizational motivations, stakeholders' expectations, and sustainability agendas. *Sustainable Development*, *25*(4), 288–298.

Purvis, B., Mao, Y., & Robinson, D. (2019). Three pillars of sustainability: In search of conceptual origins. *Sustainability Science*, *14*, 681–695.

Ranjbari, M., Esfandabadi, Z. S., Zanetti, M. C., Scagnelli, S. D., Siebers, P. O., Aghbashlo, M., … Tabatabaei, M. (2021). Three pillars of sustainability in the wake of COVID-19: A systematic review and future research agenda for sustainable development. *Journal of Cleaner Production*, *297*, 126660.

Rhee, Y. P., Park, C., & Petersen, B. (2021). The effect of local stakeholder pressures on responsive and strategic CSR activities. *Business & Society*, *60*(3), 582–613.

Ritala, P., Golnam, A., & Wegmann, A. (2014). Coopetition-based business models: The case of Amazon.com. *Industrial Marketing Management*, *43*(2), 236–249.

Ruijs, A., Vardon, M., Bass, S., & Ahlroth, S. (2019). Natural capital accounting for better policy. *Ambio*, *48*, 714–725.

Sammalisto, K., & Brorson, T. (2008). Training and communication in the implementation of environmental management systems (ISO 14001): A case study at the University of Gävle, Sweden. *Journal of Cleaner Production*, *16*(3), 299–309.

Sharma, S., & Henriques, I. (2005). Stakeholder influences on sustainability practices in the Canadian forest products industry. *Strategic Management Journal*, *26*(2), 159–180.

Unilever. (2021). Unilever celebrates 10 years of the Sustainable Living Plan. https://www.unilever.com/news/press-and-media/press-releases/2020/unilever-celebrates-10-years-of-the-sustainable-living-plan/

United Nations. (1987). *Report of the World Commission on Environment and Development: Our Common Future*. Oxford University Press.

United Nations. (2020). From Mission Zero to Climate Take Back: How Interface is Transforming its Business to Have Zero Negative Impact|Global. https://www.interface.com/GB/en-GB/sustainability/our-mission.html

Wals, A. E. (2019). Sustainability-oriented ecologies of learning: A response to systemic global dysfunction. In *Ecologies for learning and practice* (pp. 61–78). Routledge.

Wandana, L. S., Wadanambi, R. T., Preethika, D. D. P., Dassanayake, N. P., Chathumini, K. K. G. L., & Arachchige, U. S. P. R. (2021). Carbon footprint analysis: Promoting sustainable development. *International Journal for Research in Science Engineering*, *2*(1), 73–80.

WCED (1987). World commission on environment and development. *Our Common Future*, *17*(1), 1–91.

Wolf, J. (2014). The relationship between sustainable supply chain management, stakeholder pressure and corporate sustainability performance. *Journal of Business Ethics*, *119*, 317–328.

Zhu, Q., & Sarkis, J. (2016). An interplay framework of organizational factors in environmental management systems implementation. *Journal of Environmental Management*, *166*, 557–565.

10 The Effect of Organizational Green Operations and Digitalization to Promote Green Supply Chain Performance

Zia-ur-Rehman and Asghar Hayyat
Ghazi University, Dera Ghazi Khan, Pakistan

10.1 INTRODUCTION

As environmental concerns continue to gain momentum, organizations are under increasing pressure to adopt eco-friendly practices and reduce their carbon footprint. The integration of green operations involved in life cycle approaches – considering environmental impacts, optimizing resource usage to minimize waste, promoting sustainability – and green operations extended beyond the individuals and digitalization has emerged as a powerful approach to enhance supply chain sustainability and performance (Umair et al., 2023). This chapter aims to explore the intersection of green operations as a continuous improvement process, the ways to minimize waste generation, and the ways to implement effective recycling programs through digitalization, showcasing how these components can work together to drive positive environmental impacts and operational efficiency in implementing green operations, which represent a fundamental shift in how organizations approach their operational processes. By adhering to the principles of green operations, businesses can contribute positively to the environment while achieving long-term economic viability. These principles form the foundation for sustainable practices that protect natural resources, reduce environmental impact, and foster a healthier and more sustainable future for generations to come.

DOI: 10.1201/9781032616810-10

10.2 SECTION 1: UNDERSTANDING GREEN OPERATIONS AND ITS SIGNIFICANCE

10.2.1 Definition and Principles of Green Operations

Definition: Green operations, also known as sustainable operations or eco-friendly operations, refer to the adoption of environmentally responsible practices and strategies within an organization's operational processes (Khan, Yu, Umar, & Tanveer, 2022). The primary aim of green operations is to minimize negative environmental impacts, conserve natural resources, and promote sustainability while maintaining or improving operational efficiency and profitability (Baah et al., 2021). Green operations encompass various aspects of an organization's activities, ranging from production and supply chain management to waste reduction and energy consumption (Ghosh, Mandal, & Ray, 2022).

10.2.2 Principles of Green Operations

10.2.2.1 Environmental Stewardship

The core principle of green operations is the commitment to environmental stewardship. Organizations must take responsibility for their ecological footprint and actively work toward reducing negative impacts on the environment. This includes reducing greenhouse gas emissions, minimizing waste generation, and conserving natural resources (Aftab, Abid, Sarwar, & Veneziani, 2022).

10.2.2.2 Life Cycle Thinking

Green operations involve adopting a life cycle approach, considering the environmental impacts of products or services at every stage of their life cycle. This includes raw material extraction, production, distribution, use, and end-of-life disposal or recycling. By understanding the full life cycle of their offerings, companies can identify opportunities for improvement and make more sustainable decisions (Braulio-Gonzalo, Jorge-Ortiz, & Bovea, 2022).

10.2.2.3 Resource Efficiency

Green operations focus on optimizing resource usage to minimize waste and maximize efficiency. This includes reducing energy consumption, water usage, and materials wastage throughout the organization's processes. Resource efficiency helps lower operational costs and environmental impacts simultaneously (Pimenov et al., 2022).

10.2.2.4 Renewable Energy Adoption

To promote sustainability, organizations embracing green operations seek to increase their reliance on renewable energy sources such as solar, wind, or hydroelectric power. By transitioning away from fossil fuels, companies can significantly reduce their carbon footprint and contribute to combating climate change (Shaikh et al., 2017).

10.2.2.5 Green Supply Chain Management

Green operations extend beyond an individual organization and encompass its entire supply chain. Companies work collaboratively with suppliers and partners to ensure sustainable sourcing and ethical practices, and reduce environmental impacts throughout the supply chain (Lin & Zhang, 2023).

10.2.2.6 Waste Reduction and Recycling

Minimizing waste generation and implementing effective recycling programs are vital aspects of green operations. By adopting circular economy principles, organizations aim to reduce the amount of waste sent to landfills and promote the reuse of materials (Oloruntobi et al., 2023).

10.2.2.7 Continuous Improvement

Green operations are an ongoing journey of continuous improvement. Organizations set environmental targets, monitor their progress, and regularly reassess their processes to identify further opportunities for enhancing sustainability (Yang, Zhao, & Ma, 2022).

10.2.2.8 Transparency and Reporting

Transparency is essential for green operations. Companies should communicate their environmental efforts and progress to stakeholders, customers, and the public. Regular sustainability reporting allows for accountability and fosters trust with stakeholders (Adhi Santharm & Ramanathan, 2022).

10.2.2.9 Compliance and Regulation

Green operations align with environmental regulations and standards. Organizations adhere to relevant laws, certifications, and industry guidelines to ensure they are operating within sustainable and legal parameters (Kedward, Gabor, & Ryan-Collins, 2022).

10.2.2.10 Innovation and Technology

Embracing innovative technologies is key to advancing green operations. Digitalization, IoT, AI, and other technologies can enable real-time monitoring, data analysis, and smarter decision-making to improve sustainability efforts (Lazaroiu, Androniceanu, Grecu, Grecu, & Neguriță, 2022).

10.2.3 The Importance of Sustainability in Modern Supply Chain Management

The idea of sustainability has become increasingly important in all industries in the connected and quickly changing world of today (Fiksel & Fiksel, 2015). Businesses are starting to understand how essential sustainability is to building a reliable, moral, and successful supply chain. The success of an organization's supply chain is crucial, so adopting sustainability in this area has become more of a requirement than an option (Orji, Kusi-Sarpong, & Gupta, 2020). This extensive chapter focuses on the environmental, social, and economic facets of sustainability and examines its significance in contemporary supply chain management.

10.2.4 Environmental Sustainability in Supply Chain Management

Environmental sustainability refers to the responsible use of natural resources and the minimization of environmental impacts throughout the supply chain (Kazancoglu, Sagnak, Kayikci, & Kumar Mangla, 2020; Pang & Zhang, 2019; Sugandini, Susilowati, Siswanti, & Syafri, 2020). The modern supply chain faces a myriad of challenges related to resource depletion, pollution, and climate change. Embracing environmental sustainability in supply chain management brings numerous benefits.

10.2.4.1 Resource Efficiency

Sustainable supply chain practices encourage the efficient use of resources, such as energy, water, and raw materials (Grejo & Lunkes, 2022; Rajaeifar et al., 2022). By optimizing resource consumption, businesses can reduce costs and decrease their ecological footprint.

10.2.4.2 Carbon Footprint Reduction

Adopting sustainable transportation and logistics practices can significantly reduce greenhouse gas emissions. This not only contributes to mitigating climate change but also enhances a company's reputation and brand value (H. Li, Chen, & Umair, 2023; Olabi et al., 2022; Suchek et al., 2021).

10.2.4.3 Circular Economy

Integrating circular economy principles into the supply chain promotes recycling, reusing, and reducing waste. This approach fosters a more sustainable and resource-efficient system (Asthana, 2023; Hossain, Ng, Antwi-Afari, & Amor, 2020).

10.2.4.4 Compliance and Regulation

Embracing environmental sustainability enables businesses to comply with increasingly stringent environmental regulations and reduce the risk of potential fines and penalties (Hossain et al., 2020; Z. Zhang, Peng, Yang, & Lee, 2022).

10.2.5 Social Sustainability in Supply Chain Management

Social sustainability encompasses the fair treatment of workers, respect for human rights, and community engagement throughout the supply chain (Stahl, Brewster, Collings, & Hajro, 2020; Z. Zhang et al., 2022). Neglecting social sustainability can lead to reputational damage and legal consequences (Chan, Wei, Guo, & Leung, 2020; Thorisdottir & Johannsdottir, 2020). The incorporation of social sustainability in supply chain management brings the following advantages:

10.2.5.1 Ethical Labor Practices

Sustainable supply chains prioritize fair wages, safe working conditions, and the prohibition of child labor and forced labor (Antonini, Beck, & Larrinaga, 2020; Desiderio et al., 2022). By adhering to ethical labor practices, businesses can attract and retain a skilled workforce and avoid negative publicity (Santos, 2023).

10.2.5.2 Supply Chain Transparency

Transparency in the supply chain allows companies to trace the origin of their products and ensure that suppliers follow ethical standards (Brun, Karaosman, & Barresi, 2020). This fosters accountability and builds trust among consumers and stakeholders (Bai, Quayson, & Sarkis, 2022; Ebinger & Omondi, 2020; S. A. Khan et al., 2022).

10.2.5.3 Social Responsibility

By engaging in socially responsible practices, companies can positively impact the communities in which they operate, leading to improved relationships with local stakeholders and a more favorable business environment (Camilleri, 2022; Pfajfar, Shoham, Małecka, & Zalaznik, 2022; Singh & Misra, 2021).

10.2.5.4 Supplier Relationships

Building strong, collaborative relationships with suppliers based on ethical principles can enhance supply chain stability and resilience, minimizing the risk of disruptions.

10.2.6 Economic Sustainability in Supply Chain Management

Economic sustainability refers to the ability of a supply chain to remain financially viable in the long term. Integrating economic sustainability into supply chain management offers several key advantages:

10.2.6.1 Cost Efficiency

Sustainable supply chain practices often lead to cost savings through reduced waste, improved resource management, and streamlined processes (García Alcaraz et al., 2022).

10.2.6.2 Risk Management

Diversifying the supply chain and adopting sustainable practices can mitigate risks associated with price fluctuations, supply disruptions, and market uncertainties (S. Prakash et al., 2022; Zahraee, Shiwakoti, & Stasinopoulos, 2022).

10.2.6.3 Innovation and Competitive Advantage

Embracing sustainability can drive innovation, leading to the development of new products and services that meet evolving customer demands (Acciarini et al., 2022). This, in turn, can provide a competitive edge in the market (Musiello-Neto, Rua, Arias-Oliva, & Silva, 2021).

10.2.6.4 Investor Confidence

Many investors now consider a company's sustainability performance as a critical factor when making investment decisions (In, Rook, & Monk, 2019). Demonstrating strong sustainability practices can attract more investors and strengthen the company's financial position (Almeyda & Darmansya, 2019).

The importance of sustainability in modern supply chain management cannot be overstated and the businesses that integrate environmental, social, and economic sustainability principles into their supply chain operations stand to gain numerous advantages. From reducing environmental impacts and enhancing corporate social responsibility to achieving long-term economic viability and competitive advantage, sustainability is not only a moral imperative but also a strategic necessity. As we move forward into an increasingly interconnected and resource-constrained world, embracing sustainability in supply chain management becomes the key to building a better, more resilient future for businesses and society as a whole.

10.2.7 Environmental Benefits of Adopting Green Practices

In an era characterized by climate change and environmental degradation, the adoption of green practices has become imperative for individuals, businesses, and governments alike (Woo & Kang, 2020). Green practices encompass a wide range of sustainable actions that aim to minimize the impact of human activities on the environment. From renewable energy utilization and waste reduction to eco-friendly transportation and sustainable agriculture, green practices offer significant environmental benefits (Raimi, 2020; Souto, 2022). This comprehensive note explores the various environmental advantages of adopting green practices across different sectors.

10.2.7.1 Mitigating Climate Change

One of the most critical environmental benefits of green practices is their potential to mitigate climate change (Souto, 2022). Green practices, such as the use of renewable energy sources like solar, wind, and hydropower, reduce the reliance on fossil fuels, thereby lowering greenhouse gas emissions (Zhou & Li, 2022). By curbing carbon dioxide and other greenhouse gas emissions, these practices help stabilize the climate and contribute to global efforts to limit the rise in temperature (Chen, Tee, Elnahass, & Ahmed, 2023).

10.2.7.2 Reducing Air Pollution

Green practices also play a vital role in reducing air pollution. The burning of fossil fuels for energy and transportation releases harmful pollutants and particulate matter into the atmosphere, leading to smog, respiratory issues, and other health problems (Chen et al., 2023). By shifting to cleaner energy sources and promoting eco-friendly transportation options like electric vehicles (EVs) and public transit, we can significantly improve air quality and protect public health (Farid et al., 2023; Zikirillo & Ataboyev, 2023).

10.2.7.3 Conserving Natural Resources

The adoption of green practices emphasizes resource conservation. Sustainable resource management strategies, such as water-saving techniques in agriculture, efficient recycling processes, and responsible forest management, help preserve precious natural resources for future generations. Conserving resources like water, timber, and minerals contributes to the overall health of ecosystems and biodiversity

(Farrukh, Mathrani, & Sajjad, 2023; Kotsiuk, Rogova, Medvid, & Popovych, 2023; Maftouh et al., 2022; Mukhopadhyay & Thakur, 2021; Srivastav et al., 2021).

10.2.7.4 Protecting Biodiversity

Biodiversity is crucial for the health and stability of ecosystems. Green practices that prioritize biodiversity conservation, such as sustainable land use planning, habitat restoration, and protected areas, safeguard endangered species and promote ecological balance (Maftouh et al., 2022). Preserving biodiversity not only ensures the continuation of various plant and animal species but also enhances the resilience of ecosystems against environmental threats (Gao, Zou, Zhang, & Xu, 2020; Hoffmann, 2022).

10.2.7.5 Minimizing Waste Generation

The implementation of green practices emphasizes waste reduction and efficient waste management. Through practices like recycling, composting, and waste-to-energy technologies, the volume of waste sent to landfills is minimized (Gao et al., 2020; Levaggi, Levaggi, Marchiori, & Trecroci, 2020). This reduces the environmental burden of waste disposal, lowers methane emissions from landfills, and conserves resources that would otherwise be used to manufacture new products from virgin materials (Badgett & Milbrandt, 2020; Ghozatfar, Yaghoubi, & Bahrami, 2023; Mukherjee et al., 2020).

10.2.7.6 Improving Water Quality

Water is a precious and limited resource, and green practices contribute to improving water quality. Sustainable agricultural practices, such as organic farming and precision irrigation, reduce the use of harmful chemicals and prevent agricultural runoff from contaminating water bodies (Adedibu, 2023). Additionally, green infrastructure, like rainwater harvesting systems and permeable pavements, helps to manage stormwater runoff, preventing pollutants from entering waterways (Gupta & Shashidhara, 2023; McLennon, Dari, Jha, Sihi, & Kankarla, 2021).

10.2.7.7 Promoting Sustainable Agriculture

Green practices in agriculture prioritize sustainable farming methods that protect soil health, conserve water, and minimize the use of synthetic fertilizers and pesticides (Kumar et al., 2023; Sultan, & El-Qassem, 2021). By adopting practices like crop rotation, agroforestry, and integrated pest management, farmers can achieve higher yields while preserving the long-term fertility of the land and reducing negative impacts on surrounding ecosystems (Manna et al., 2021).

10.2.7.8 Encouraging Eco-Friendly Transportation

Transportation is a significant source of greenhouse gas emissions and air pollution. Green practices in transportation involve promoting the use of public transit, cycling, and walking as alternatives to single-occupancy vehicles (Patil, 2021). Additionally, the adoption of EVs and the integration of renewable energy sources in transportation systems contribute to reducing carbon emissions and fostering cleaner air (Bartle & Chatterjee, 2019; Whillans et al., 2021).

10.2.7.9 Supporting Renewable Energy

Transitioning to renewable energy sources is a cornerstone of green practices. Wind, solar, hydro, geothermal, and biomass energy offer sustainable alternatives to fossil fuels, which deplete finite resources and contribute to climate change (Osman et al., 2023). Embracing renewable energy not only reduces greenhouse gas emissions but also helps create a more resilient and decentralized energy grid (Kumar & Rathore, 2023).

The environmental benefits of adopting green practices cannot be overstated. From mitigating climate change and reducing air pollution to conserving natural resources and promoting biodiversity, green practices offer practical and effective solutions to the environmental challenges we face today. Governments, businesses, and individuals all have a role to play in embracing sustainable practices and making environmentally conscious choices. By working together to prioritize environmental stewardship, we can build a more sustainable and resilient future for generations to come (Callaghan & Colton, 2008; Meacham, 2019; Zhou, Ren, Du, Tang, & Li, 2023).

10.2.8 Economic and Reputational Advantages of Green Operations

Green operations, also known as sustainable or eco-friendly practices, have emerged as a strategic imperative for businesses in the modern world (Gill, Ahmad, & Kazmi, 2021; Mukonza et al., 2021). These practices involve integrating environmentally responsible measures into various aspects of an organization's operations, supply chain, and overall business strategy. The adoption of green operations offers a multitude of benefits, including both economic advantages and enhanced reputational standing (Ye & Dela, 2023; Zhou et al., 2023). This comprehensive note explores the economic and reputational advantages of green operations and highlights the value they bring to businesses.

10.2.8.1 Economic Advantages of Green Operations

10.2.8.1.1 Cost Savings

Contrary to the belief that sustainability incurs higher costs, green operations can actually lead to substantial cost savings over time (Cocco, Pinna, & Marchesi, 2017). By implementing energy-efficient technologies, optimizing resource consumption, and reducing waste, businesses can lower operational expenses. For example, energy-efficient lighting, heating, and cooling systems reduce utility bills, while waste reduction and recycling programs can result in reduced waste disposal costs (Deegan, 2013). Additionally, green practices that prioritize resource efficiency can extend the lifespan of equipment and assets, leading to lower maintenance and replacement expenses (Clarkson, Li, Richardson, & Vasvari, 2011).

10.2.8.1.2 Energy Efficiency

One of the primary economic advantages of green operations is improved energy efficiency. Businesses that invest in energy-efficient technologies and practices can significantly reduce their energy consumption and associated costs (Bharany et al., 2022). Energy audits, smart building systems, and renewable energy integration help organizations optimize energy usage, leading to long-term savings and a smaller carbon footprint (Kapp, Choi, & Kissock, 2022).

10.2.8.1.3 Supply Chain Resilience

Green operations extend beyond an organization's boundaries and encompass the entire supply chain (Alonso-Muñoz, González-Sánchez, Siligardi, & García-Muiña, 2021). By promoting sustainable practices among suppliers and partners, businesses can build a more resilient supply chain (Nayal et al., 2022). This resilience is particularly valuable in mitigating the risks associated with resource scarcity, price fluctuations, and disruptions caused by climate-related events (Siagian, Tarigan, & Jie, 2021). A robust and sustainable supply chain contributes to consistent operations and enhances the organization's ability to respond to market challenges (Trabucco & De Giovanni, 2021).

10.2.8.1.4 Regulatory Compliance

With increasing environmental regulations worldwide, businesses that adopt green operations can stay ahead of compliance requirements (Tu & Wu, 2021). Proactively integrating sustainability practices ensures that an organization is better prepared to meet existing and future environmental regulations (Kuo, Fang, & LePage, 2022). By avoiding potential fines and penalties, businesses can protect their financial health and reputation (Baah, Opoku-Agyeman, Acquah, Issau, & Abdoulaye, 2020).

10.2.8.1.5 Access to Green Markets and Incentives

As sustainability gains prominence, many industries are witnessing the emergence of green markets and eco-conscious consumers. Adopting green operations allows businesses to access these markets and appeal to environmentally conscious consumers who prefer eco-friendly products and services (J.-E. Park & E. J. S. Kang, 2022a; J. Park & E. Kang, 2022b; RISTESKA, 2023). Additionally, governments and municipalities often provide incentives, grants, or tax breaks to encourage businesses to adopt sustainable practices, further supporting the economic advantages of going green.

10.2.8.2 Reputational Advantages of Green Operations

10.2.8.2.1 Enhanced Brand Image and Value

Embracing green operations can significantly enhance a company's brand image and value. A commitment to sustainability demonstrates corporate responsibility and a dedication to environmental stewardship. Consumers are increasingly making purchasing decisions based on a company's social and environmental impact (Tsai et al., 2020). By adopting green operations, businesses can build a positive brand perception and attract more environmentally conscious customers, contributing to brand loyalty and differentiation in the market.

10.2.8.2.2 Improved Stakeholder Relations

Green operations resonate positively with stakeholders, including customers, employees, investors, and local communities. Customers are more likely to support companies that prioritize sustainability, and employees often feel a stronger sense of purpose and satisfaction when working for environmentally responsible organizations (Gabler, Landers, & Rapp, 2020). Investors are increasingly factoring environmental, social, and governance criteria into their decision-making, making sustainable businesses more attractive for investment. Additionally, local communities tend to view environmentally responsible businesses favorably, fostering better relations and support (Afshar Jahanshahi, Maghsoudi, & Shafighi, 2021; Karatepe, Rezapouraghdam, & Hassannia, 2021).

10.2.8.2.3 Crisis Resilience and Risk Management

Companies that prioritize sustainability through green operations tend to be more resilient during times of crisis. Environmental disasters or incidents of non-compliance can have severe reputational repercussions, leading to public backlash and loss of trust. Green operations help organizations mitigate environmental risks and demonstrate proactive risk management, thus reducing the likelihood of crises and protecting their reputation in the face of adversity (Obrenovic et al., 2020; Woo & Kang, 2020).

10.2.8.2.4 Attracting and Retaining Talent

Younger generations of employees, in particular, are increasingly seeking job opportunities with companies that align with their values, including a commitment to sustainability. Organizations with green operations can attract and retain top talent, fostering a more engaged and motivated workforce. Employees are more likely to be proud of their association with environmentally responsible companies and contribute positively to the organization's reputation (Boğan & Dedeoğlu, 2020; Raza et al., 2021; Zafar, Ho, Cheah, & Mohamed, 2023).

The economic and reputational advantages of green operations make them a compelling strategy for businesses of all sizes and sectors. From cost savings and energy efficiency to enhanced brand image and stakeholder relations, green operations deliver tangible benefits that positively impact the bottom line and long-term sustainability (Khan, Yu, & Farooq, 2023). By embracing green practices, businesses not only contribute to a healthier environment but also position themselves for success in an increasingly environmentally conscious marketplace. The combination of economic prosperity and a strong reputational standing makes green operations a win–win approach for organizations seeking to thrive in a dynamic and sustainable future (Berrone, Rousseau, Ricart, Brito, & Giuliodori, 2023).

10.2.9 Key Challenges in Implementing Green Initiatives in Organizations

In the face of mounting environmental concerns, organizations worldwide are increasingly recognizing the importance of adopting green initiatives to promote sustainability and reduce their ecological footprint. Green initiatives encompass a wide range of environmentally conscious actions, such as energy conservation, waste reduction, sustainable sourcing, and carbon footprint reduction. However, despite the evident benefits, implementing green initiatives in organizations comes with its fair share of challenges.

This note explores some of the key hurdles faced by businesses in their journey toward sustainability, along with relevant industrial examples.

10.2.9.1 Financial Constraints and Return on Investment Concerns

One of the primary challenges organizations face in implementing green initiatives is the perceived financial burden and concerns about the return on investment (Ghadge et al., 2021). Many eco-friendly technologies and practices require upfront capital investments, and organizations may hesitate to allocate funds for sustainability projects if the benefits are not immediately evident (Asghar et al., 2022; Ye & Dela, 2023). However, it is essential to consider the long-term financial gains, such as energy cost savings and reduced operational expenses, that green initiatives can deliver. Industrial Example: The adoption of solar energy systems is a common green initiative for companies seeking to reduce their dependence on fossil fuels and lower energy costs (Mutezo, & Mulopo, 2021). Despite the initial investment in solar panels and infrastructure, the long-term benefits often outweigh the upfront costs, leading to significant energy savings over time (Gawusu et al., 2022).

10.2.9.2 Lack of Awareness and Education

A lack of awareness and understanding of sustainable practices and their benefits can hinder the successful implementation of green initiatives. Employees and management may not fully comprehend the importance of environmental stewardship or the specific actions they can take to contribute to sustainability. Proper education and training programs are crucial to creating a culture of environmental responsibility within the organization (Perron, Côté, & Duffy, 2006). Industrial Example: In the hospitality industry, hotels face challenges in implementing green initiatives when their staff is not fully aware of the significance of water and energy conservation (Wan, Chan, & Huang, 2017). By providing comprehensive training and education on sustainable practices, hotels can improve employee engagement and encourage more proactive participation in green initiatives (Ali Ababneh, Awwad, & Abu-Haija, 2021).

10.2.9.3 Resistance to Change

Resistance to change is a common barrier to adopting green initiatives. Employees may resist altering established practices and processes, especially if they perceive sustainability measures as inconvenient or disruptive to their daily routines (Yeatts et al., 2000). Overcoming this resistance requires effective communication and involvement of employees in the decision-making process (Rayner, 2016). Industrial Example: Manufacturing companies aiming to implement energy-efficient production processes may encounter resistance from workers accustomed to traditional, energy-intensive methods (De Groot, Verhoef, & Nijkamp, 2001). In such cases, involving employees in the planning and decision-making and highlighting the benefits of energy conservation can help overcome resistance (Cagno, Trianni, Spallina, & Marchesani, 2017).

10.2.9.4 Complex Regulatory Landscape

Navigating the complex and ever-changing regulatory landscape related to environmental compliance can pose significant challenges for organizations. Different regions and industries have varying environmental regulations, and businesses must stay informed and compliant to avoid legal repercussions and financial penalties (Karneli, 2023). Industrial Example: The automotive industry faces stringent emissions standards in various countries. Implementing green initiatives, such as developing electric or hybrid vehicles, can help automobile manufacturers meet regulatory requirements and align with global efforts to reduce carbon emissions (Lukin, Krajnović, & Bosna, 2022).

10.2.9.5 Supply Chain Complexity

Green initiatives must often extend beyond an organization's boundaries to encompass its supply chain. Ensuring that suppliers and partners adhere to sustainable practices can be challenging, especially in global supply chains with diverse stakeholders (Paulraj, 2011). Industrial Example: The fashion industry faces complexities in ensuring sustainable sourcing and ethical labor practices throughout the supply chain (Y. Li, Zhao, Shi, & Li, 2014). Brands that prioritize green initiatives must collaborate closely with suppliers and conduct audits to verify compliance with sustainable standards (Kashmanian, 2017).

10.2.9.6 Technological Limitations

The availability and maturity of green technologies can impact the feasibility of certain green initiatives. Some industries may struggle to find suitable and cost-effective green alternatives for specific processes or operations (Matuštík & Kočí, 2021). Industrial Example: The aviation industry faces challenges in adopting sustainable aviation fuels (SAFs) due to limited production capacity and higher costs compared to conventional aviation fuels (L. Zhang, Butler, & Yang, 2020). Research and development efforts are ongoing to overcome these technological limitations and make SAFs more viable.

10.2.9.7 Reporting and Measurement

Accurately measuring and reporting the impact of green initiatives can be a challenge for organizations. Setting meaningful sustainability targets and tracking progress requires robust data collection and analysis, which can be resource-intensive (C. Li et al., 2023). Industrial Example: Retail chains implementing waste reduction initiatives may find it challenging to track and quantify the reduction in waste generated across multiple store locations. Investing in data management systems and engaging with waste management partners can facilitate accurate reporting (Asiaei, Bontis, Alizadeh, & Yaghoubi, 2022).

Implementing green initiatives in organizations is crucial for building a sustainable future, but it comes with several challenges. Overcoming financial constraints, educating stakeholders, addressing resistance to change, complying with complex regulations, and managing supply chain complexities are some of the key hurdles organizations face. By learning from industrial examples and understanding the unique challenges of their specific sectors, businesses can develop effective strategies to navigate these hurdles successfully. Embracing green initiatives not only contributes to environmental preservation but also enhances a company's reputation, attracts environmentally conscious consumers, and fosters long-term profitability and competitiveness. With collaborative efforts, innovative solutions, and a commitment to sustainability, organizations can overcome these challenges and make significant strides toward a greener and more sustainable future.

10.3 SECTION 2: ROLE OF DIGITALIZATION IN GREEN SUPPLY CHAIN PERFORMANCE

10.3.1 Digital Technologies Transforming the Supply Chain Landscape

In the ever-evolving world of business, digital technologies have emerged as a powerful force, revolutionizing various industries, and significantly impacting the supply chain landscape. The traditional supply chain model, once characterized by manual processes, is now being reshaped by innovative digital solutions, leading to increased efficiency, transparency, and flexibility (Helmold & Terry, 2021; Souza, Valamede, & Akkari, 2020; Veile, Schmidt, Müller, & Voigt, 2021).

In this transformative era, businesses are adopting cutting-edge technologies to streamline their supply chain operations. One of the key drivers of this transformation is the Internet of Things (Allioui & Mourdi, 2023; Kumar, Payal, Dixit, & Chatterjee, 2020; Peter, Pradhan, & Mbohwa, 2023). IoT-enabled devices are now embedded throughout the supply chain, providing real-time data on inventory levels, temperature, humidity, and other critical factors. With this wealth of data, companies can optimize their supply chain processes, anticipate disruptions, and make data-driven decisions to improve overall performance.

Furthermore, the implementation of artificial intelligence (AI) and machine learning (ML) has proven to be a game-changer in the supply chain arena. These technologies enable advanced demand forecasting, dynamic route planning, and predictive maintenance, reducing lead times and cutting down on operational costs. AI-powered algorithms can analyze vast amounts of historical data to identify patterns, ensuring that inventory levels are optimized, and supply meets demand with precision (SOHRABI, 2023).

The rise of cloud computing has also contributed to the digital transformation of supply chains. Cloud-based platforms offer secure and accessible data storage and collaboration, allowing stakeholders across the supply chain to share real-time information seamlessly. This enhanced visibility enables smoother coordination among suppliers, manufacturers, distributors, and retailers, resulting in a more agile and responsive supply chain ecosystem (Kouhizadeh, Saberi, & Sarkis, 2021; Tay & Loh, 2021).

Moreover, blockchain technology is gaining traction as a transformative force in supply chain management. By creating an immutable and transparent ledger of transactions, blockchain enhances traceability and authenticity across the supply chain. It helps mitigate the risks of counterfeiting and ensures compliance with regulatory standards, fostering trust among all stakeholders (Kouhizadeh et al., 2021).

As the digital transformation continues to unfold, companies must embrace these digital technologies to remain competitive in today's dynamic marketplace. Embracing digital solutions not only streamlines supply chain processes but also facilitates sustainability efforts, reduces waste, and enhances overall customer experience.

10.3.2 Internet of Things and Its Applications in Green Operations

In the realm of sustainability and environmental consciousness, the Internet of Things has emerged as a game-changing technology, driving significant advancements in green operations across various industries. IoT, a network of interconnected devices embedded with sensors and software, enables seamless communication and data exchange, leading to more efficient and eco-friendly practices (Peter et al., 2023).

One of the primary applications of IoT in green operations is in smart energy management (Matuštík & Kočí, 2021). IoT-enabled devices, such as smart meters and sensors, are integrated into buildings and industrial facilities to monitor and optimize energy consumption. Real-time data on electricity, water, and gas usage allow businesses to identify energy inefficiencies, pinpoint areas of excessive consumption, and implement energy-saving measures. By making informed decisions based on IoT-generated insights, companies can reduce their carbon footprint and achieve substantial cost savings (Kaur & Kaur, 2022).

The term "smart agriculture," also known as "precision agriculture" or "precision farming," refers to a new method of farming that makes use of data analysis and cutting-edge technology to optimize many areas of agricultural production (Bhattacharyay et al., 2020). It entails integrating cutting-edge technologies like sensors, GPS navigation, IoT gadgets, robotics, big data analytics, and AI to improve the effectiveness, productivity, and sustainability of agricultural methods.

In the agricultural sector, IoT is revolutionizing sustainable farming practices. Smart agriculture involves deploying IoT sensors in fields to monitor soil moisture, temperature, and nutrient levels. This data-driven approach enables farmers to employ precision irrigation and fertilization techniques, reducing water and chemical usage while maximizing crop yield. Moreover, IoT-powered autonomous farming machinery optimizes planting and harvesting schedules, minimizing fuel consumption and soil compaction (Manne & Kantheti, 2021).

IoT's impact on transportation and logistics is also pivotal in promoting greener operations. Fleet managers leverage IoT to monitor vehicle performance, fuel efficiency, and route optimization. Real-time tracking of cargo shipments allows for efficient logistics planning, reducing the number of empty or half-loaded trucks on the road. This optimization not only reduces greenhouse gas emissions but also cuts operational costs and delivery times (A. Kumar et al., 2020).

In the context of sustainable cities, IoT applications have transformative effects. Smart city initiatives leverage IoT technology to manage traffic flow, monitor air quality, and optimize street lighting. By dynamically adjusting traffic signals based on real-time data, cities can reduce congestion, minimize idling, and lower vehicle emissions. Similarly, intelligent street lighting systems dim or brighten lights based on pedestrian and vehicular traffic, enhancing energy efficiency and reducing light pollution.

The convergence of IoT and green operations promises a more sustainable and environmentally responsible future. By harnessing the power of IoT to monitor, analyze, and optimize resource usage, businesses and industries can drive positive change, reducing their environmental impact while achieving economic benefits. The continued evolution of IoT technology and its widespread adoption hold the key to a greener, more sustainable world for generations to come.

10.3.3 Big Data Analytics for Sustainability and Waste Reduction

In the quest for a more sustainable future, big data analytics has emerged as a powerful tool, revolutionizing efforts to reduce waste and promote sustainability. The exponential growth of data from various sources has presented an opportunity to gain valuable insights into consumption patterns, resource utilization, and waste generation. Leveraging big data analytics, businesses and organizations are making significant strides in identifying opportunities for waste reduction and implementing eco-friendly practices (Mondejar et al., 2021).

At the heart of big data analytics for sustainability is the ability to collect and process vast amounts of data from diverse sources. Sensors, Internet of Things devices, social media, and other digital platforms generate data points that provide a comprehensive view of operations and consumer behaviors. Through advanced data analytics tools and techniques, this data is analyzed to detect patterns, trends, and anomalies, shedding light on areas where waste occurs and resources can be utilized more efficiently.

In the context of waste reduction, big data analytics offers precise insights into production processes and supply chains. By monitoring production lines and logistics in real time, businesses can identify inefficiencies and bottlenecks that lead to waste generation (Javaid, Haleem, Singh, Rab, & Suman, 2021). This allows for targeted improvements and adjustments, reducing waste at its source and optimizing resource allocation.

Beyond waste reduction, big data analytics empowers sustainable decision-making across various industries. In agriculture, for example, data-driven insights enable precision farming practices, minimizing the use of water, pesticides, and fertilizers while maximizing crop yields. In energy management, data analytics aids in optimizing energy consumption and identifying opportunities for renewable energy integration (Yadav et al., 2022).

In essence, big data analytics is a transformative force in the pursuit of sustainability and waste reduction. Its ability to analyze vast and complex datasets provides organizations with valuable information to make data-driven decisions, leading to more efficient resource management and reduced environmental impact. As data analytics tools continue to evolve and become more accessible, the potential for promoting sustainability and minimizing waste becomes even more promising. Embracing big data analytics is not just a competitive advantage but a crucial step toward building a greener and more sustainable world for generations to come.

10.3.4 Cloud Computing for Real-Time Monitoring and Collaboration

In the fast-paced and interconnected world of today, cloud computing has emerged as a game-changing technology, revolutionizing real-time monitoring and collaboration across various industries (Aryal, Liao, Nattuthurai, & Li, 2020). The traditional model of on-premises data storage and isolated communication has given way to cloud-based solutions that offer enhanced accessibility, flexibility, and efficiency (Vance et al., 2019).

Real-time monitoring is one of the most significant advantages offered by cloud computing. Through the integration of Internet of Things devices and sensors, data is collected in real time and transmitted to the cloud for analysis. This enables organizations to monitor and track various parameters, such as production processes, supply chain logistics, environmental conditions, and more. Decision-makers can access this data from anywhere and at any time, empowering them to make informed decisions promptly.

In addition to real-time monitoring, cloud computing facilitates collaborative efforts among teams and stakeholders, irrespective of their geographical locations. Cloud-based collaboration tools, such as project management platforms, document sharing, and video conferencing solutions, enable seamless communication and cooperation (Yadav et al., 2022). This fosters a culture of teamwork and innovation, bringing together experts from diverse backgrounds to collectively address challenges and find solutions.

Moreover, cloud computing enhances data security and resilience. Reputed cloud service providers invest heavily in state-of-the-art security measures, ensuring that data is protected from cyber threats and unauthorized access (Tabrizchi & Kuchaki Rafsanjani, 2020). Additionally, cloud backups and redundancies guarantee data continuity, even in the face of hardware failures or natural disasters.

The impact of cloud computing is particularly evident in industries like healthcare, finance, and manufacturing. In healthcare, cloud-based electronic health records enable healthcare professionals to access patient information instantly, leading to more efficient diagnosis and treatment (Phillips, 2021).

In conclusion, cloud computing has become an indispensable tool for real-time monitoring and collaboration, driving transformative changes in how businesses and organizations operate. The ability to access, process, and share data seamlessly from the cloud offers unprecedented opportunities for efficiency, innovation, and sustainability. As cloud technologies continue to advance, the possibilities for real-time monitoring and collaboration are bound to expand, paving the way for a more connected and agile world.

10.3.5 Artificial Intelligence and Machine Learning in Eco-friendly Decision-making

In the pursuit of a more sustainable and eco-friendlier world, AI and ML have emerged as powerful tools, revolutionizing decision-making processes across various industries. These technologies, fueled by vast amounts of data and advanced algorithms, have the capacity to analyze, predict, and optimize eco-friendly initiatives with unprecedented precision and efficiency.

One of the primary applications of AI and ML in eco-friendly decision-making lies in the realm of energy management (Matuštík & Kočí, 2021). Smart grids, powered by AI-driven analytics, continuously monitor energy consumption patterns, enabling real-time adjustments to optimize energy distribution and usage. Through predictive modeling, AI algorithms anticipate demand fluctuations, facilitating the integration of renewable energy sources like solar and wind power into the grid. This not only reduces reliance on fossil fuels but also ensures a stable and sustainable energy supply.

AI and ML are also making significant contributions to waste reduction efforts. Intelligent waste management systems employ sensors and data analytics to monitor fill levels in bins and dumpsters, optimizing waste collection routes and schedules. By minimizing unnecessary trips and maximizing capacity utilization, these systems reduce fuel consumption and carbon emissions associated with waste disposal. Additionally, ML algorithms are used to identify patterns in waste generation and guide companies and municipalities in developing targeted recycling and waste reduction programs (Cioffi, Travaglioni, Piscitelli, Petrillo, & De Felice, 2020).

In essence, AI and ML are driving a paradigm shift in eco-friendly decision-making. By harnessing the potential of data and advanced algorithms, businesses and organizations can make informed choices that align with sustainability goals (Abdullah & Lim, 2023). From energy management and waste reduction to agriculture and transportation, these technologies empower a more resource-efficient and environmentally conscious approach to decision-making. As AI and ML continue to evolve, the possibilities for achieving a greener and more sustainable future become increasingly promising. Embracing these technologies is not just an opportunity but a responsibility toward building a better world for generations to come.

10.3.6 Blockchain for Traceability and Transparency in Sustainable Sourcing

In the ever-evolving landscape of sustainable sourcing, blockchain technology has emerged as a transformative force, providing unprecedented levels of traceability and transparency throughout supply chains (Navaneethakrishnan, Melam, Bothra, Haldar, & Pal, 2023). With its decentralized and immutable ledger, blockchain offers a secure and efficient way to track the journey of products from their origin to the end consumer, ensuring ethical and sustainable practices are upheld.

At the core of blockchain's impact on sustainable sourcing is its ability to create an unalterable record of every transaction and movement within the supply chain. Each step, from raw material extraction to manufacturing, transportation, and distribution, is securely recorded on the blockchain. This transparency allows all stakeholders, including consumers, to trace the provenance of products, ensuring they are sourced ethically and sustainably (S. A. Khan et al., 2022).

Additionally, blockchain combats issues like illegal logging and wildlife trafficking (Pullman, McCarthy, & Mena, 2023). By recording the origin and movement of timber and wildlife products, blockchain helps authorities and consumers identify and eliminate products sourced through illegal or environmentally harmful means.

Blockchain's impact on sustainable sourcing extends to other industries as well. In the energy sector, blockchain can verify the provenance of renewable energy sources, guaranteeing that consumers are using clean and sustainable power. In the electronics industry, blockchain can ensure responsible mineral sourcing, preventing the use of conflict minerals in manufacturing processes (Núñez-Merino, Maqueira-Marín, Moyano-Fuentes, & Martínez-Jurado, 2020).

In conclusion, blockchain technology is a game-changer in promoting traceability and transparency in sustainable sourcing. By providing an immutable record of supply chain activities, blockchain empowers consumers to make ethical choices, supports businesses in showcasing their sustainable practices, and fosters a culture of responsible sourcing and environmental stewardship (Kayikci, Durak Usar, & Aylak, 2022). As blockchain continues to evolve and gain wider adoption, the journey toward a more sustainable and ethically conscious future becomes increasingly tangible. Embracing blockchain for sustainable sourcing is not just a step toward responsible business practices but a significant stride toward building a better world for current and future generations.

10.4 SECTION 3: INTEGRATING GREEN OPERATIONS AND DIGITALIZATION IN SUPPLY CHAIN MANAGEMENT

10.4.1 Developing a Green Supply Chain Strategy: From Sourcing to Logistics

In today's rapidly evolving business landscape, the integration of green operations and digitalization has become a pivotal aspect of supply chain management. Organizations are increasingly recognizing the importance of adopting environmentally sustainable practices while leveraging digital technologies to optimize their supply chain processes. Developing a comprehensive green supply chain strategy that encompasses every stage, from sourcing to logistics, has become a strategic imperative for businesses committed to sustainability and efficiency (Feng, Lai, & Zhu, 2022).

The journey toward a green supply chain begins with responsible sourcing practices. Sustainable sourcing entails selecting suppliers and partners who align with environmental standards and ethical guidelines. This includes considering factors like carbon emissions, waste management, and adherence to eco-friendly certifications. Digitalization plays a crucial role in this process, enabling organizations to gather and analyze vast amounts of data related to supplier performance, environmental impact, and compliance (Kuusisto, 2017) with advanced data analytics, companies can make informed decisions about their sourcing partners, ensuring that the entire supply chain operates in harmony with sustainable principles (Ebinger & Omondi, 2020).

As the green supply chain strategy progresses, logistics management emerges as a critical area for optimization. The integration of digital technologies such as the IoT, AI, and blockchain enhances transparency, traceability, and efficiency in logistics operations. Digitalization facilitates seamless communication and data exchange between partners, supporting joint initiatives for eco-friendly sourcing, waste reduction, and energy efficiency (Akbari & Hopkins, 2022).

In conclusion, integrating green operations and digitalization in supply chain management is essential for organizations committed to environmental sustainability and long-term success. From responsible sourcing to optimized logistics, a green supply chain strategy leverages digital technologies to enhance efficiency, transparency, and collaboration. By embracing this transformation, businesses can contribute positively to the environment while creating value for all stakeholders involved in the journey toward a greener and more sustainable future.

10.4.2 Assessing Environmental Impacts and Conducting Life Cycle Assessments

Assessing environmental impacts and conducting life cycle assessments (LCAs) have become integral components of sustainability initiatives for businesses and industries worldwide. These practices involve a comprehensive evaluation of the environmental implications associated with products, services, or processes throughout their entire life cycle, from raw material extraction to end-of-life disposal (Ferrari, Volpi, Settembre-Blundo, & García-Muiña, 2021; Najjar et al., 2022; Teh, Khan, Corbitt, & Ong, 2020).

The process of assessing environmental impacts begins with data collection and analysis. A wide range of environmental indicators, such as greenhouse gas emissions, water usage, energy consumption, and waste generation, are considered to understand the full scope of the environmental footprint (Matuštík & Kočí, 2021). This data can be collected from internal sources, supply chain partners, and third-party databases, providing a holistic view of the ecological consequences of the subject under assessment. One of the key tools employed in this process is the LCA, which follows a systematic methodology to quantify the environmental impacts at each stage of the product's life cycle (Passon, Galante, & Ogliari, 2023). The life cycle typically includes stages like raw material acquisition, manufacturing, transportation, product use, and end-of-life disposal or recycling. LCA involves evaluating the inputs (resources and energy) and outputs (emissions, waste) associated with each stage, enabling a comprehensive understanding of the product's environmental profile.

Environmental impacts and LCA assessments facilitate informed decision-making for sustainable practices (Barbhuiya & Das, 2023). They help businesses identify hotspots of high environmental impact, enabling them to focus on improving the most critical areas. This may involve reducing energy consumption during manufacturing, optimizing transportation routes, or using more eco-friendly materials. By identifying opportunities for improvement, companies can develop targeted strategies to minimize their ecological footprint (Lee, 2011).

Moreover, digital technologies have significantly enhanced the efficiency and accuracy of environmental assessments. Advanced data analytics, ML, and AI algorithms can process vast amounts of environmental data, enabling real-time monitoring and continuous improvement. This digital transformation empowers businesses to make data-driven decisions, respond rapidly to emerging environmental concerns, and adapt their sustainability strategies as needed (Attaran, 2020).

In conclusion, assessing environmental impacts and conducting LCAs are indispensable practices for businesses striving to achieve sustainability goals. By understanding the full scope of their environmental footprint, organizations can identify areas for improvement, implement targeted strategies, and enhance their overall environmental performance. As businesses embrace digital technologies and the importance of sustainable practices grows, the adoption of robust environmental assessments becomes essential in ensuring a greener and more sustainable future for generations to come.

10.4.3 Supplier Selection and Collaboration for Sustainable Sourcing

Supplier selection and collaboration for sustainable sourcing have emerged as crucial pillars in the pursuit of environmental and social responsibility within supply chain management. As businesses increasingly recognize the importance of sustainability, they are reshaping their supplier selection criteria to ensure that partners align with eco-friendly practices and ethical standards (Macchion, Toscani, & Vinelli, 2023).

The process of supplier selection for sustainable sourcing begins with a comprehensive evaluation of potential partners (Sarkis & Talluri, 2002). Beyond traditional considerations like cost and quality, companies now assess suppliers based on their environmental impact (Bras, 1997), commitment to sustainability (Rao & Holt, 2005), and adherence to ethical labor practices (Marshall, McCarthy, McGrath, & Claudy, 2015). This involves conducting audits, requesting sustainability reports, and engaging in transparent dialogue with suppliers to gain insights into their environmental policies and practices (Khokhar et al., 2022).

Collaboration is key to promoting sustainable sourcing within the supply chain (Mishra, Singh, & Rana, 2022). Businesses seek to build strong partnerships with suppliers who share their sustainability vision. By fostering open communication and joint goal-setting, companies can work together with suppliers to develop and implement sustainable practices throughout the supply chain (Park, Kim, & Lee, 2022). Incorporating sustainable sourcing into supplier selection and collaboration also involves considering the entire life cycle of products or materials. This means evaluating not only the environmental impact of the supplier's operations but also the sustainability of the raw materials they use and the end-of-life disposal or recycling processes. Emphasis is placed on minimizing waste, reducing carbon emissions, and promoting the circular economy (Negri, Cagno, Colicchia, & Sarkis, 2021).

Digital technologies play a vital role in supplier selection and collaboration for sustainable sourcing (Ghadimi, Wang, Lim, & Heavey, 2019). Advanced data analytics and supply chain management platforms enable businesses to gather and analyze vast amounts of data, allowing for a more informed assessment of supplier performance and environmental impact. This data-driven approach empowers companies to make strategic decisions that align with their sustainability goals.

In conclusion, supplier selection and collaboration for sustainable sourcing are instrumental in fostering responsible and ethical supply chain practices. By partnering with suppliers who share their sustainability vision and leveraging digital technologies to promote transparency and data-driven decision-making, businesses can drive positive change, reduce environmental impact, and contribute to a more sustainable future for both society and the planet.

10.4.4 Implementing Green Transportation and Distribution Practices

Implementing green transportation and distribution practices has become a critical aspect of sustainable supply chain management. With the increasing awareness of environmental concerns, businesses are striving to reduce their carbon footprint and promote eco-friendly alternatives in the transportation and distribution of goods.

At the heart of green transportation practices lies the adoption of low-carbon and energy-efficient modes of transport (Shah et al., 2021). Companies are exploring alternatives such as EVs, hybrid vehicles, and biodiesel-powered trucks to replace traditional fossil fuel-dependent fleets. These greener alternatives not only reduce greenhouse gas emissions but also pave the way for a more sustainable future. In addition to upgrading their vehicle fleets, businesses are optimizing transportation routes and logistics operations to reduce energy consumption and minimize emissions. Advanced analytics and AI-powered algorithms are utilized to identify the most efficient routes, considering factors like traffic patterns, weather conditions, and delivery schedules (Kozlov, 2022). By streamlining transportation processes, companies can achieve significant fuel savings and contribute to cleaner air and reduced congestion.

Furthermore, green transportation practices involve the consolidation and collaboration of shipments to minimize the number of trips and empty spaces in transportation vehicles. Sharing transportation resources through collaborative networks enables businesses to achieve economies of scale and maximize efficiency while reducing overall environmental impact. In the realm of distribution practices, companies are focusing on implementing sustainable packaging solutions. The move toward recyclable and biodegradable packaging materials helps reduce waste and supports the principles of a circular economy. Additionally, optimizing packaging sizes and materials minimizes the use of resources and lowers transportation costs, contributing to overall sustainability goals (Dekker, Bloemhof, & Mallidis, 2012).

Digital technologies also play a vital role in green transportation and distribution practices. Internet of Things devices, RFID tags, and tracking systems enable real-time monitoring of shipments, allowing businesses to proactively manage logistics and respond to potential issues promptly. This not only enhances efficiency but also reduces the risk of spoilage and waste during transportation. Embracing green transportation and distribution practices is not only a moral imperative but also a strategic advantage for businesses. As consumers increasingly prioritize sustainability, companies that prioritize eco-friendly transportation and distribution are likely to gain a competitive edge and enhance brand reputation (Yan & Yazdanifard, 2014).

In conclusion, the implementation of green transportation and distribution practices is a significant step toward building a more sustainable supply chain. By adopting energy-efficient vehicles, optimizing transportation routes, and embracing sustainable packaging, businesses can reduce their environmental impact while improving operational efficiency. Digital technologies further enhance these efforts, enabling data-driven decision-making and promoting transparency throughout the supply chain. As businesses continue to integrate green practices into their operations, they contribute to a greener, more environmentally responsible future for all.

10.4.5 Eco-friendly Packaging and Waste Reduction Techniques

In the global push toward sustainability and environmental responsibility, eco-friendly packaging, and waste reduction techniques have emerged as critical strategies to minimize the impact of consumer goods on the environment.

Eco-friendly packaging focuses on reducing the ecological footprint of product packaging throughout its life cycle (G. Prakash & Pathak, 2017). Businesses are increasingly opting for materials that are recyclable, biodegradable, or made from renewable resources. By moving away from traditional single-use plastics and excessive packaging, companies are striving to minimize waste generation and promote a more circular economy.

Innovative packaging solutions include compostable materials, plant-based plastics, and reusable containers. Compostable packaging breaks down naturally, returning to the earth without leaving harmful residues. Plant-based plastics, derived from renewable resources, offer a more sustainable alternative to petroleum-based plastics. Reusable containers, such as refillable bottles and containers, encourage consumers to reduce single-use packaging and actively participate in waste reduction efforts (Jadhav, Fulke, & Giripunje, 2023).

Moreover, businesses are employing lightweight packaging designs that not only save resources but also reduce transportation emissions. Lighter packages require less energy to produce and transport, thereby contributing to lower carbon emissions throughout the supply chain. Waste reduction techniques further complement eco-friendly packaging efforts. Companies are adopting practices like source reduction, which involves redesigning products to use fewer materials, and encouraging consumers to make more conscious choices. By promoting sustainable consumption and responsible disposal practices, businesses actively engage consumers in the journey toward waste reduction (Vachon, 2007).

Circular economy initiatives are gaining traction, encouraging businesses to design products with end-of-life considerations in mind (De Angelis, Howard, & Miemczyk, 2018). This involves developing products that are easily repairable, upgradable, or recyclable. By promoting product stewardship, businesses aim to extend the life of products and minimize waste generation (Genovese, Acquaye, Figueroa, & Koh, 2017).

Digital technologies also play a vital role in waste reduction. Advanced analytics and supply chain management platforms enable companies to optimize inventory levels, reducing the risk of excess stock and potential waste. Moreover, data-driven insights help businesses forecast demand more accurately, preventing overproduction and unnecessary waste. Eco-friendly packaging and waste reduction techniques not only align with environmental goals but also positively impact a company's brand reputation. Consumers are increasingly seeking products from businesses that demonstrate a genuine commitment to sustainability. By embracing these practices, companies can attract environmentally conscious consumers and build a loyal customer base (Majeed, Aslam, Murtaza, Attila, & Molnár, 2022).

In conclusion, eco-friendly packaging and waste reduction techniques are instrumental in creating a more sustainable future. By choosing environmentally responsible packaging materials and implementing waste reduction strategies, businesses can significantly reduce their ecological footprint and contribute to a cleaner, healthier planet. As industries continue to prioritize sustainability, these practices will continue to evolve and drive positive change, promoting a greener world for generations to come.

10.4.6 Monitoring and Reporting Sustainability Metrics with Digital Tools

In the realm of sustainability, monitoring and reporting metrics play a pivotal role in measuring the environmental and social impact of businesses and organizations (Gong, Simpson, Koh, & Tan, 2018). With the increasing focus on corporate responsibility and transparency, digital tools have become indispensable in efficiently collecting, analyzing, and reporting sustainability data (Venkatesh et al., 2020).

Digital tools have revolutionized the way sustainability metrics are monitored, making the process more streamlined and accurate (Saner, Yiu, & Nguyen, 2020). IoT devices, sensors, and data loggers are deployed to collect real-time data on energy consumption, water usage, waste generation, and other key sustainability indicators. These devices continuously monitor operations, enabling businesses to identify trends and patterns that may have otherwise gone unnoticed (Baron, 2006).

By leveraging advanced data analytics and ML algorithms, digital tools process the vast amounts of sustainability data, offering deeper insights and actionable information (Guo et al., 2020). The ability to crunch numbers, detect anomalies, and generate predictive models empowers organizations to make data-driven decisions, optimize resource usage, and identify opportunities for sustainable improvements. Digital dashboards and customized reporting tools facilitate effective communication of sustainability metrics. Businesses can generate visual reports and interactive dashboards that showcase progress toward sustainability goals, providing stakeholders with a clear understanding of their environmental impact. These reports can be shared internally with employees, management, and board members, as well as externally with investors, customers, and regulatory bodies (Frick, 2023).

Moreover, digital tools enhance the accuracy and reliability of sustainability reporting. Manual data entry and spreadsheets are prone to errors and can be time-consuming. By automating data collection and analysis, digital tools minimize the risk of inaccuracies and ensure that the reported metrics are consistent and trustworthy. Cloud-based platforms enable easy access to sustainability data from anywhere and at any time. This fosters collaboration among different departments and locations within an organization, facilitating coordinated efforts toward sustainability goals. Additionally, cloud-based solutions offer a secure and centralized repository for sustainability data, ensuring data integrity and compliance with data privacy regulations (Xing, Qian, & Zaman, 2016).

The integration of digital tools into sustainability monitoring and reporting also promotes transparency and accountability. Businesses can share their sustainability metrics and progress with external stakeholders through dedicated sustainability portals or websites. By being transparent about their environmental and social impact, organizations build trust and credibility with customers, investors, and the wider community.

In conclusion, monitoring and reporting sustainability metrics with digital tools have become essential components of modern sustainability efforts. Through IoT devices, data analytics, and cloud-based platforms, businesses can effectively collect, analyze, and report sustainability data, making informed decisions and demonstrating their commitment to environmental responsibility. As technology continues to advance, digital tools will continue to evolve, empowering organizations to proactively address sustainability challenges and contribute to a more sustainable and responsible future.

10.5 SECTION 4: BENEFITS AND CHALLENGES OF IMPLEMENTING GREEN SUPPLY CHAIN INITIATIVES

10.5.1 Environmental Benefits: Reduction in Greenhouse Gas Emissions, Waste, and Pollution

Implementing green supply chain initiatives brings forth a host of environmental benefits that pave the way toward a more sustainable future. At the forefront of these advantages lies a significant reduction in greenhouse gas emissions, waste, and pollution.

One of the most profound impacts of green supply chain initiatives is the substantial reduction in greenhouse gas emissions (Wang & Gupta, 2011). By adopting eco-friendly practices throughout the supply chain, businesses minimize their reliance on fossil fuels, which are major contributors to greenhouse gas emissions. Utilizing energy-efficient transportation, optimizing logistics routes, and embracing renewable energy sources in operations all contribute to lowering the carbon footprint. As a result, the overall impact of the supply chain on the environment is lessened, and progress is made toward mitigating the effects of climate change (Sanders, Boone, Ganeshan, & Wood, 2019).

Green supply chain initiatives also play a crucial role in waste reduction. By embracing sustainable packaging solutions (Mwaura, Letting, Ithinji, & Bula, 2016; Zsidisin & Siferd, 2001) and source reduction techniques (Kitazawa & Sarkis, 2000), companies minimize the amount of waste generated during production, transportation, and distribution processes (Kozlov, 2022; Romero-Hernández & Romero, 2018). This includes the reduction of single-use plastics, opting for recyclable or compostable materials, and promoting responsible consumption. As a result, less waste ends up in landfills or oceans, reducing environmental degradation and promoting a cleaner, healthier ecosystem.

Furthermore, green supply chain initiatives contribute to curbing pollution levels. By adopting cleaner production methods and adhering to stringent environmental regulations, businesses minimize the release of harmful pollutants into the air, water, and soil (Panigrahi & Sahu, 2018). Sustainable practices in waste management and responsible disposal also prevent pollution of natural resources (Fedotkina, Gorbashko, & Vatolkina, 2019; Kurniawan, Liang, et al., 2022). Consequently, the overall environmental impact of the supply chain is reduced, benefiting both the planet and the communities in which businesses operate.

In conclusion, green supply chain initiatives yield a range of environmental benefits, including a reduction in greenhouse gas emissions, waste, and pollution. By embracing sustainability practices, businesses contribute to a cleaner and greener world, while inspiring positive changes throughout the supply chain network. While challenges may arise during the implementation process, the rewards of a more environmentally responsible supply chain are invaluable for the well-being of the planet and future generations.

10.5.2 Cost Savings and Efficiency Gains through Resource Optimization

Implementing green supply chain initiatives not only brings environmental benefits but also results in significant cost savings and efficiency gains through resource optimization (Esmaeilian, Sarkis, Lewis, & Behdad, 2020). By embracing sustainability practices and streamlining operations, businesses can achieve a more cost-effective and resource-efficient supply chain.

One of the primary areas of cost savings is in energy consumption. Green supply chain initiatives focus on reducing energy usage through various means, such as adopting energy-efficient technologies, optimizing transportation routes, and implementing smart energy management practices (Marchi & Zanoni, 2017). These measures lead to reduced energy bills and lower operational expenses, contributing to overall cost savings.

Resource optimization also extends to materials and waste management (Van Engeland, Beliën, De Boeck, & De Jaeger, 2020). By adopting sustainable packaging solutions and source reduction techniques, businesses can minimize material usage and waste generation (Guillard et al., 2018; Romero-Hernández & Romero, 2018). This not only reduces the cost of raw materials but also lowers disposal and waste management expenses. Additionally, businesses can explore recycling and upcycling opportunities, turning waste into valuable resources and reducing the need for new materials.

Optimizing transportation and logistics operations contributes significantly to efficiency gains (Agyabeng-Mensah, Ahenkorah, Afum, Dacosta, & Tian, 2020; Ranieri, Digiesi, Silvestri, & Roccotelli, 2018). By using data analytics and advanced technologies, companies can identify the most efficient transportation routes, minimize delivery times, and reduce fuel consumption (Gružauskas, Baskutis, & Navickas, 2018). This optimization streamlines the supply chain, leading to faster delivery times and improved customer satisfaction.

In conclusion, green supply chain initiatives provide cost savings and efficiency gains through resource optimization. By reducing energy consumption, minimizing waste, optimizing transportation, and streamlining processes, businesses can achieve a more resource-efficient supply chain. These sustainable practices not only lead to financial benefits but also strengthen the company's reputation and positioning in a market increasingly driven by environmental consciousness. As businesses continue to embrace sustainability, the rewards of cost savings and efficiency gains will undoubtedly reinforce the commitment to a greener and more prosperous future.

10.5.3 Enhanced Brand Image and Customer Loyalty

In the pursuit of sustainable and responsible practices, green supply chain initiatives offer businesses more than just environmental benefits and cost savings. They also lead to an enhanced brand image and foster strong customer loyalty, establishing a lasting and positive impact on the company's reputation.

Embracing green supply chain initiatives demonstrates a commitment to environmental stewardship and corporate social responsibility (Akbari & McClelland, 2020). As consumers increasingly prioritize sustainability, businesses that adopt eco-friendly practices gain a competitive advantage by resonating with environmentally conscious customers (Rathore, 2018). Such a commitment to sustainability communicates that the company is mindful of its impact on the planet and the community, earning the trust and admiration of consumers.

An enhanced brand image emerges as businesses showcase their sustainability efforts through transparent and effective communication (Kim, Song, Lee, & Lee, 2017). Companies can leverage digital platforms, social media, and sustainability reports to share their green initiatives, accomplishments, and progress toward sustainability goals. This open dialogue helps build a genuine and positive perception of the brand, showcasing the company as a responsible and caring corporate citizen.

Moreover, a strong brand image is reinforced by customer experiences that align with sustainable values (Cambra-Fierro, Fuentes-Blasco, Huerta-Álvarez, & Olavarría, 2021). Sustainable products and packaging appeal to environmentally conscious consumers who seek products with reduced environmental impact. The sustainable choices offered by businesses become an extension of their brand, fostering a sense of loyalty among customers who align with these values (Maspul, 2023).

Green supply chain initiatives also create opportunities for businesses to engage customers in the journey toward sustainability. Companies can involve consumers in recycling programs, circular economy initiatives, or eco-friendly campaigns, making customers active participants in the company's sustainability mission (Kurniawan, Othman, Hwang, & Gikas, 2022). Such involvement deepens the connection between customers and the brand, nurturing a sense of loyalty and belonging.

Customer loyalty is further strengthened when companies prioritize transparency and accountability. Openly sharing information about sustainable sourcing, ethical practices, and progress toward sustainability goals builds trust with customers (Guerreiro & Pacheco, 2021). When customers know that a company is actively striving to minimize its environmental impact, they are more likely to remain loyal to the brand.

In conclusion, green supply chain initiatives contribute to an enhanced brand image and customer loyalty. By showcasing a commitment to sustainability, engaging in transparent communication, and offering eco-friendly products and experiences, businesses can attract environmentally conscious consumers and foster a loyal customer base. As the importance of sustainability continues to grow in consumer decision-making, businesses that prioritize green practices will undoubtedly reap the rewards of an esteemed brand image and enduring customer loyalty.

10.5.4 Regulatory Compliance and Risk Mitigation

Green supply chain initiatives not only bring about environmental and economic benefits but also play a crucial role in ensuring regulatory compliance and mitigating potential risks for businesses (Xiao, Wilhelm, van der Vaart, & Van Donk, 2019). As governments and international bodies increasingly prioritize sustainability and environmental protection, adhering to green supply chain practices becomes essential for avoiding legal and reputational risks.

Regulatory compliance is a key driver for the adoption of green supply chain initiatives (Hsu, Choon Tan, Hanim Mohamad Zailani, & Jayaraman, 2013). Governments around the world have implemented strict environmental regulations and standards aimed at reducing carbon emissions, promoting resource conservation, and preventing environmental degradation (Lin & Zhang, 2023). Moreover, some industries are subject to specific environmental certifications or eco-labeling requirements (Gulbrandsen, 2005). By incorporating green practices into their supply chain, businesses can achieve these certifications, enhancing their credibility and marketability. Customers and stakeholders are increasingly seeking products and services from companies that demonstrate their commitment to regulatory compliance and sustainability.

Green supply chain initiatives also contribute to risk mitigation by reducing vulnerabilities and exposure to environmental risks. Climate change, natural disasters, and resource scarcity are potential risks that can disrupt supply chain operations and impact business continuity. By adopting sustainable practices, businesses build resilience and strengthen their ability to withstand these challenges (DiBella et al., 2023). In addition to regulatory compliance and risk mitigation, green supply chain initiatives help businesses stay ahead of emerging sustainability requirements (Hasan, 2013). As consumers and investors demand greater transparency and accountability, businesses that proactively implement sustainable practices are better positioned to navigate evolving market expectations.

In conclusion, green supply chain initiatives are vital for regulatory compliance and risk mitigation in an increasingly sustainability-focused world. By adhering to environmental regulations, achieving certifications, and adopting eco-friendly practices, businesses avoid legal liabilities and reputational risks. Additionally, sustainability practices enhance resilience, enabling businesses to better withstand environmental challenges and uncertainties. Embracing green supply chain initiatives is not only a responsible business choice but also a strategic approach to safeguarding a company's long-term success in a rapidly changing and sustainability-driven business landscape (Rattalino, 2018).

10.6 CHALLENGES OF ADOPTING GREEN PRACTICES: COST BARRIERS, LACK OF AWARENESS, AND RESISTANCE TO CHANGE

The adoption of green practices in supply chains is a transformative journey, but it comes with its fair share of challenges (Rattalino, 2018). Three significant obstacles that businesses encounter when embracing sustainability are cost barriers, lack of awareness, and resistance to change.

One of the primary challenges is the perception of higher costs associated with green practices (Mariani & Vastola, 2015). While sustainable initiatives can lead to long-term cost savings, some businesses may face initial investments and operational adjustments that seem financially burdensome (Amankwah-Amoah, & Syllias, 2020). Upgrading to energy-efficient technologies, sourcing eco-friendly materials, and implementing waste reduction measures might require additional capital upfront (Shrivastava, 1995). Overcoming this challenge requires a strategic approach, demonstrating the long-term benefits of sustainability, including reduced energy expenses, waste management savings, and enhanced brand reputation.

A lack of awareness and understanding about green practices can also hinder their widespread adoption. Many companies, particularly small and medium-sized enterprises, may not be fully aware of the potential environmental and economic benefits that sustainability can offer. This lack of awareness might result in the overlooking of opportunities to improve resource efficiency and reduce environmental impact. Educating businesses about the advantages of green practices and providing accessible resources can help bridge this knowledge gap, encouraging more companies to embrace sustainability.

In conclusion, while adopting green practices in supply chains offers numerous benefits, there are challenges to overcome. Cost barriers, lack of awareness, and resistance to change can impede progress. Nevertheless, with proactive measures, collaboration, and effective leadership, businesses can address these challenges and successfully integrate sustainability into their supply chain operations, reaping the rewards of environmental responsibility and long-term economic viability.

10.7 CONCLUSION

The integration of green operations and digitalization has become imperative for organizations seeking to achieve sustainable and competitive supply chain performance. By leveraging digital technologies, companies can optimize their processes, reduce environmental impacts, and gain a competitive edge. Embracing green supply chain practices not only benefits the planet but also contributes to long-term business success.

10.8 DISCUSSION QUESTIONS

1. How can digital technologies like IoT and AI contribute to reducing waste and enhancing sustainability in supply chain operations?
2. What are some of the key challenges organizations might face when implementing green supply chain initiatives, and how can they overcome them?
3. Discuss the role of blockchain technology in ensuring transparency and traceability in sustainable sourcing.
4. How can organizations strike a balance between achieving green objectives and maintaining cost-effectiveness in their supply chain?
5. In your opinion, what factors are critical for ensuring the successful adoption of green operations and digitalization in supply chain management?

10.9 ANSWERS TO DISCUSSION QUESTIONS

1. Digital technologies like IoT enable real-time monitoring of energy consumption and process efficiency, allowing organizations to identify areas

for optimization and waste reduction. AI can analyze vast amounts of data to optimize transportation routes and inventory management, leading to reduced fuel usage and greenhouse gas emissions.

2. Some key challenges in implementing green supply chain initiatives include high initial investment costs, resistance to change from stakeholders, and the need for supplier collaboration. To overcome these challenges, organizations can conduct thorough cost-benefit analyses, promote a culture of sustainability, and incentivize suppliers to adopt eco-friendly practices.
3. Blockchain technology provides an immutable and transparent ledger, making it ideal for ensuring traceability and transparency in sustainable sourcing. It can be used to track the origin of raw materials, verify certifications, and monitor supply chain activities to ensure compliance with environmental standards.
4. Organizations can achieve a balance between green objectives and cost-effectiveness by conducting a LCA of their products, optimizing processes to reduce waste and resource consumption, and exploring sustainable alternatives that may initially cost more but yield long-term benefits.
5. Successful adoption of green operations and digitalization requires strong leadership commitment, employee engagement, continuous monitoring and improvement, and collaboration with suppliers and other stakeholders. Additionally, a clear communication strategy can help create awareness and support for sustainability initiatives throughout the organization.

10.10 SUMMARY

This chapter delved into the concept of green operations and their significance in promoting sustainability within organizations. It explored how digitalization plays a crucial role in enhancing green supply chain performance by leveraging technologies such as IoT, AI, and blockchain. The integration of green practices and digital technologies allows companies to achieve environmental objectives while improving operational efficiency and reducing costs. The chapter also discussed real-world case studies, providing practical insights into successful green supply chain initiatives. By adopting green operations and digitalization, organizations can not only contribute to a greener world but also secure a competitive advantage in the market.

REFERENCES

Abdullah, N., Lim, A. (2023). The incorporating sustainable and green IT practices in modern IT service operations for an environmentally conscious future. *Journal of Sustainable Technologies and Infrastructure Planning*, 7(3), 17–47.

Acciarini, C., Borelli, F., Capo, F., Cappa, F., Sarrocco, C. (2022). Can digitalization favour the emergence of innovative and sustainable business models? A qualitative exploration in the automotive sector. *Journal of Strategy and Management*, 15(3), 335–352.

Adedibu, P. A. (2023). *Ecological problems of agriculture: impacts and sustainable solutions.* ScienceOpen Preprints.

Adhi Santharm, B., & Ramanathan, U. (2022). Supply chain transparency for sustainability–an intervention-based research approach. *International Journal of Operations & Production Management*, 42(7), 995–1021.

Afshar Jahanshahi, A., Maghsoudi, T., Shafighi, N. (2021). Employees' environmentally responsible behavior: The critical role of environmental justice perception. *Sustainability: Science, Practice and Policy*, 17(1), 1–14.

Aftab, J., Abid, N., Sarwar, H., & Veneziani, M. (2022). Environmental ethics, green innovation, and sustainable performance: Exploring the role of environmental leadership and environmental strategy. *Journal of Cleaner Production*, 378, 134639.

Agyabeng-Mensah, Y., Ahenkorah, E., Afum, E., Dacosta, E., & Tian, Z. (2020). Green warehousing, logistics optimization, social values and ethics and economic performance: The role of supply chain sustainability. *The International Journal of Logistics Management*, 31(3), 549–574.

Akbari, M., & Hopkins, J. L. (2022). Digital technologies as enablers of supply chain sustainability in an emerging economy. *Operations Management Research*, 15(3–4), 689–710.

Akbari, M., & McClelland, R. (2020). Corporate social responsibility and corporate citizenship in sustainable supply chain: A structured literature review. *Benchmarking: An International Journal*, 27(6), 1799–1841.

Ali Ababneh, O. M., Awwad, A. S., Abu-Haija, A. (2021). The association between green human resources practices and employee engagement with environmental initiatives in hotels: The moderation effect of perceived transformational leadership. *Journal of Human Resources in Hospitality & Tourism*, 20(3), 390–416.

Allioui, H., Mourdi, Y. (2023). Unleashing the potential of AI: Investigating cutting-edge technologies that are transforming businesses. *International Journal of Computer Engineering and Data Science (IJCEDS)*, 3(2), 1–12.

Almeyda, R., & Darmansya, A. (2019). The influence of environmental, social, and governance (ESG) disclosure on firm financial performance. *IPTEK Journal of Proceedings Series*, (5), 278–290.

Alonso-Muñoz, S., González-Sánchez, R., Siligardi, C., & García-Muiña, F. E. (2021). New circular networks in resilient supply chains: An external capital perspective. *Sustainability*, 13(11), 6130.

Amankwah-Amoah, J., Syllias, J. (2020). Can adopting ambitious environmental sustainability initiatives lead to business failures? An analytical framework. *Business Strategy and the Environment*, 29(1), 240–249.

Antonini, C., Beck, C., Larrinaga, C. (2020). Subpolitics and sustainability reporting boundaries. The case of working conditions in global supply chains. *Accounting, Auditing & Accountability Journal*, 33(7), 1535–1567.

Aryal, A., Liao, Y., Nattuthurai, P., & Li, B. (2020). The emerging big data analytics and IoT in supply chain management: A systematic review. *Supply Chain Management: An International Journal*, 25(2), 141–156.

Asghar, R., Sulaiman, M. H., Mustaffa, Z., Ullah, N., Hassan, W. (2022). The important contribution of renewable energy technologies in overcoming Pakistan's energy crisis: Present challenges and potential opportunities. *Energy & Environment*, 34(8), 3450–3494.

Asiaei, K., Bontis, N., Alizadeh, R., Yaghoubi, M. (2022). Green intellectual capital and environmental management accounting: Natural resource orchestration in favor of environmental performance. *Business Strategy and the Environment*, 31(1), 76–93.

Asthana, A. N. (2023). Wastewater management through circular economy: A pathway towards sustainable business and environmental protection. *Advances in Water Science*, 34(3), 87–98.

Attaran, M. (2020). Digital technology enablers and their implications for supply chain management. Paper presented at *the Supply Chain Forum: An International Journal*.

Baah, C., Opoku-Agyeman, D., Acquah, I. S. K., Agyabeng-Mensah, Y., Afum, E., Faibil, D., & Abdoulaye, F. A. M. (2021). Examining the correlations between stakeholder pressures, green production practices, firm reputation, environmental and financial performance: Evidence from manufacturing SMEs. *Sustainable Production and Consumption*, 27, 100–114.

Baah, C., Opoku-Agyeman, D., Acquah, I. S. K., Issau, K., & Abdoulaye, F. A. M. (2020). Understanding the influence of environmental production practices on firm performance: A proactive versus reactive approach. *Journal of Manufacturing Technology Management*, 32(2), 266–289.

Badgett, A., & Milbrandt, A. (2020). A summary of standards and practices for wet waste streams used in waste-to-energy technologies in the United States. *Renewable and Sustainable Energy Reviews*, 117, 109425.

Bai, C., Quayson, M., & Sarkis, J. (2022). Analysis of Blockchain's enablers for improving sustainable supply chain transparency in Africa cocoa industry. *Journal of Cleaner Production*, 358, 131896.

Barbhuiya, S., & Das, B. B. (2023). Life Cycle Assessment of construction materials: Methodologies, applications and future directions for sustainable decision-making. *Case Studies in Construction Materials*, 19, e02326.

Baron, R. A. (2006). Opportunity recognition as pattern recognition: How entrepreneurs "connect the dots" to identify new business opportunities. *Academy of Management Perspectives*, 20(1), 104–119.

Bartle, C., Chatterjee, K. (2019). Employer perceptions of the business benefits of sustainable transport: A case study of peri-urban employment areas in South West England. *Transportation Research Part A: Policy and Practice*, 126, 297–313.

Berrone, P., Rousseau, H. E., Ricart, J. E., Brito, E., & Giuliodori, A. (2023). How can research contribute to the implementation of sustainable development goals? An interpretive review of SDG literature in management. *International Journal of Management Reviews*, 25(2), 318–339.

Bharany, S., Badotra, S., Sharma, S., Rani, S., Alazab, M., & Jhaveri, R. H. (2022). Energy efficient fault tolerance techniques in green cloud computing: A systematic survey and taxonomy. *Sustainable Energy Technologies and Assessments*, 53, 102613.

Bhattacharyay, D., Maitra, S., Pine, S., Shankar, T., & Pedda Ghouse Peera, S. K. (2020). Future of precision agriculture in India. *Protected Cultivation and Smart Agriculture*, 1, 289–299.

Boğan, E., Dedeoğlu, B. B. (2020). Hotel employees' corporate social responsibility perception and organizational citizenship behavior: Perceived external prestige and pride in organization as serial mediators. *Corporate Social Responsibility and Environmental Management*, 27(5), 2342–2353.

Bras, B. (1997). Incorporating environmental issues in product design and realization. *Industry and Environment*, 20(1), 7–13.

Braulio-Gonzalo, M., Jorge-Ortiz, A., & Bovea, M. D. (2022). How are indicators in Green Building Rating Systems addressing sustainability dimensions and life cycle frameworks in residential buildings? *Environmental Impact Assessment Review*, 95, 106793.

Brun, A., Karaosman, H., & Barresi, T. (2020). Supply chain collaboration for transparency. *Sustainability*, 12(11), 4429.

Cagno, E., Trianni, A., Spallina, G., & Marchesani, F. (2017). Drivers for energy efficiency and their effect on barriers: Empirical evidence from Italian manufacturing enterprises. *Energy Efficiency*, 10, 855–869.

Callaghan, E. G., Colton, J. (2008). Building sustainable & resilient communities: A balancing of community capital. *Environment, Development and Sustainability*, 10, 931–942.

Cambra-Fierro, J. J., Fuentes-Blasco, M., Huerta-Álvarez, R., Olavarría, A. (2021). Customer-based brand equity and customer engagement in experiential services: Insights from an emerging economy. *Service Business*, 15, 467–491.

Camilleri, M. A. (2022). Strategic attributions of corporate social responsibility and environmental management: The business case for doing well by doing good! *Sustainable Development*, 30(3), 409–422.

Guo, H., Nativi, S., Liang, D., Craglia, M., Wang, L., Schade, S., ... Li, J. (2020). Big Earth Data science: An information framework for a sustainable planet. *International Journal of Digital Earth*, 13(7), 743–767.

Gupta, I., & Shashidhara, Y. (2023). Enhancing crop optimization in organic agriculture: Insights from a polynomial regression model and the grain bot innovation.

Hasan, M. (2013). Sustainable supply chain management practices and operational performance. *American Journal of Industrial and Business Management*, 3(1), 42–48. https://doi.org/10.4236/ajibm.2013.31006

Helmold, M., & Terry, B. (2021). *Operations and supply management 4.0: Industry insights, case studies and best practices*. Springer Nature.

Hoffmann, S. (2022). Challenges and opportunities of area-based conservation in reaching biodiversity and sustainability goals. *Biodiversity and Conservation*, 31(2), 325–352.

Hossain, M. U., Ng, S. T., Antwi-Afari, P., & Amor, B. (2020). Circular economy and the construction industry: Existing trends, challenges and prospective framework for sustainable construction. *Renewable and Sustainable Energy Reviews*, 130, 109948.

Hsu, C. C., Choon Tan, K., Hanim Mohamad Zailani, S., Jayaraman, V. (2013). Supply chain drivers that foster the development of green initiatives in an emerging economy. *International Journal of Operations & Production Management*, 33(6), 656–688.

In, S. Y., Rook, D., & Monk, A. (2019). Integrating alternative data (also known as ESG data) in investment decision making. *Global Economic Review*, 48(3), 237–260.

Jadhav, H., Fulke, A., & Giripunje, M. (2023). Recent global insight into mitigation of plastic pollutants, sustainable biodegradable alternatives, and recycling strategies. *International Journal of Environmental Science and Technology*, 20(7), 8175–8198.

Javaid, M., Haleem, A., Singh, R. P., Rab, S., & Suman, R. (2021). Upgrading the manufacturing sector via applications of industrial internet of things (IIoT). *Sensors International*, 2, 100129.

Kapp, S., Choi, J.-K., & Kissock, K. (2022). Toward energy-efficient industrial thermal systems for regional manufacturing facilities. *Energy Reports*, 8, 1377–1387.

Karatepe, O. M., Rezapouraghdam, H., & Hassannia, R. (2021). Sense of calling, emotional exhaustion and their effects on hotel employees' green and non-green work outcomes. *International Journal of Contemporary Hospitality Management*, 33(10), 3705–3728.

Karneli, O. (2023). The Role of Adhocratic Leadership in Facing the Changing Business Environment. *Journal of Contemporary Administration and Management (ADMAN)*, 1(2), 77–83.

Kashmanian, R. M. (2017). Building greater transparency in supply chains to advance sustainability. *Environmental Quality Management*, 26(3), 73–104.

Kaur, M., & Kaur, S. (2022). Internet of Things in the Smart Automotive Sector: A Review. Paper presented at the *2022 11th International Conference on System Modeling & Advancement in Research Trends (SMART)*.

Kayikci, Y., Durak Usar, D., & Aylak, B. L. (2022). Using blockchain technology to drive operational excellence in perishable food supply chains during outbreaks. *The International Journal of Logistics Management*, 33(3), 836–876.

Kazancoglu, Y., Sagnak, M., Kayikci, Y., & Kumar Mangla, S. (2020). Operational excellence in a green supply chain for environmental management: A case study. *Business Strategy and the Environment*, 29(3), 1532–1547.

Kedward, K., Gabor, D., & Ryan-Collins, J. (2022). Aligning finance with the green transition: From a risk-based to an allocative green credit policy regime. Available at SSRN 4198146.

Khan, S. A., Mubarik, M. S., Kusi-Sarpong, S., Gupta, H., Zaman, S. I., & Mubarik, M. (2022). Blockchain technologies as enablers of supply chain mapping for sustainable supply chains. *Business Strategy and the Environment*, 31(8), 3742–3756.

Khan, S. A. R., Yu, Z., & Farooq, K. (2023). Green capabilities, green purchasing, and triple bottom line performance: Leading toward environmental sustainability. *Business Strategy and the Environment*, 32(4), 2022–2034.

Khan, S. A. R., Yu, Z., Umar, M., & Tanveer, M. (2022). Green capabilities and green purchasing practices: A strategy striving towards sustainable operations. *Business Strategy and the Environment*, 31(4), 1719–1729.

Khokhar, M., Zia, S., Islam, T., Sharma, A., Iqbal, W., & Irshad, M. (2022). Going green supply chain management during covid-19, assessing the best supplier selection criteria: A triple bottom line (tbl) approach. *Problemy Ekorozwoju*, 17(1), 36–51.

Kim, J. S., Song, H., Lee, C.-K., & Lee, J. Y. (2017). The impact of four CSR dimensions on a gaming company's image and customers' revisit intentions. *International Journal of Hospitality Management*, 61, 73–81.

Kitazawa, S., & Sarkis, J. (2000). The relationship between ISO 14001 and continuous source reduction programs. *International Journal of Operations & Production Management*, 20(2), 225–248.

Kotsiuk, V., Rogova, N., Medvid, L., & Popovych, O. (2023). Ecological rural tourism: Sustainable use of natural resources and ecosystems.

Kouhizadeh, M., Saberi, S., & Sarkis, J. (2021). Blockchain technology and the sustainable supply chain: Theoretically exploring adoption barriers. *International Journal of Production Economics*, 231, 107831.

Kozlov, I. P. (2022). Optimizing Public Transport Services using AI to Reduce Congestion in Metropolitan Area. *International Journal of Intelligent Automation and Computing*, 5(2), 1–14.

Kumar, A., Payal, M., Dixit, P., & Chatterjee, J. M. (2020). Framework for realization of green smart cities through the Internet of Things (IoT). *Trends in Cloud-based IoT*, 85–111. https://doi.org/10.1007/978-3-030-40037-8_6

Kumar, M., Kaushik, K., Singh, S., Kumar, S., Rai, A., Kumar, R., … Kumar, A. (2023). Sustainable horticulture practices: An environmental-friendly approaches.

Kumar, S., & Rathore, K. (2023). Renewable Energy for Sustainable Development Goal of Clean and Affordable Energy. *International Journal of Materials Manufacturing and Sustainable*, 1, 1–15. https://doi.org/10.56896/ijmmst

Kuo, F.-I., Fang, W.-T., & LePage, B. A. (2022). Proactive environmental strategies in the hotel industry: Eco-innovation, green competitive advantage, and green core competence. *Journal of Sustainable Tourism*, 30(6), 1240–1261.

Kurniawan, T. A., Liang, X., O'Callaghan, E., Goh, H., Othman, M. H. D., Avtar, R., & Kusworo, T. D. (2022). Transformation of solid waste management in China: Moving towards sustainability through digitalization-based circular economy. *Sustainability*, 14(4), 2374.

Kurniawan, T. A., Othman, M. H. D., Hwang, G. H., & Gikas, P. (2022). Unlocking digital technologies for waste recycling in Industry 4.0 era: A transformation towards a digitalization-based circular economy in Indonesia. *Journal of Cleaner Production*, 357, 131911.

Kuusisto, M. (2017). Organizational effects of digitalization: A literature review. *International Journal of Organization Theory and Behavior*, 20(3), 341–362.

Lazaroiu, G., Androniceanu, A., Grecu, I., Grecu, G., & Neguriță, O. (2022). Artificial intelligence-based decision-making algorithms, Internet of Things sensing networks, and sustainable cyber-physical management systems in big data-driven cognitive manufacturing. *Oeconomia Copernicana*, 13(4), 1047–1080.

Lee, K.-H. (2011). Integrating carbon footprint into supply chain management: The case of Hyundai Motor Company (HMC) in the automobile industry. *Journal of Cleaner Production*, 19(11), 1216–1223.

Levaggi, L., Levaggi, R., Marchiori, C., & Trecroci, C. (2020). Waste-to-Energy in the EU: The effects of plant ownership, waste mobility, and decentralization on environmental outcomes and welfare. *Sustainability*, 12(14), 5743.

Li, C., Deng, Z., Wang, Z., Hu, Y., Wang, L., & Yu, S. (2023). Responses to the COVID-19 pandemic have impeded progress towards the Sustainable Development Goals. *Communications Earth & Environment*, 4(1), 252.

Li, H., Chen, C., & Umair, M. (2023). Green finance, enterprise energy efficiency, and green total factor productivity: Evidence from China. *Sustainability*, 15(14), 11065.

Li, Y., Zhao, X., Shi, D., & Li, X. (2014). Governance of sustainable supply chains in the fast fashion industry. *European Management Journal*, 32(5), 823–836.

Lin, B., & Zhang, A. (2023). Can government environmental regulation promote low-carbon development in heavy polluting industries? Evidence from China's new environmental protection law. *Environmental Impact Assessment Review*, 99, 106991.

Lukin, E., Krajnović, A., & Bosna, J. (2022). Sustainability strategies and achieving SDGs: A comparative analysis of leading companies in the automotive industry. *Sustainability*, 14(7), 4000.

Macchion, L., Toscani, A. C., & Vinelli, A. (2023). Sustainable business models of small and medium-sized enterprises and the relationships to be established within the supply chain to support these models. *Corporate Social Responsibility and Environmental Management*, 30(2), 563–573.

Maftouh, A., El Fatni, O., Fayiah, M., Liew, R., Lam, S., Bahaj, T., & Butt, M. (2022). The application of water–energy nexus in the Middle East and North Africa (MENA) region: A structured review. *Applied Water Science*, 12(5), 83.

Majeed, M. U., Aslam, S., Murtaza, S. A., Attila, S., & Molnár, E. (2022). Green marketing approaches and their impact on green purchase intentions: Mediating role of green brand image and consumer beliefs towards the environment. *Sustainability*, 14(18), 11703.

Manna, M. C., Rahman, M. M., Naidu, R., Bari, A. F., Singh, A., Thakur, J., … Subbarao, A. (2021). Organic farming: A prospect for food, environment and livelihood security in Indian agriculture. *Advances in Agronomy*, 170, 101–153.

Manne, R., & Kantheti, S. C. (2021). Green IoT Towards Environmentally Friendly, Sustainable and Revolutionized Farming. *Green Internet of Things and Machine Learning: Towards a Smart Sustainable World*, 113–139.

Marchi, B., & Zanoni, S. (2017). Supply chain management for improved energy efficiency: Review and opportunities. *Energies*, 10(10), 1618.

Mariani, A., & Vastola, A. (2015). Sustainable winegrowing: Current perspectives. *International Journal of Wine Research*, 37–48. https://doi.org/10.2147/IJWR.S68003

Marshall, D., McCarthy, L., McGrath, P., & Claudy, M. (2015). Going above and beyond: How sustainability culture and entrepreneurial orientation drive social sustainability supply chain practice adoption. *Supply Chain Management: An International Journal*, 20(4), 434–454.

Maspul, K. A. (2023). The Emergence of Local Coffee Brands: A Paradigm Shift in Jakarta Coffee Culture. *EKOMA: Jurnal Ekonomi, Manajemen, Akuntansi*, 3(1), 135–149.

Matuštík, J., & Kočí, V. (2021). What is a footprint? A conceptual analysis of environmental footprint indicators. *Journal of Cleaner Production*, 285, 124833.

McLennon, E., Dari, B., Jha, G., Sihi, D., & Kankarla, V. (2021). Regenerative agriculture and integrative permaculture for sustainable and technology driven global food production and security. *Agronomy Journal*, 113(6), 4541–4559.

Meacham, B. J. (2019). Sustainability and resiliency objectives in performance building regulations. In *Building Governance and Climate Change* (pp. 8–23). Routledge.

Mishra, R., Singh, R. K., & Rana, N. P. (2022). Developing environmental collaboration among supply chain partners for sustainable consumption & production: Insights from an auto sector supply chain. *Journal of Cleaner Production*, 338, 130619.

Mondejar, M. E., Avtar, R., Diaz, H. L. B., Dubey, R. K., Esteban, J., Gómez-Morales, A., … Prasad, K. A. (2021). Digitalization to achieve sustainable development goals: Steps towards a Smart Green Planet. *Science of the Total Environment*, 794, 148539.

Mukherjee, C., Denney, J., Mbonimpa, E., Slagley, J., & Bhowmik, R. (2020). A review on municipal solid waste-to-energy trends in the USA. *Renewable and Sustainable Energy Reviews*, 119, 109512.

Mukhopadhyay, D., & Thakur, P. (2021). Natural resource management and conservation for smallholder farming in India: Strategies and challenges. In *Soil Science: Fundamentals to Recent Advances*, 731–749. https://doi.org/10.1007/978-981-16-0917-6_36

Mukonza, C., Hinson, R. E., Adeola, O., Adisa, I., Mogaji, E., & Kirgiz, A. C. (2021). Green marketing: An introduction. *Green marketing in Emerging Markets: Strategic and Operational Perspectives*, 3–14. https://doi.org/10.1007/978-3-030-74065-8_1

Musiello-Neto, F., Rua, O. L., Arias-Oliva, M., & Silva, A. F. (2021). Open innovation and competitive advantage on the hospitality sector: The role of organizational strategy. *Sustainability*, 13(24), 13650.

Mutezo, G., & Mulopo, J. (2021). A review of Africa's transition from fossil fuels to renewable energy using circular economy principles. *Renewable and Sustainable Energy Reviews*, 137, 110609.

Mwaura, A. W., Letting, N., Ithinji, G. K., & Bula, H. O. (2016). Green distribution practices and competitiveness of food manufacturing firms in Kenya. *International Journal of Economics, Commerce and Management*, 4(3), 189–207.

Najjar, M. K., Figueiredo, K., Evangelista, A. C. J., Hammad, A. W., Tam, V. W., & Haddad, A. (2022). Life cycle assessment methodology integrated with BIM as a decision-making tool at early-stages of building design. *International Journal of Construction Management*, 22(4), 541–555.

Navaneethakrishnan, S. R., Melam, D. P. K., Bothra, N., Haldar, B., & Pal, S. (2023). Blockchain and machine learning integration: Enhancing transparency and traceability in supply chains. *Remittances Review*, 8(4), 2396–2409.

Nayal, K., Raut, R. D., Yadav, V. S., Priyadarshinee, P., & Narkhede, B. E. (2022). The impact of sustainable development strategy on sustainable supply chain firm performance in the digital transformation era. *Business Strategy and the Environment*, 31(3), 845–859.

Negri, M., Cagno, E., Colicchia, C., & Sarkis, J. (2021). Integrating sustainability and resilience in the supply chain: A systematic literature review and a research agenda. *Business Strategy and the Environment*, 30(7), 2858–2886.

Núñez-Merino, M., Maqueira-Marín, J. M., Moyano-Fuentes, J., & Martínez-Jurado, P. (2020). Information and digital technologies of Industry 4.0 and Lean supply chain management: A systematic literature review. *International Journal of Production Research*, 58(16), 5034–5061.

Obrenovic, B., Du, J., Godinic, D., Tsoy, D., Khan, M. A. S., & Jakhongirov, I. (2020). Sustaining enterprise operations and productivity during the COVID-19 pandemic: "Enterprise Effectiveness and Sustainability Model". *Sustainability*, 12(15), 5981.

Olabi, A., Obaideen, K., Elsaid, K., Wilberforce, T., Sayed, E. T., & Maghrabie, H. M. (2022). Assessment of the pre-combustion carbon capture contribution into sustainable development goals SDGs using novel indicators. *Renewable and Sustainable Energy Reviews*, 153, 111710.

Oloruntobi, O., Mokhtar, K., Rozar, N. M., Gohari, A., Asif, S., & Chuah, L. F. (2023). Effective technologies and practices for reducing pollution in warehouses-a review. *Cleaner Engineering and Technology*, 100622. https://doi.org/10.1016/j.clet.2023.100622

Orji, I. J., Kusi-Sarpong, S., & Gupta, H. (2020). The critical success factors of using social media for supply chain social sustainability in the freight logistics industry. *International Journal of Production Research*, 58(5), 1522–1539.

Osman, A. I., Chen, L., Yang, M., Msigwa, G., Farghali, M., Fawzy, S., … Yap, P.-S. (2023). Cost, environmental impact, and resilience of renewable energy under a changing climate: A review. *Environmental Chemistry Letters*, 21(2), 741–764.

Pang, R., & Zhang, X. (2019). Achieving environmental sustainability in manufacture: A 28-year bibliometric cartography of green manufacturing research. *Journal of Cleaner Production*, 233, 84–99.

Panigrahi, S. S., & Sahu, B. (2018). Analysis of interactions among the enablers of green supply chain management using interpretive structural modelling: An Indian perspective. *International Journal of Comparative Management*, 1(4), 377–399.

Park, J.-E., & Kang, E. (2022a). The mediating role of eco-friendly artwork for urban hotels to attract environmental educated consumers. *Sustainability*, 14(7), 3784.

Park, J., & Kang, E. (2022b). The Mediating Role of Eco-Friendly Art-work for Urban Hotels to Attract Environmental Educated Consumers. Sustainability 2022, 14, x. In: Note: MDPI stays neutral with regard to jurisdictional claims in ….

Park, S. R., Kim, S. T., & Lee, H.-H. (2022). Green supply chain management efforts of first-tier suppliers on economic and business performances in the electronics industry. *Sustainability*, 14(3), 1836.

Passon, B. C., Galante, E. B., & Ogliari, A. (2023). A systematic approach to assist in life-cycle assessment of ammunition demilitarization process: A case study with the 105-mm HE M1 ammunition. *The International Journal of Life Cycle Assessment*, 28(4), 398–428.

Patil, P. (2021). Sustainable transportation planning: strategies for reducing greenhouse gas emissions in urban areas. *Empirical Quests for Management Essences*, 1(1), 116–129.

Paulraj, A. (2011). Understanding the relationships between internal resources and capabilities, sustainable supply management and organizational sustainability. *Journal of Supply Chain Management*, 47(1), 19–37.

Perron, G. M., Côté, R. P., & Duffy, J. F. (2006). Improving environmental awareness training in business. *Journal of Cleaner Production*, 14(6–7), 551–562.

Peter, O., Pradhan, A., & Mbohwa, C. (2023). Industrial internet of things (IIoT): Opportunities, challenges, and requirements in manufacturing businesses in emerging economies. *Procedia Computer Science*, 217, 856–865.

Pfajfar, G., Shoham, A., Małecka, A., & Zalaznik, M. (2022). Value of corporate social responsibility for multiple stakeholders and social impact–Relationship marketing perspective. *Journal of Business Research*, 143, 46–61.

Phillips, A. (2021). Artificial intelligence-enabled healthcare delivery and digital epidemiological surveillance in the remote treatment of patients during the COVID-19 pandemic. *American Journal of Medical Research*, 8(1), 40–49.

Pimenov, D. Y., Mia, M., Gupta, M. K., Machado, Á. R., Pintaude, G., Unune, D. R., … Wojciechowski, S. (2022). Resource saving by optimization and machining environments for sustainable manufacturing: A review and future prospects. *Renewable and Sustainable Energy Reviews*, 166, 112660.

Prakash, G., & Pathak, P. (2017). Intention to buy eco-friendly packaged products among young consumers of India: A study on developing nation. *Journal of Cleaner Production*, 141, 385–393.

Prakash, S., Kumar, S., Soni, G., Jain, V., Dev, S., & Chandra, C. (2022). Evaluating approaches using the Grey-TOPSIS for sustainable supply chain collaboration under risk and uncertainty. *Benchmarking: An International Journal*, 30, 3124–3149.

Pullman, M., McCarthy, L., & Mena, C. (2023). Breaking bad: How can supply chain management better address illegal supply chains? *International Journal of Operations & Production Management*. https://doi.org/10.1108/IJOPM-02-2023-0079

Raimi, M. O. (2020). A review of environmental, social and health impact assessment (Eshia) practice in Nigeria: A panacea for sustainable development and decision making. *MOJ Public Health*, 9(3-2020).

Rajaeifar, M. A., Ghadimi, P., Raugei, M., Wu, Y., & Heidrich, O. (2022). Challenges and recent developments in supply and value chains of electric vehicle batteries: A sustainability perspective. *Resources, Conservation and Recycling*, 180, 106144.

Ranieri, L., Digiesi, S., Silvestri, B., & Roccotelli, M. (2018). A review of last mile logistics innovations in an externalities cost reduction vision. *Sustainability*, 10(3), 782.

Rao, P., & Holt, D. (2005). Do green supply chains lead to competitiveness and economic performance? *International Journal of Operations & Production Management*, 25(9), 898–916.

Rathore, B. (2018). Navigating the Green Marketing Landscape: Best Practices and Future Trends. *International Journal of New Media Studies: International Peer Reviewed Scholarly Indexed Journal*, 5(2), 1–9.

Rattalino, F. (2018). Circular advantage anyone? Sustainability-driven innovation and circularity at Patagonia, Inc. *Thunderbird International Business Review*, 60(5), 747–755.

Rayner, S. (2016). Uncomfortable knowledge: The social construction of ignorance in science and environmental policy discourses. In *An introduction to the sociology of ignorance* (pp. 107–125). Routledge.

Raza, A., Farrukh, M., Iqbal, M. K., Farhan, M., & Wu, Y. (2021). Corporate social responsibility and employees' voluntary pro-environmental behavior: The role of organizational pride and employee engagement. *Corporate Social Responsibility and Environmental Management*, 28(3), 1104–1116.

Risteska, L. J. V. (2023). Green Marketing (41).

Romero-Hernández, O., & Romero, S. (2018). Maximizing the value of waste: From waste management to the circular economy. *Thunderbird International Business Review*, 60(5), 757–764.

Sanders, N. R., Boone, T., Ganeshan, R., & Wood, J. D. (2019). Sustainable supply chains in the age of AI and digitization: Research challenges and opportunities. *Journal of Business Logistics*, 40(3), 229–240.

Saner, R., Yiu, L., & Nguyen, M. (2020). Monitoring the SDGs: Digital and social technologies to ensure citizen participation, inclusiveness and transparency. *Development Policy Review*, 38(4), 483–500.

Santos, E. (2023). From neglect to progress: Assessing social sustainability and decent work in the tourism sector. *Sustainability*, 15(13), 10329.

Sarkis, J., & Talluri, S. (2002). A model for strategic supplier selection. *Journal of Supply Chain Management*, 38(4), 18–28.

Shah, K. J., Pan, S.-Y., Lee, I., Kim, H., You, Z., Zheng, J.-M., & Chiang, P.-C. (2021). Green transportation for sustainability: Review of current barriers, strategies, and innovative technologies. *Journal of Cleaner Production*, 326, 129392.

Shaikh, P. H., Nor, N. B. M., Sahito, A. A., Nallagownden, P., Elamvazuthi, I., & Shaikh, M. (2017). Building energy for sustainable development in Malaysia: A review. *Renewable and Sustainable Energy Reviews*, 75, 1392–1403.

Shrivastava, P. (1995). Environmental technologies and competitive advantage. *Business Ethics and Strategy*, 16(S1), 183–200.

Siagian, H., Tarigan, Z. J. H., & Jie, F. (2021). Supply chain integration enables resilience, flexibility, and innovation to improve business performance in COVID-19 era. *Sustainability*, 13(9), 4669.

Singh, K., & Misra, M. (2021). Linking corporate social responsibility (CSR) and organizational performance: The moderating effect of corporate reputation. *European Research on Management and Business Economics*, 27(1), 100139.

Sohrabi, M. J. I. O. D. T. I. S. C. M. (2023). *Artificial Intelligence in Logistic Industry*. 73–86.

Souto, J. E. (2022). Organizational creativity and sustainability-oriented innovation as drivers of sustainable development: Overcoming firms' economic, environmental and social sustainability challenges. *Journal of Manufacturing Technology Management*, 33(4), 805–826.

Souza, G. R., Valamede, L. S., & Akkari, A. C. S. (2020). Characterization of Digital Supply Chain: A Focus on Configuration and Resources. Paper presented at *the Brazilian Technology Symposium*.

Srivastav, A. L., Dhyani, R., Ranjan, M., Madhav, S., & Sillanpää, M. (2021). Climate-resilient strategies for sustainable management of water resources and agriculture. *Environmental Science and Pollution Research*, 28(31), 41576–41595.

Stahl, G. K., Brewster, C. J., Collings, D. G., & Hajro, A. (2020). Enhancing the role of human resource management in corporate sustainability and social responsibility: A multi-stakeholder, multidimensional approach to HRM. *Human Resource Management Review*, 30(3), 100708.

Suchek, N., Fernandes, C. I., Kraus, S., Filser, M., & Sjögrén, H. (2021). Innovation and the circular economy: A systematic literature review. *Business Strategy and the Environment*, 30(8), 3686–3702.

Sugandini, D., Susilowati, C., Siswanti, Y., & Syafri, W. (2020). Green supply management and green marketing strategy on green purchase intention: SMEs cases. *Journal of Industrial Engineering and Management (JIEM)*, 13(1), 79–92.

Sultan, S., & El-Qassem, A. (2021). Future prospects for sustainable agricultural development. *International Journal of Modern Agriculture and Environment*, 1(2), 54–82.

Tabrizchi, H., & Kuchaki Rafsanjani, M. (2020). A survey on security challenges in cloud computing: Issues, threats, and solutions. *The Journal of Supercomputing*, 76(12), 9493–9532.

Tay, H. L., & Loh, H. S. (2021). Digital transformations and supply chain management: A Lean Six Sigma perspective. *Journal of Asia Business Studies*, 16(2), 340–353.

Teh, D., Khan, T., Corbitt, B., & Ong, C. E. (2020). Sustainability strategy and blockchain-enabled life cycle assessment: A focus on materials industry. *Environment Systems and Decisions*, 40, 605–622.

Thorisdottir, T. S., & Johannsdottir, L. (2020). Corporate social responsibility influencing sustainability within the fashion industry. A systematic review. *Sustainability*, 12(21), 9167.

Trabucco, M., & De Giovanni, P. (2021). Achieving resilience and business sustainability during COVID-19: The role of lean supply chain practices and digitalization. *Sustainability*, 13(22), 12369.

Tsai, P.-H., Lin, G.-Y., Zheng, Y.-L., Chen, Y.-C., Chen, P.-Z., & Su, Z.-C. (2020). Exploring the effect of Starbucks' green marketing on consumers' purchase decisions from consumers' perspective. *Journal of Retailing and Consumer Services*, 56, 102162.

Tu, Y., & Wu, W. (2021). How does green innovation improve enterprises' competitive advantage? The role of organizational learning. *Sustainable Production and Consumption*, 26, 504–516.

Umair, S., Waqas, U., & Mrugalska, B. (2023). Cultivating sustainable environmental performance: The role of green talent management, transformational leadership, and employee engagement with green initiatives. *Work* 78(4), 1–13.

Vachon, S. (2007). Green supply chain practices and the selection of environmental technologies. *International Journal of Production Research*, 45(18–19), 4357–4379.

Van Engeland, J., Beliën, J., De Boeck, L., & De Jaeger, S. (2020). Literature review: Strategic network optimization models in waste reverse supply chains. *Omega*, 91, 102012.

Vance, T. C., Wengren, M., Burger, E., Hernandez, D., Kearns, T., Medina-Lopez, E., … Potemra, J. T. (2019). From the oceans to the cloud: Opportunities and challenges for data, models, computation and workflows. *Frontiers in Marine Science*, 6, 211.

Veile, J. W., Schmidt, M.-C., Müller, J. M., & Voigt, K.-I. (2021). Relationship follows technology! How Industry 4.0 reshapes future buyer-supplier relationships. *Journal of Manufacturing Technology Management*, 32(6), 1245–1266.

Venkatesh, V., Kang, K., Wang, B., Zhong, R. Y., & Zhang, A. (2020). System architecture for blockchain based transparency of supply chain social sustainability. *Robotics and Computer-Integrated Manufacturing*, 63, 101896.

Wan, Y. K. P., Chan, S. H. J., & Huang, H. L. W. (2017). Environmental awareness, initiatives and performance in the hotel industry of Macau. *Tourism Review*, 72(1), 87–103.

Wang, H.-F., & Gupta, S. M. (2011). *Green supply chain management: Product life cycle approach*. McGraw-Hill Education.

Whillans, A., Sherlock, J., Roberts, J., O'Flaherty, S., Gavin, L., & Dykstra, H. (2021). Nudging the commute: Using behaviorally informed interventions to promote sustainable transportation. *Behavioral Science & Policy*, 7(2), 27–49.

Woo, E.-J., & Kang, E. (2020). Environmental issues as an indispensable aspect of sustainable leadership. *Sustainability*, 12(17), 7014.

Xiao, C., Wilhelm, M., van der Vaart, T., & Van Donk, D. P. (2019). Inside the buying firm: Exploring responses to paradoxical tensions in sustainable supply chain management. *Journal of Supply Chain Management*, 55(1), 3–20.

Xing, K., Qian, W., & Zaman, A. U. (2016). Development of a cloud-based platform for footprint assessment in green supply chain management. *Journal of Cleaner Production*, 139, 191–203.

Yadav, V. S., Singh, A., Raut, R. D., Mangla, S. K., Luthra, S., & Kumar, A. (2022). Exploring the application of Industry 4.0 technologies in the agricultural food supply chain: A systematic literature review. *Computers & Industrial Engineering*, 169, 108304.

Yan, Y. K., & Yazdanifard, R. (2014). The concept of green marketing and green product development on consumer buying approach. *Global Journal of Commerce & Management Perspective*, 3(2), 33–38.

Yang, Q., Zhao, Y., & Ma, D. (2022). Cu-mediated Ullmann-type cross-coupling and industrial applications in route design, process development, and scale-up of pharmaceutical and agrochemical processes. *Organic Process Research & Development*, 26(6), 1690–1750.

Ye, J., & Dela, E. (2023). The effect of green investment and green financing on sustainable business performance of foreign chemical industries operating in Indonesia: The mediating role of corporate social responsibility. *Sustainability*, 15(14), 11218.

Zafar, H., Ho, J. A., Cheah, J. H., & Mohamed, R. (2023). Promoting pro-environmental behavior through organizational identity and green organizational climate. *Asia Pacific Journal of Human Resources*, 61(2), 483–506.

Zahraee, S. M., Shiwakoti, N., & Stasinopoulos, P. (2022). Agricultural biomass supply chain resilience: COVID-19 outbreak vs. sustainability compliance, technological change, uncertainties, and policies. *Cleaner Logistics and Supply Chain*, 4, 100049.

Zhang, L., Butler, T. L., & Yang, B. (2020). Recent trends, opportunities and challenges of sustainable aviation fuel. *Green Energy to Sustainability: Strategies for Global Industries*, 85–110. https://doi.org/10.1002/9781119152057.ch5

Zhang, Z., Peng, X., Yang, L., & Lee, S. (2022). How does Chinese central environmental inspection affect corporate green innovation? The moderating effect of bargaining intentions. *Environmental Science and Pollution Research*, 29(28), 42955–42972.

Zhou, L., Ren, S., Du, L., Tang, F., & Li, R. (2023). Is environmental labeling certification a "green passport" for firm exports in emerging economies? Evidence from China. *International Business Review*, 32, 102171.

Zhou, M., & Li, X. (2022). Influence of green finance and renewable energy resources over the sustainable development goal of clean energy in China. *Resources Policy*, 78, 102816.

Zikirillo, S., & Ataboyev, I. (2023). Air pollution and control engineering and technology. Paper presented at *the Proceedings of International Conference on Modern Science and Scientific Studies*.

Zsidisin, G. A., & Siferd, S. P. (2001). Environmental purchasing: A framework for theory development. *European Journal of Purchasing & Supply Management*, 7(1), 61–73.

11 Innovative Transport *Environmental, Social, and Economic Aspects*

Magdalena Dalewska and Beata Mrugalska
Poznan University of Technology, Poznan, Poland

11.1 INTRODUCTION

Transport plays a significant role in the life of society as it stimulates development, communication, and coordination of all other sectors. It enables accessibility to opportunities through movement of people and goods. It is also associated with a number of direct and indirect externalities such as traffic congestion, air pollution, and road accidents. However, consumption of raw materials, noise, and emission of pollutants are some of the main problems related to the transport industry. Therefore, in recent decades transport ecology has started to develop significantly focusing on investigation of all correlations between humans, transport, and the environment, including various systems and rebound effects (Sudowski & Mrugalska, 2017; Umair et al., 2023a, 2023b).

Transport is responsible for about a quarter of the EU's total greenhouse gas emissions and causes air pollution, noise pollution, and habitat fragmentation (Bakker, 2023). Its contribution to global warming and climate change is significant as its emissions continue to grow, achieving 26% of the global CO_2 emissions (Zhang & Witlox, 2019). It results from the fact that in recent years, the amount of cargo transport and the number of means of transport have increased significantly. It leads to vehicle operational pollution (transport, loading, and unloading of means of transport), breakdowns, and illegal activities. During transport, exhaust gases, noise, and other waste are emitted into the environment. However, most exhaust gases are emitted in land transport, among which road transport generates a significant amount. Carbon dioxide emission into the environment is approximately 30% of the carbon dioxide emissions from all modes of transport. However, it has to be underlined that pollution emitted into the environment is mainly produced as a result of unforeseen events. Transport failures are quite rare, but their effects have a huge impact on the environment. For example, in water transport, they can be such as fuel leaks, fires, cargo falling, and ship collisions In the case of land transport, the main accidents involve other means of transport, people, or animals, as well as loss of fuel. The greatest consequences of failures are generated by air transport. However, they occur the least frequently in reference to all means of transport. Illegal activities mainly involve non-compliance with regulations specifying exhaust gas emission standards. Comparing all means of transport, it is estimated that land transport has the largest

 DOI: 10.1201/9781032616810-11

share in environmental pollution. In Europe, EU emission standards have to be applied to all vehicles. There are differentiated two separate standards for light-duty vehicles (categories: M_1, M_2, N_1, and N_2 not exceeding 2,610 kg) and heavy-duty vehicles (categories M_1, M_2, N_1, and N_2, mass exceeding 2,610 kg, and all vehicles of category M_3 and N_3). The numbering method used by these two distinct standards allows you to tell them apart: the heavy-duty standards use Roman numbers (Euro I, II, III, etc.), whereas the light-duty standards use Arabic numbers (Euro 1, 2, 3, etc.). Since January 1, 2021, the European Union has been operating under the new emission standard, Euro 6D ISC-FCM. It encompasses not just the problem of cleanliness but also the duty to monitor the degree of combustion. According to the European Commission in 2050 only zero-emission vehicles will be driving in the EU, but in the meantime another standard: Euro 7 is planned to be established (Valverde et al., 2023).

In order to reduce the intensification of the greenhouse effect and emissions of harmful substances, producers increasingly turn to alternative energy sources, such as renewable electricity, biofuels, hydrogen, synthetic, or paraffin fuels. Renewable electricity is one of the sources which is inexhaustible as it replenishes naturally and does not cause damage to the environment. These renewable energy sources are dependable, affordable, and environmentally friendly; generate little secondary waste; and promote sustainable development. These energy sources can meet two-thirds of the global energy demand, which means they can make a substantial contribution to reducing greenhouse gas emissions and, ultimately, global warming (Agrawal & Soni, 2021).

One of the fuels increasingly used by producers is bio-LPG, which is produced from processed waste and vegetable oils. Vehicles, which are powered by natural gas, are mainly used in heavy road transport to reduce emissions of harmful substances, such as nitrogen oxides and particulate matter. Means of transport using biogas have lower carbon dioxide emissions than traditional combustion engines. In 2022, renewable energy sources accounted for 41.2% of the EU's gross electricity consumption, which is 3.4 percentage points more than in 2021 (37.8%) and far ahead of other electricity-generating sources like nuclear (less than 22%), gas (less than 20%) or coal (less than 17%). Between 2021 and 2022, the overall amount of renewable energy sources grew by 5.7%. More than two-thirds of the electricity produced from renewable sources came from wind and hydropower (37.5% and 29.9%, respectively). Solid biofuels (6.9%), other renewable sources (7.5%), and solar (18.2%) accounted for the remaining one-third of the electricity generated. The fastest-growing energy source is solar power, which was only 1% of the EU's total electricity consumption in 2008 (Eurostat, 2024).

Hydrogen is also an alternative transport solution to conventional fuels. It is considered to be the fuel of the future in transport, which may revolutionize this industry. It works on traditional combustion engines as well as in electric vehicles. Instead of gasoline or diesel fuel, hydrogen is injected and combined with oxygen in the fuel cell. Vehicles using hydrogen are zero-emission because the byproducts are water and heat (SES Hydrogen, 2022). However, there are several types of technologies for producing this element. These technologies affect greenhouse gas emissions to diverse degree. Moreover, in transport, ecological fuel equivalents are being created to use existing combustion engines. Gasoline and diesel oil are replaced with synthetic and paraffin fuels, which are produced as a result of chemical reactions. These fuels can be obtained from materials such as natural gas, coal, biomass, and plastics. The production of synthetic fuels is currently more expensive than the extraction of fossil fuels. It involves the expenditure of a large amount of energy and the complexity of the production process (Łodygowski, 2013).

11.2 THE IMPACT OF INNOVATIVE TRANSPORT ON THE ENVIRONMENT

Electromobility is a modern solution for eco-innovative transport management. It covers fully electric vehicles, hybrid electric vehicles, and vehicles using hydrogen fuel cell technology. These vehicles are introduced as alternatives to conventional drive units. They are the main driving force for sustainable transport and efficient mobility as they are introduced to the market mainly to reduce exhaust emissions. Zero emission transport is particularly important in cities and urban agglomerations where there is heavy traffic and an increased population. Therefore, electromobility is expected to revolutionize the automotive market in the future. (Yılmaz, Özceylan & Mrugalska, 2023; Kenger et al., 2023). The use of electric vehicles can reduce greenhouse gas emissions by up to 60%. However, the concept of electromobility brings many difficulties. The limitations are the high costs that must be incurred when purchasing vehicles. Moreover, the cities' transport infrastructure is not yet adapted. Electric vehicles have a short range, and the power grid in cities is not developed. A very important aspect affecting the environment is the disposal of batteries used in electric vehicles. Currently, there are no solutions for the disposal or recycling of such batteries. The service life of such batteries is approximately five to ten years. After this time, the batteries lose their properties and the capacity becomes very low and cannot be reused. The environmental impact of electric vehicles also depends on the type of energy used to power the battery.

Using renewable energy to charge means of transport, for example, from wind, solar, or water power plants, has a positive impact on the environment. However, energy produced from fossil fuels does not mean that an electric vehicle has a beneficial impact on climatic conditions (Babicz & Nowakowicz-Dębek, 2021).

Autonomous means of transport become a modern technology used to reduce the negative effects of transport on the environment. Autonomous vehicles are intended to reduce carbon dioxide emissions and reduce the use of all types of fuels, especially fossil fuels. It is estimated that the introduction of autonomous vehicles to the market will result in a downward trend in pollution by up to 5% by 2050 (Ranft et al., 2016).

The implementation of autonomous vehicles will largely minimize the utility of fuels. Eco-driving is expected to contribute to this, the idea of which is to reduce emissions and fuel consumption while driving efficiently and dynamically. Appropriate braking and acceleration systems, maintaining a constant speed, and controlling the route are some of the optimization functions used in automatic means of transport. Moreover, the implemented systems allowing several vehicles to move in columns may have a relatively high impact on improving the quality of nature. Eco-driving can reduce emissions by 10% to 15%. The development of transportation technologies resulting from the Environmental Protection Agency (EPA) research plays a key role in increasing the economics of fuel use. Autonomous vehicles (AV), thanks to their light construction and small size, will have less air resistance, which will result in a reduction in the fuel used. Innovative technologies in autonomous means of transport intensify the use of alternative energy sources. The latest AV systems show productivity of up to 90%. In autonomous vehicles, modern technology is to ensure the use of batteries that will have competitive prices compared to traditional batteries and are just as effective (Neumann, 2018).

Requirements related to reducing emissions in transport are regulated largely by the European Union. It takes action to reduce carbon dioxide emissions in road transport. It introduces standards limiting emissions for new heavy vehicles. From 2025, newly registered vehicles will be subject to stricter emission requirements. All road transport modes will have to reduce emissions by 15%. In 2030, carbon dioxide emissions from trucks will be reduced by 31% and from passenger cars by 37.5% (Stowarzyszenie Prawników…, 2021).

11.3 ECONOMIC, SOCIAL, AND ENVIRONMENTAL ASPECTS IN TRANSPORT

Transport is a sector of great importance in economic development and influences the economy as it drives the entire economy. Thanks to it, it is possible to move goods and passengers over significant distances. Trade exchanges generate high revenues for private and public enterprises. In 2021, the transport, forwarding, and logistics industry had an approximately 6% share in Poland's Gross Domestic Production (Polski Instytut Ekonomiczny, 2023).

In literature, we can find many connections between economic and social systems, and ecology and transport. The systems interact with each other to a significant extent. Its interactions can be noticed when we analyze the impact of transport on the environment, which causes economic problems, such as negative impacts on the agricultural industry, as well as social problems in terms of health. Interdependencies between systems can be grouped as follows:

- economic
- economic and social
- ecological

When we analyze the economic correlations of transport, special attention should be paid to ecology. Economical route planning, tracking orders, and using vehicle cargo spaces can make a significant contribution to improving the climate. Means of transport should cover the shortest routes while carrying out the largest number of orders. Additionally, it is important to optimize travel time. Vehicle routes should be planned to avoid cities and urban agglomerations during peak hours.

The Internet of Things, artificial intelligence, and data science are used for appropriate transport optimization. Thanks to them, it is possible to design systems that will implement innovative ecological solutions for the transport industry, which will directly translate into reductions in emissions of harmful substances into the environment. These systems collect data using the Internet of Things and then use artificial intelligence and data science algorithms. It is estimated that the collected data allows you to reduce carbon footprint by up to 50%, thus saving 80% of the energy. Moreover, it allows for fuel reduction, which contributes to lower transport costs. Economic and social interdependencies between ecology and transport are noticed with the increasing availability of public transport and the expansion of transport infrastructure. The investments in public transport increase the dynamics of economic systems. In addition, projects related to changes in transport infrastructure generate new jobs. All economic and social activities contribute to reducing traffic on the roads. Users are giving up private vehicles in favor of public transport, which is more economical and ecological. In line with ecological aspects of transport, CarSharing will be popularized in the future and will be used to share vehicles. The means of transport owned by society are not used 90% of the time. Only 10% of these vehicles are used. Therefore, in the future, vehicles will be shared with other road users. Looking at the ecological aspect, a future solution to reduce the emission of harmful substances will be to share rides in vehicles with other users moving in the same direction. It is expected that by 2050 approximately 70% of the global population will live in urban agglomerations. The developed infrastructure is not able to accommodate all means of transport during urban peaks. Therefore, all kinds of applications and programs will be introduced to the market that will plan the route and enable the use of common means of transport. Ecological correlations between other systems are seen in their impact on the environment. Economic and social activity in transport affects the deteriorating climate. Toxins enter the water, soil, and air. Environmental pollution by society is inevitable; therefore, efforts are made to reduce greenhouse gas emissions by introducing eco-innovative transport solutions (Latuszyńska & Strulak-Wójcikiewicz, 2011; Mambiznes, 2021).

Currently, the idea of sustainable development in the context of transport is being emphasized. The essence of sustainable transport is that the movement does not need to cause negative effects and inconvenience. This idea strives for effective and economical communication, limiting the harmful impact on the environment (Dovramadjiev & Mrugalska, 2023). Sustainable transport enables economic development and local development; supports efficient operation; and reduces the consumption of natural resources while minimizing the seizure of goods and noise. The implementation of sustainable transport has a direct long-term impact on the condition of the planet, as well as individual economies, life, and health. In the context of the transport industry, it is primarily about the optimization of routes, fuel consumption, types of fuel used, and the use of intermodal transport. The growing demand for various types of goods and transport services is one of the main challenges of sustainable transport. In practice, sustainable transport also means popularizing innovative transport solutions and improvements in the drive units. An important aspect is the implementation of appropriate solutions at the grassroots level, especially taking into account the role of the automotive industry in generating greenhouse gases. Public awareness and education also play a key role in sustainable transport (Wyszomirski, 2017, pp. 1–6).

11.4 INNOVATIVE TRANSPORT SOLUTIONS – CASE STUDIES

The future transport solution means transport using renewable energy sources and ecological fuels, which will reduce emissions of harmful substances to zero or reduce the release of toxins.

11.4.1 Land Transport

An innovative concept in road transport that limits the negative impact on the environment is the Toyota i-ROAD vehicle. It is a fully electric, zero-emission means of transport that can transport two people. The car is equipped with a three-wheeled chassis and is powered by two electric motors. The built-in lithium-ion batteries allow you to cover a distance of 50 km, reaching a maximum speed of 45 km/h. It takes three hours to charge the battery. In addition, the built-in Active Lean system allows you to tilt during a turning maneuver. The small design makes the vehicle an ideal means of transport in cities and urban agglomerations. A three-wheeled car could revolutionize urban transport and contribute to the popularization of car-sharing (Szczepański, Wiśniewska & Zajkowski, 2016).

The concept of an autonomous car from Renault EZ-GO may contribute to the revolution in the transport sector. The vehicle enables connection to urban infrastructure, which is a key solution for smart cities of the future. The car was designed to promote the sharing of means of transport for passengers traveling in one direction. Renault EZ-GO is equipped with self-management systems, so it does not require driver intervention. The built-in eclectic drive allows you to reach a speed of 50 km/h. The modern design allows passengers to enter through the floating rear doors (Factum-info, 2023).

An example of a Dutch innovative and ecological solution is the car Lightyear 0. The vehicle moves using renewable solar energy, thanks to built-in solar panels in the body. Lightyear 0 is the first passenger car of its kind with very low energy consumption. The streamlined design and relatively low weight result in lower air resistance on the vehicle. During testing, the solar vehicle showed a range of 560 km at a constant speed of 110 km/h (Brzeziński, 2022).

The future of transport may be modular vehicles consisting of capsules that can be combined into platoons. Passengers will be able to change capsules while traveling, which will increase mobility and flexibility. The modular transport concept was proposed by Scania in cooperation with NEXT. An autonomous vehicle consists of biodegradable components and technologies that reduce harmful environmental impacts (Cats et al., 2023).

A modern means of public transport proposed by China is the TEB (Transit Elevated Bus). The transit bus runs on two rails and thus allows other vehicles to move underneath it on two lanes. It is powered by renewable electricity, which does not contribute to environmental pollution. The concept of such a project is to reduce traffic load on the road without reducing passenger traffic. It is estimated that the bus design will reduce traffic jams by up to 35% (Fox News, 2016).

Olli's autonomous invention may contribute to reducing the intensification of the greenhouse effect. It is an independent, electric bus printed in 3D technology by Local Motors. This vehicle is equipped with one of the most advanced technologies in the world.

The IBM Watson system allows integration with the vehicle via voice commands. This is made possible by 30 built-in sensors. Passengers will be able to communicate with Olli during their journey to communicate their destination preferences. Olli is an innovative transport solution that can accommodate 12 people (Geekweek, 2016).

Automatic robots using alternative energy sources may contribute to the revolution in the transport industry. They can be found in agriculture, industry, logistics, and the military, but they are not yet widespread solutions. An example of a mobile robot is MOBOT, designed by WObit. It serves as a replacement for traditional logistics trains. The robot's design is small and light, making it more flexible. The robots are equipped with intelligent terrain mapping systems. They can cooperate with each other and communicate with other devices via WiFi. Moreover, the built-in Mecalac wheels enable movement in different directions (Mobot, 2023).

An innovative transport solution is the Iveco Z Truck. The Z truck is a vehicle powered by biomethane, but it will also be adapted to refuel with liquid gas. The estimated range of this vehicle will be approximately 2,200 km. It has a 400 HP engine with 2,000 N/m torque. This vehicle uses 33% less natural gas, which contributes to a significant reduction in carbon dioxide emissions. Additionally, the truck has a 16-speed PowerShift automatic transmission and uses low-viscosity oil. In addition, the Iveco Z Truck is equipped with Michelin X Line Energy tires, which reduce fuel consumption by up to 1 liter per 100 km, thanks to reduced rolling resistance. The exhaust gases produced by this means of transport are converted into useful thermal energy. In the future, the Iveco truck manufacturer will strive to retrain truck tractor drivers into onboard logistics operators who will plan routes, and the driver's functions will be taken over by electronics (Złoty, 2016).

The future of heavy transport may also be the Freightliner Inspiration Truck. This vehicle uses the Highway Pilot autonomous driving system. It includes front radar, cameras, adaptive cruise control, and an active braking system. This system does not replace the driver but facilitates driving. Using the collected data, it can maintain a constant speed and, if necessary, accelerate or slow down. Highway Pilot has a Start-Stop mode, which is useful when there is traffic congestion. The use of the Highway Pilot system allows you to reduce fuel consumption by up to 5%, which will contribute to reducing exhaust emissions (Freightliner, 2023).

A modern solution in rail transport is the magnetic railway. Maglev is a train that uses superconducting magnets to move. When moving, it levitates at a height of 10 cm above the guide, and when standing still, it lowers itself and stands on its wheels. This is the fastest railway so far, which will ultimately be able to reach speeds of up to 700 km/h (Stawska-Magdziak, 2022).

Hyperloop is a future-proof rail transport concept capable of performing high-speed operations (up to 1,200 km/h). The means of transport is run through a low-pressure pipe to reduce resistance and shorten travel time. A single capsule can accommodate 28 people. Due to the speed achieved, passengers will not be able to move during the journey (Grasso Macola, 2021).

11.4.2 Air Transport

Unmanned aerial vehicles will also be a modern application in air transport. Drones are intelligent, reliable devices that do not emit toxins through the use of electricity.

They are commonly used as entertainment machines, but they are increasingly used in industry. An example of an unmanned vessel in the agricultural industry is the DJI Agras T16 drone. Used for observation purposes, and also for watering and fertilizing green areas, the design of the drone allows it to transport up to 16 liters of liquids. DJI Agras T16 can spray at a rate of approximately 5 liters per minute. The implemented systems allow for precise discharge of materials in a designated place. The technology of unmanned ships can significantly contribute to the development of the agricultural economy and exclude traditional agricultural solutions. In addition, drones equipped with pumps or water tanks can be used to extinguish fires or clean facades (Solectric, 2023).

Unmanned ships are also found in the TSL industry. The use of drones in warehouses is at the testing stage. The first companies are conducting commercial research in which drones could be used to conduct warehouse inventories. These devices would be equipped with autonomous navigation systems inside buildings, thanks to which they would not require access to the GPS satellite. They would save time, minimize the human factor, and enable smooth distribution. It is estimated that implementing unmanned ships in warehouses brings 80% savings compared to standard inventories (Logistics manager, 2021).

The introduction of drones in the TSL industry would also revolutionize outside warehouse transport. The first prototypes of unmanned ships enabling the transport of cargo are being created in the world. One of them is the RDSX delivery drone developed by the American company A2Z. This device has eight propellers that enable it to transport loads weighing four kilograms over a distance of 30 kilometers. It draws its power from built-in batteries, which can be quickly replaced. The built-in boom uses precise sensors to lower the goods in the designated location. RDSX can be controlled automatically or manually by the operator, thanks to two optical cameras. The widespread introduction of unmanned ships to the market would reduce delivery times while reducing transport costs (Łysoń, 2021).

The latest invention is self-flying passenger drones. The first passenger road prototype is the Ehang 184 produced by the Chinese company Eheng. The Ehang 184 electric vehicle is operated by an automatic flight system, so passengers do not have to interfere with the flight. Users only need to enter destination data via tablet. In addition, the machine has special safety systems that take control of the drone in the event of a failure. The use of such systems allows the use of the machine without pilot permissions. Ehang 184 uses electricity. When fully charged, the drone can travel 30 kilometers at a maximum speed of 130 kilometers per hour (Electric VTOL News, 2023).

Renewable energy sources are also the future of air transport. More and more electric aircraft prototypes are appearing on the market. One of them is proposed by the Israeli company Eviation Alice. It is an aircraft adapted for both cargo and freight transport. It can transport 1.2 tons of cargo or nine passengers on board. Built-in batteries allow a range of up to 815 km. The Eviation Alice is fully powered by electricity, making it an environmentally friendly and zero-emission of carbon dioxide (Walków, 2021).

11.4.3 Water Transport

Yara Birkeland is the world's first zero-emission and autonomous merchant ship designed by Norwegian engineers. It runs entirely on batteries, which means it doesn't need a captain to control it. This ship also does not need any diesel fuel to run as everything depends on automatic systems. The main goal of designing the ship was to reduce nitrogen oxide and carbon dioxide emissions in the aquatic environment (Sąsiada, 2019).

Turanor PlanetSolar is a type of ship that uses renewable energy. It is a Swiss design by Raphael Domjan. The ship is 31 meters long and 15 meters wide. To power the Turanor PlanetSolar engines, 537 square meters of solar panels are needed, which store energy in 12-ton batteries. The aerodynamic hull allows you to reach speeds of up to 17 kilometers per hour. Currently, the ship is used as a mobile laboratory that collects and analyzes the amount of plastic in the water and combats pollution in the oceans (Ekosun, 2016).

Oto Ocenbrid is an ecological cargo ship with a length of 200 meters and a width of 40 meters. Powered by wind energy, thanks to built-in telescopic wing sails, the sails can rotate around the vertical axis and adjust their length depending on needs and weather conditions.

They are made of composite materials and steel. The design and modern technology of the ship enable reductions in carbon dioxide emissions by 90% compared to traditional ships. The first cruise is estimated for 2024 (Geekweek, 2020).

Boats powered by warm hydrogen are an increasingly common means of transport replacing conventional fuels. One example is the fully autonomous Energy Observer ship. Initially, it was used for hobby purposes but was later modified for green technology.

The catamaran is energy self-sufficient by using hydrogen from seawater through electrolysis. The generated energy is stored in aluminum tanks with carbon fiber. The method of obtaining energy is fully ecological. In addition, the ship has two built-in wind generators, solar panels, and a generation drive. Energy Observer for movement can obtain energy from several renewable energy sources regardless of the season and place (Energy-observer, 2023).

11.5 CONCLUSIONS

The potential of eco-innovative means of transport depends largely on technological aspects but also on adapted infrastructure. The introduction of hydrogen stations and the development of charging networks may intensify the adaptation of eco-innovation transport solutions. Therefore, the engagement of both private and public entities is crucial to expand such infrastructure.

The analysis of eco-innovative means of transport showed many challenges related to the economic aspect. Production and operation of modern transport solutions cost currently more than traditional means of transport. Therefore, you should strive to reduce the costs associated with production of this type of means of transport, as well as improvement of technologies that would allow for an increase in the life of batteries and hydrogen cells. The study showed that social awareness is equally important. The implementation process of eco-innovative means of transport requires adequate education on their benefits.

REFERENCES

Agrawal, S., & Soni, R. (2021). Renewable energy: Sources, importance and prospects for sustainable future. *Energy: Crises, Challenges and Solutions*, 131–150. https://doi.org/10.1002/9781119741503.ch7

Babicz, M., & Nowakowicz-Dębek, B. (eds.) (2021). *Wybrane zagadnienia z zakresu ochrony i zagrożeń środowiska*. Lublin: Uniwersytet Przyrodniczy w Lublinie.

Bakker, S. (2023). Transport and mobility. European Environment Agency. Retrieved on 6 November 2023 from: https://www.eea.europa.eu/en/topics/in-depth/transport-and-mobility

Brzeziński, M. (2022). Lightyear 0. Energia ze słońca. Samochód trafia do produkcji. Retrieved on 12 June 2023 from: https://automotyw.com/lightyear-0-energia-ze-slonca-samochod-trafia-do-produkcji/

Cats, O., Gidofalvi, G., Jenelius, E., & Hatzenbuhler, J. (2023). Modular vehicle routing for combined passenger freight transport, *Transportation Research Part A* 173, 1–2.

Dovramadjiev, T., & Mrugalska, B. (2023). Real-time planning and monitoring of the steel pipes towards life cycle sustainability management. *Annals of Operations Research* 324(1–2), 1485–1499.

Ekosun (2016). Katamaran zasilany energią paneli fotowoltaicznych. Retrieved on 12 June 2023 from: https://ekosun.pl/blog/oze/katamaran-zasilany-energia-paneli-fotowoltaicznych/

Electric VTOL News (2023). EHang 184 (defunct). Retrieved on 12 June 2023 from: https://evtol.news/ehang/

Energy-observer (2023). Our vessel. Retrieved on 12 June 2023 from: https://www.energy-observer.org/about/vessel

Eurostat (2024). Electricity from renewable sources up to 41% in 2022. Retrieved on 7 April 2024 from: https://ec.europa.eu/eurostat/web/products-eurostat-news/w/ddn-20240221-1

Factum-info (2023). Renault EZ-GO to pojazd współużytkowany przyszłości. Retrieved on 10 June 2023 from: https://factum-info.net/pl/interesnoe/avtomobili/444-renault-ez-go-robomobil-budushchego-dlya-rajdsheringa

Fox News (2016). China begins tests of giant elevated bus. Retrieved on 15 June 2023 from: https://www.foxnews.com/auto/china-begins-tests-of-giant-elevated-bus

Freightliner (2023). The Freightliner Inspiration Truck. Retrieved on 12 June 2023 from: https://roadstars.mercedes-benz-trucks.com/en_GB/events/2015/may/freightliner-inspiration-truck.html

Geekweek (2016). Olli - autonomiczny autobus przyszłości. Retrieved on 12 June 2023 from: https://geekweek.interia.pl/newsroom,nDate,2016-06-20

Geekweek (2020). Ten ekologiczny statek przewiezie tysiące samochodów pomiędzy kontynentami. Retrieved on 12 June 2023 from: https://geekweek.interia.pl/ekotechnologia/news-ten-ekologiczny-statek-przewiezie-tysiace-samochodow-pomiedz,nId,5546430

Grasso Macola, I. (2021). Timeline: tracing the evolution of hyperloop rail technology. Retrieved on 12 June 2023 from: https://www.railway-technology.com/features/timeline-tracing-evolution-hyperloop-rail-technology/

Kenger, Ö. N., Kenger, Z., Özceylan, E., & Mrugalska, B. (2023). Clustering of Cities based on their Smart Performances: A Comparative Approach of Fuzzy C-Means, K-Means, and K-Medoids. *IEEE Access*.

Latuszyńska, M., & Strulak-Wójcikiewicz, R. (2011). Ekologiczne aspekty rozwoju infrastruktury transportu. In: Kryk, B. (ed.), *Trendy i wyzwania zrównoważonego rozwoju*. Szczecin: Uniwersytet Szczeciński, 189–199.

Łodygowski, K. (2013). Paliwa syntetyczne do zasilania silników spalinowych z zapłonem samoczynnym, *Technika Transportu Szynowego* 10, 655–663.

Logistics manager (2021). Pierwsza komercyjna inwentaryzacja rojem autonomicznych dronów. Retrieved on 12 June 2023 from: https://www.logistics-manager.pl/2021/06/24/pierwsza-komercyjna-inwentaryzacja-rojem-autonomicznych-dronow/

Łysoń, M. (2021). Drony dostawcze osiągnęły właśnie nowy poziom dzięki RDSX. Retrieved on 12 June 2023 from: https://whatnext.pl/539550-rdsx-dron-dostawczy-nowej-generacji/

Mambiznes (2021). Internet rzeczy i sztuczna inteligencja przyspieszą odchodzenie od węgla w przemyśle. Retrieved on 15 June 2023 from: https://mambiznes.pl/wlasny-biznes/internet-rzeczy-sztuczna-inteligencja-przyspiesza-odchodzenie-wegla-przemysle-105678

Mobot (2023). Roboty mobilne versus pociągi logistyczne. Retrieved on 12 June 2023 from: http://www.mobot.pl/artykul/5051/roboty-mobilne-versus-pociagi-logistyczne/

Neumann, T. (2018). Perspektywy wykorzystania pojazdów autonomicznych w transporcie drogowym w Polsce. *Autobusy* 12, 788–789.

Polski Instytut Ekonomiczny (2023), Polska branżą TSL tworzy 6 proc. Polskiego PKB i jest liderem wśród krajów UE. Retrieved on 10 June 2023 from: https://pie.net.pl/polska-branza-tsl-tworzy-6-proc-polskiego-pkb-i-jest-liderem-wsrod-krajow-ue/

Ranft, F., Adler, M., Diamond, P., Guerrero, E., & Laza, M. (2016). Freeing the Road. Shaping the future for autonomous vehicles, Policy network special report.

Sąsiada, T. (2019). Norwegowie budują bezzałogowy kontenerowiec z napędem elektrycznym. Retrieved on 7 April 2024 from: https://seshydrogen.com/en/objective-2-use-hydrogen-as-an-alternative-fuel-for-transportation/

SES Hydrogen (2022). Objective. 2 Use hydrogen as an alternative fuel for transportation.

Solectric (2023). Agras T16 – najnowsze drony rolnicze. Retrieved on 10 June 2023 from: https://solectric.pl/agras-t16/

Stawska-Magdziak, K. (2022). Maglev - najszybszy pociąg świata. Najciekawsze ma w środku. Retrieved on 10 June 2023 from: https://gadzetomania.pl/maglev-najszybszy-pociag-swiata-najciekawsze-ma-w-srodku-wideo,6705085918218369a

Stowarzyszenie Prawników Rynku Motoryzacyjnego (2021). Unia Europejska wprowadza Zielony Ład. Retrieved on 15 June 2023 from: https://sprm.org.pl/aktualnosci/rynek-motoryzacyjny/unia-europejska-wprowadza-zielony-lad-zakazujac-rejestracji-pojazdow-spalinowych

Sudowski, M., Mrugalska, B. (2017). Assurance of road transport safety and manipulation of professional drivers working time. *Zeszyty Naukowe Politechniki Poznańskiej. Organizacja i Zarządzanie* 73, 245–251.

Szczepański, W., Wiśniewska, J., Zajkowski, K. (2016). Nowoczesne technologie w systemach transportowych przyszłości. *Autobusy* 8, 349.

Umair, S., Waqas, U., Al Shamsi, I. R., Kamran, H., & Mrugalska, B. (2023a). Impact of Strategic Orientation and Supply Chain Integration on Firm's Innovation Performance: A Mediation Analysis. In Mrugalska, B., Ahram, T., and Karwowski, W. (eds), *Human Factors in Engineering* (pp. 85–103). CRC Press, Boca Raton.

Umair, S., Waqas, U., & Mrugalska, B. (2023b). Cultivating sustainable environmental performance: The role of green talent management, transformational leadership, and employee engagement with green initiatives. *Work* 78(4), 1–13.

Valverde, V., Kondo, Y., Otsuki, Y., Krenz, T., Melas, A., Suarez-Bertoa, R., & Giechaskiel, B. (2023). Measurement of Gaseous Exhaust Emissions of Light-Duty Vehicles in Preparation for Euro 7: A Comparison of Portable and Laboratory Instrumentation. *Energies*, 16(6), 2561.

Walków, M. (2021). DHL zamawia elektryczne samoloty. Pierwszy lot planuje jeszcze w tym roku. Retrieved on 22 June 2023 from: https://businessinsider.com.pl/technologie/nowe-technologie/dhl-pierwsze-elektryczne-samoloty-towarowe-eviation-alice/x0hdek0

Wyszomirski, O. (2017). Zrównoważony rozwój transportu w miastach, a jakość życia. *Transport miejski i regionalny*, 12, 1–6.

Yılmaz, I., Eren Özceylan, E., & Mrugalska, B. (2023). A framework for evaluating electrical vehicle charging station location decisions in a spherical fuzzy environment: A case of shopping malls, *Annals of Operations Research* (in print). https://doi.org/10.1007/s10479-023-05772-x

Zhang, S., & Witlox, F. (2019). Analyzing the impact of different transport governance strategies on climate change. *Sustainability* 12(1), 200.

Złoty, P. (2016). IAA 2016 - Iveco Z Truck z technologią LNG. Retrieved on 12 June 2023 from: https://cng-lng.pl/motoryzacja/pojazdy/IAA-2016-Iveco-Z-Truck-z-technologia-LNG,artykul,8918.html

12 Managing Career Attitudes in the Era of the 21st Century

Muhammad Latif Khan
Global College of Engineering & Technology, Ruwi, Sultanate of Oman

Hyder Kamran
University of Buraimi, Al Buraimi, Oman

Rohani Salleh
Universiti Teknologi PETRONAS, Seri Iskandar, Malaysia

Pooyan Rahmanivahid
Department of Mechanical Engineering Global College of Engineering & Technology, Ruwi, Sultanate of Oman

Waseem Fatima
Tawam International School AL Buraimi, Sultanate of Oman

12.1 CAREER REDEFINING IN A COMPLEX WORK ENVIRONMENT

The change in the workplace over the past few decades has led to research on modern job types. Increasing self-direction, flexibility, and the pursuit of subjective professional success characterize these careers. However, the theoretical foundations and conceptual frameworks of the field of career studies are fragmented, which presents a challenge to researchers in the area (Steindórsdóttir *et al.*, 2023). A volatile and complicated corporate climate leads to career signals that are incredibly muddled and inconsistent. In response, people are becoming skeptical about their goals and intentions for advancement. A new psychological contract based on continuous learning and identity change has replaced the traditional psychological contract. It is the "path with a heart" (Herb Shepard, 2010), where an employee enters a company, works hard, performs well, is loyal and committed, and receives ever-greater rewards and job security (Tomprou & Lee, 2022).

Traditional careers depend on upward mobility, salary increases, and responsibilities across a few organizations. Contrary to that, under new ideas of career pursuit, an employee is the agent who determines personal career goals, and these may be independent of organizational boundaries (Arthur, Khapova, & Wilderom, 2005).

DOI: 10.1201/9781032616810-12

The capacity to manage connections with intra- and inter-organizational contacts is one area where experts have begun to highlight the importance of the link between the person and the external setting in promoting protean career orientation (PCO) (Waters et al., 2015; Tee, Cham, Low, & Lau, 2022). We explore the interaction between the individual and the situation in line with the protean career paradigm by adopting a self-actor approach. This approach holds that people may achieve their inherent goals and values by influencing their surroundings (Kim, Hood, Creed, & Bath, 2022).

A considerable degree of ambiguity exists nowadays about career trajectories and expectations due to the fluctuating and unstable career environment. Because of the corporate environment's extreme turbulence and complexity, career signals are incredibly murky and conflicting. People are becoming similarly hesitant about their goals and ambitions for job advancement, maybe as a form of self-defense (Briscoe, Henagan, Burton, & Murphy, 2012; Hofstetter & Rosenblatt, 2017). In response to this ambiguity, individuals are increasingly managing their careers instead of organizations. In addition, intrinsic factors influence them more than extrinsic factors (Quigley & Tymon 2006). PCO, a novel strategy introduced by Hall (1976), is an individual attitude in which job decisions are individualized and serve as the foundation for the pursuit of self-fulfillment (Hall, 2004; Khan, Salleh, Shamim, & Hemdi, 2023b).

Career redefinitions are often required in complex work environments. Adaptability and openness to change can help individuals stay current and grow in a changing economy. Therefore, continual learning, collaboration, and adaptability are essential prerequisites to pursue new directions in career. Embracing change and seeking innovative solutions is what makes professionals thrive. Adapting to emerging challenges and opportunities is essential for a successful career path.

12.2 PROTEAN CAREER

According to Briscoe et al. (2006), the word "protean" comes from the Latin word "Proteus," which denotes the exceptional capacity of a human to modify the form of anything in order to cope with uncertainty. Value drive and self-direction are two characteristics that characterize prototypic career attitudes. When someone speaks of self-direction, it relates to how much they proactively take charge of their own profession (Briscoe, Hall, & DeMuth, 2006; Mirvis & Hall, 1994). In contrast, a value-driven approach describes an individual's heightened consciousness of their own unique priorities and serves as a benchmark for decision-making and evaluation (Hall, 1996). Therefore, different from traditional approach to career, the concept of protean career highlights individual values and aspirations which are flexible, self-directed, and adaptable to changing goals and skills. A protean career emphasizes individual autonomy and fulfillment through self-directed goal setting (Hirschi et al., 2017). Continuous learning, adaptability to change, and values-driven decision-making pave the way to pursue new career orientation.

The protean career concept was first proposed in the 1970s but has received little empirical study (Gubler, Arnold, & Coombs, 2014; Rodrigues, Butler, & Guest, 2019). Hence, researchers started looking into the factors that influence PCO, examining, for example, the significance of each person's positive core self-evaluations and contact networks (Khan, Salleh, & Hemdi, 2016; Rodrigues *et al.*, 2019). These contributions have brought attention to the important role that personal characteristics, including interpersonal skills, play in PCO. PCO emphasizes on behavioral competencies as antecedents of PCO. These abilities relate to the capacity to comprehend and regulate oneself and the environment (Emmerling & Boyatzis, 2012; Khan, Salleh, Javaid, *et al.*, 2023a).

12.3 ANTECEDENTS OF PROTEAN CAREER ORIENTATION

A person's individual personality, including autonomy, self-direction, and a desire to continuously learn and develop skills, is also important. Values, preferences, and psychological characteristics play an important role in the development of a PCO. An individual may also desire work–life balance, be autonomous, be intrinsically motivated, and be open to learning new things. In addition, increased job mobility and societal shifts toward a knowledge-based economy have contributed to the rise of PCOs. An organizational culture that fosters a PCO also accepts influences from external factors, such as labor market shifts and organizational practices. As a result, career paths are flexible and adaptable (Sullivan, 2011; Joo, Park & Oh, 2013).

Psychological factors, personality traits, career goals, and role models influence a PCO along with external factors (McDonald et al., 2005; Lin, 2015; Hall, Yip, & Doiron, 2018). PCO is a result of the interaction and influence of these factors. Personal fulfillment and career success are often in balance for people with a PCO.

12.4 BEHAVIORAL COMPETENCIES AND PROTEAN CAREER ORIENTATION

This idea includes both behaviors (i.e., a range of potential actions depending on the circumstance) and the motivations behind those acts. In seminal contributions on this topic, research identifies self-awareness and flexibility as essential qualities for cultivating a PCO (Boyatzis, Rochford, & Cavanagh, 2017). According to the attributes and skills an individual possesses, their behavioral competencies determine how they behave at work. PCOs emphasize flexibility, adaptability, and pursuing personal interests and values based on career goals. However, subsequent research (Gubler *et al.*, 2014; Hall, Yip, & Doiron, 2018) has shown that there is little empirical evidence for this association. There has been widespread recognition in the previous decades of the research on behavioral abilities' beneficial effects on effective individual performance across industries and professional vocations. According to earlier research (Gianakos, 1999; Santos, Wang, & Lewis, 2018), a person's skills also play a role in determining their orientation and inclination to behave in the context of professional choices (Chang *et al.*, 2023; Khan *et al.*, 2016; Rowe, 2013). Since behavioral competencies can be developed by educational institutions, as shown by prior studies (Barker, Moore, Olmi, & Rowsey, 2019; Purwanto, 2020), exploring their role as PCO predictors is particularly important. This has important implications for preparing people for modern careers (Chui, Li, & Ngo, 2022).

Individuals always learn to attain their particular goals (Hall, 2004). Under the aspirations of motivational behavior, the pursuit of individual learning objectives, accomplishment orientation, and the view of failure as constructs of constructive criticism. The interaction between an individual and their surroundings has a significant impact on how their careers turn out. Therefore, learning capabilities contribute significantly to PCO.

12.5 SELF-DIRECTED CAREER ATTITUDE

Self-directed career attitudes emphasize proactive and autonomous decision-making when it comes to career development and decision-making. This attitude indicates motivation, initiative, and a sense of personal responsibility when it comes to planning, pursuing, and adapting to career goals (Cortellazzo *et al.*, 2020). The self-directed individual constantly assesses and adjusts their career path to align with their values and ambitions.

Protean careers emphasize self-direction, adaptability, and continuous learning. Morel (2019) points out that self-regulation and self-motivation are essential elements of career decision-making. Individuals who manage their career based on their values and subjective success criteria use the protean career approach because they follow a career path based on their career objectives. Individual's values and beliefs shape career orientation rather than the organization's values and beliefs (Ahmed, 2019), thus making it self-directed and value-driven. Personal values and self-direction are the distinguishing characteristics of protean individuals (Orpen, 1994). Therefore, having a self-directed career attitude has become increasingly important as individuals navigate complex and ever-changing career landscapes. Taking control of one's career is empowering for individuals personally and professionally.

12.6 EMPLOYABILITY AND CAREER SUCCESS

An individual's career success depends on their employability. Combining technical skills (technical expertise) and soft skills (communication, teamwork) with adaptability, learning capabilities, and problem-solving abilities makes an individual employable in the job market (Fajaryati *et al.*, 2020). Those with high employability get hired more effectively, secure rewarding jobs, and progress in their careers. Developing skills, networking actively, and improving oneself are essential to career success. In general, career success is defined as achieving one's career goals, being satisfied and fulfilled at work and developing professionally. Having a sense of personal satisfaction can manifest itself in the form of a promotion, an increase in income, more stability in the workplace, or an increase in income (Seibert *et al.*, 2001). To be employable, one needs to acquire relevant knowledge, skills, and certifications, and remain current with industry trends. Building professional relationships and networking are also important for improving employability. When career decisions are guided by personal values and passions, a value-driven attitude enhances job satisfaction and fulfillment (Kundi *et al.*, 2021). Therefore, an individual who has a self-directed career attitude fosters flexibility and adaptability through proactive goal setting and pursuit. Successful careers are driven by proactive, purposeful actions aligned with personal values.

12.7 VALUE-DRIVEN CAREER ATTITUDE

PCO is value-driven in addition, "in the sense that the person's internal values provide the guidance and measure of success for the individual's career" (Briscoe *et al.*, 2006; Chang, Guo, Cai, & Guo, 2023). The tendency is to seek employment that meets personal requirements as well as professional demands (Reitman & Schneer, 2008), giving the term "career" a more expansive definition of self-realization.

Scholars have mainly examined the effects of PCO on career success in adult workers when considering protean career outcomes (De Bruin & Buchner, 2010; Rastgar, Ebrahimi, & Hessan, 2014). There hasn't been much discussion on how PCO may help young people who are just starting out in their careers, particularly those who attempt to get into the job market (Baluku, Löser, Otto, & Schummer, 2018; Rodrigues *et al.*, 2019). In times of economic, organizational, and employment difficulty, PCO may aid people in acquiring and maintaining employability. According to earlier studies (Waters, Briscoe, Hall, & Wang, 2014), the traits linked to the protean career attitude make people with this attitude more appealing to potential employers and more proactive in their search for employment opportunities.

12.8 THEORETICAL BACKGROUND

The concept of PCO is based on pioneer studies that employed a multi-dimensional approach to career explanation based on many environmental elements (De Vos & Soens, 2008; Volmer & Spurk, 2011). Proliferating professional attitudes place an emphasis on values, independence, flexibility, skill development, job success, and career management based on a thorough examination of the literature. Using the social cognitive theory (SCCT) (Bandura 1986; Lent *et al.*, 1994), followed by the social exchange theory (Blau, 1964) and the protean career theory (Hall, 1976), this study seeks to identify the elements affecting attitudes toward protean careers. Therefore, proliferating professional attitudes place an emphasis on values, independence, flexibility, skill development, job success, and career management. The following conceptual framework explains the relation between protean career attitudes and individual career outcomes like employability (Figure 12.1).

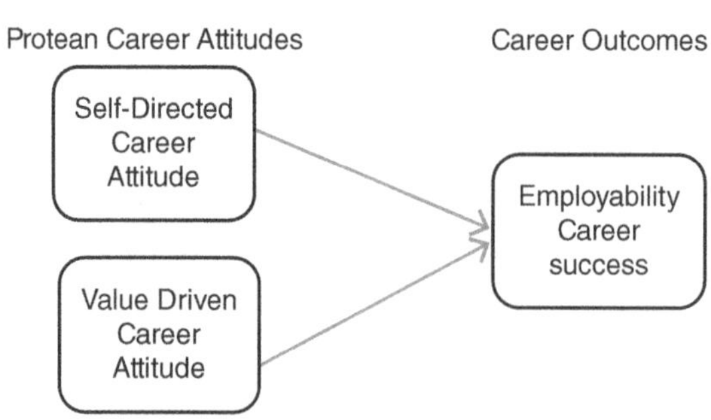

FIGURE 12.1 Conceptual framework.

12.9 EMERGING TRENDS IN CAREER MANAGEMENT

Emerging career management trends hanging employee and employer expectations reflect changes in career paths. In the ever-changing workplace, people are adopting these trends to succeed (Clarke, 2013; Baruch, 2015; Dachner *et al.*, 2021).

- Flexibility and Remote Work: Career management now requires setting boundaries, adapting to virtual collaboration, and balancing work and personal life.
- A skills-based approach to hiring: As companies emphasize the importance of continuous learning and adaptability, they are increasingly paying more attention to skills rather than traditional qualifications.
- Equity, Diversity, and Inclusion: Companies, resulting in more inclusive career management, mentoring, and leadership development, prioritize diverse talent.
- Work–Life Balance and Mental Health: Employers are now focusing a great deal on mental health, self-care, and work–life balance as part of career management.
- The Gig Economy: Building a diverse set of skills and income streams is becoming increasingly common, requiring individuals to manage a portfolio career or gig work.
- Artificial Intelligence: Individuals can use artificial intelligence to plan their career paths and learn about opportunities tailored to their strengths and goals.
- Continuous Learning: To remain competitive in the job market, continuous learning and upskilling are essential.

12.10 CONCLUSION

Employees and employers need to have a better knowledge of one another, given the differences between people's behavior and career. Pressure from technical advancements, globalization, and communication, along with these crucial job market changes in the late 20th and early 21st centuries, and adjusting to them, many employees' perceptions toward their workplaces have also altered. Employees seek greater professional autonomy; based on that, they tend to set their own benefits and levels of satisfaction. Revolution has altered the way jobs are designed, which has become essential for many employees (Greenhaus et al. 2009).

The protean mindset is a creative strategy to improve employability perceptions of an employee in an increasingly competitive labor market. People who have a strong propensity for change perceive themselves as in control of their job chances and are self-directed. These multifaceted skills function independently and are adaptable to the shifting work environment. Thus, by employing a suitable technique, the firms might affect a prodigious talent's perception of their internal employability, perhaps keeping them with the existing company. Career attitudes in the 21st century require an adaptable and forward-looking approach. Changing economic landscapes, technological advances, and workplace dynamics require a shift in mindset. Organizations can encourage career self-management behavior by implementing a wide range of human resource development strategies, such as employee engagement initiatives, HRD initiatives, and fostering an employable culture inside the company.

REFERENCES

Ahmed, N. O. A. (2019). Career commitment: the role of self-efficacy, career satisfaction and organizational commitment. *World Journal of Entrepreneurship, Management and Sustainable Development*, (just accepted).

Arthur, Michael B., Khapova, Svetlana N., & Wilderom, Celeste P. M. (2005). Career success in a boundaryless career world. Journal of Organizational Behavior: *The International Journal of Industrial, Occupational and Organizational Psychology and Behavior*, 26(2), 177–202.

Baluku, Martin Mabunda, Löser, Dorothee, Otto, Kathleen, & Schummer, Steffen Erik. (2018). Career mobility in young professionals: How a protean career personality and attitude shapes international mobility and entrepreneurial intentions. *Journal of Global Mobility*, 6(1), 102–122.

Bandura, Albert. (1986). *Social foundations of thought and action*. Englewood Cliffs, NJ, pp. 23–28.

Barker, Laura-Katherine, Moore, James W, Olmi, D Joe, & Rowsey, Kyle. (2019). A comparison of immediate and post-session feedback with behavioral skills training to improve interview skills in college students. *Journal of Organizational Behavior Management*, 39(3–4), 145–163.

Baruch, Y. (2015). Organizational and labor markets as career ecosystem. In *Handbook of research on sustainable careers* (pp. 364–380). Edward Elgar Publishing.

Blau, P. M. (1964). *Exchange and power in social life*. Wiley.

Boyatzis, Richard, Rochford, Kylie, & Cavanagh, Kevin V. (2017). Emotional intelligence competencies in engineer's effectiveness and engagement. *Career Development International*, 22(1), 70–86.

Briscoe, Jon P., Hall, Douglas T., & DeMuth, Rachel L. Frautschy. (2006). Protean and boundaryless careers: An empirical exploration. *Journal of Vocational Behavior*, 69(1), 30–47.

Briscoe, Jon P., Henagan, Stephanie C., Burton, James P., & Murphy, Wendy M. (2012). Coping with an insecure employment environment: The differing roles of protean and boundaryless career orientations. *Journal of Vocational Behavior*, 80(2), 308–316.

Chang, Po-Chien, Guo, Yuanli, Cai, Qihai, & Guo, Hongchi. (2023). Proactive career orientation and subjective career success: A perspective of career construction theory. *Behavioral Sciences*, 13(6), 503.

Chui, Hazel, Li, Hui, & Ngo, Hang-Yue. (2022). Linking protean career orientation with career optimism: Career adaptability and career decision self-efficacy as mediators. *Journal of Career Development*, 49(1), 161–173.

Clarke, M. (2013). The organizational career: Not dead but in need of redefinition. *The International Journal of Human Resource Management*, 24(4), 684–703.

Cortellazzo, L., Bonesso, S., Gerli, F., & Batista-Foguet, J. M. (2020). Protean career orientation: Behavioral antecedents and employability outcomes. *Journal of Vocational Behavior*, 116, 103343.

Dachner, A. M., Ellingson, J. E., Noe, R. A., & Saxton, B. M. (2021). The future of employee development. *Human Resource Management Review*, 31(2), 100732.

De Bruin, Gideon P., & Buchner, Morné. (2010). Factor and item response theory analysis of the Protean and Boundaryless Career Attitude Scales. *SA Journal of Industrial Psychology*, 36(2), 1–11.

De Vos, Ans, & Soens, Nele. (2008). Protean attitude and career success: The mediating role of self-management. *Journal of Vocational Behavior*, 73(3), 449–456.

Emmerling, Robert J., & Boyatzis, Richard E. (2012). Emotional and social intelligence competencies: Cross cultural implications. *Cross Cultural Management: An International Journal*, 19(1), 4–18.

Fajaryati, N., Budiyono, Akhyar M., & Wiranto. (2020). The employability skills needed to face the demands of work in the future: Systematic literature reviews. *Open Engineering*, 10(1), 595–603.

Gianakos, Irene. (1999). Patterns of career choice and career decision-making self-efficacy. *Journal of Vocational Behavior*, 54(2), 244–258.

Greenhaus, Jeffrey H., Callanan, Gerard A., & Godshalk, Veronica M. (2009). *Career management: Sage.*

Gubler, Martin, Arnold, John, & Coombs, Crispin. (2014). Reassessing the protean career concept: Empirical findings, conceptual components, and measurement. *Journal of Organizational Behavior*, 35(S1), S23–S40.

Hall, D. T., Yip, J., & Doiron, K. (2018). Protean careers at work: Self-direction and values orientation in psychological success. *Annual Review of Organizational Psychology and Organizational Behavior*, 5, 129–156.

Hall, Douglas T. (1976). *Careers in organizations*. Goodyear Pub. Co.

Hall, Douglas T. (1996). The Career Is Dead--Long Live the Career. A Relational Approach to Careers. *The Jossey-Bass Business & Management Series*: ERIC.

Hall, Douglas T. (2004). The protean career: A quarter-century journey. *Journal of Vocational Behavior*, 65(1), 1–13.

Hirschi, Andreas, Jaensch, Vanessa K., & Herrmann, Anne. (2017). Protean career orientation, vocational identity, and self-efficacy: an empirical clarification of their relationship. *European Journal of Work and Organizational Psychology*, 26(2), 208–220.

Hofstetter, Hila, & Rosenblatt, Zehava. (2017). Predicting protean and physical boundaryless career attitudes by work importance and work alternatives: Regulatory focus mediation effects. *The International Journal of Human Resource Management*, 28(15), 2136–2158.

Joo, B. K., Park, S., & Oh, J. R. (2013). The effects of learning goal orientation, developmental needs awareness and self-directed learning on career satisfaction in the Korean public sector. *Human Resource Development International*, 16(3), 313–329.

Khan, Muhammad Latif, Salleh, Rohani, & Hemdi, Mohamad Abdullah Bin. (2016). Effect of protean career attitudes on organizational commitment of employees with moderating role of organizational career management. *International Review of Management and Marketing*, 6(4), 155–160.

Khan, Muhammad Latif, Salleh, Rohani, Javaid, Muhammad Umair, Arshad, Muhammad Zulqarnain, Saleem, Muhammad Shoaib, & Younas, Samia. (2023a). Managing butterfly career attitudes: The moderating interplay of Organisational Career Management. *Sustainability*, 15(6), 5099.

Khan, Muhammad Latif, Salleh, Rohani, Shamim, Amjad, & Hemdi, Mohamad Abdullah. (2023b). Role-play of employees' protean career and career success in affective organizational commitment. *Asia-Pacific Journal of Business Administration*. https://doi.org/10.1108/APJBA-07-2021-0337

Kim, Sujin, Hood, Michelle, Creed, Peter, & Bath, Debra. (2022). "New career" profiles for young adults incorporating traditional and protean career orientations and competencies. *Career Development International*, 27(5), 493–510.

Kundi, Y. M., Hollet-Haudebert, S., & Peterson, J. (2021). Linking protean and boundaryless career attitudes to subjective career success: A serial mediation model. *Journal of Career Assessment*, 29(2), 263–282.

Lent, Robert W., Brown, Steven D., & Hackett, Gail. (1994). Toward a unifying social cognitive theory of career and academic interest, choice, and performance. *Journal of Vocational Behavior*, 45(1), 79–122.

Lin, Y. C. (2015). Are you a protean talent? The influence of protean career attitude, learning-goal orientation and perceived internal and external employability. *Career Development International*, 20(7), 753–772.

McDonald, Paula, Brown, Kerry, & Bradley, Lisa. (2005). Have traditional career paths given way to protean ones? Evidence from senior managers in the Australian public sector. *Career Development International*, 10(2), 109–129.

Mirvis, Philip H., & Hall, Douglas T. (1994). Psychological success and the boundaryless career. *Journal of Organizational Behavior*, 15(4), 365–380.

Morel, S. A. (2019). *Exploring a career path towards well-being: How parental behaviors, career values awareness, and career decision-making self-efficacy impact well-being in undergraduate college students* (Doctoral dissertation, Purdue University Graduate School).

Orpen, Christopher. (1994). The effects of organizational and individual career management on career success. *International Journal of Manpower*, 15(1), 27–37.

Purwanto, Agus. (2020). Effect of hard skills, soft skills, organizational learning and innovation capability on Islamic University lecturers' performance. Systematic Reviews in Pharmacy.

Quigley, Narda R., & Tymon Jr, Walter G. (2006). Toward an integrated model of intrinsic motivation and career self-management. *Career Development International*, 11(6), 522–543.

Rastgar, Abbas Ali, Ebrahimi, Elham, & Hessan, Maryam. (2014). The effects of personality on protean and boundaryless career attitudes. *International Journal of Business Management and Economics*, 1(1), 1–5.

Reitman, Frieda, & Schneer, Joy A. (2008). Enabling the new careers of the 21st century. *Organization Management Journal*, 5(1), 17–28.

Rodrigues, Ricardo, Butler, Christina L., & Guest, David. (2019). Antecedents of protean and boundaryless career orientations: The role of core self-evaluations, perceived employability and social capital. *Journal of Vocational Behavior*, 110, 1–11.

Rowe, Kate Penelope. (2013). Psychological capital and employee loyalty: The mediating role of protean career orientation.

Santos, Angeli, Wang, Weiwei, & Lewis, Jenny. (2018). Emotional intelligence and career decision-making difficulties: The mediating role of career decision self-efficacy. *Journal of Vocational Behavior*, 107, 295–309.

Seibert, S. E., Kraimer, M. L., & Crant, J. M. (2001). What do proactive people do? A longitudinal model linking proactive personality and career success. *Personnel Psychology*, 54(4), 845–874.

Shepard, H. (2010). A path with a heart: The Cultural Context of Learning about Careers. Retrieved from https://appreciativeinquiry.case.edu/uploads/choosingapathwithheart.pdf

Steindórsdóttir, Bryndís D., Sanders, Karin, Arnulf, Jan Ketil, & Dysvik, Anders. (2023). Career transitions and career success from a lifespan developmental perspective: A 15-year longitudinal study. *Journal of Vocational Behavior*, 140, 103809.

Sullivan, S. E. (2011). Self-direction in the boundaryless career era.

Tee, Poh Kiong, Cham, Tat-Huei, Low, Mei Peng, & Lau, Teck-Chai. (2022). The role of perceived employability in the relationship between protean career attitude and career success. *Australian Journal of Career Development*, 31(1), 66–76.

Tomprou, Maria, & Lee, Min Kyung. (2022). Employment relationships in algorithmic management: A psychological contract perspective. *Computers in Human Behavior*, 126, 106997.

Volmer, Judith, & Spurk, Daniel. (2011). Protean and boundaryless career attitudes: Relationships with subjective and objective career success. *Zeitschrift für ArbeitsmarktForschung*, 43(3), 207–218.

Waters, Lea, Briscoe, Jon P, Hall, Douglas T., & Wang, Lan. (2014). Protean career attitudes during unemployment and reemployment: A longitudinal perspective. *Journal of Vocational Behavior*, 84(3), 405–419.

Waters, Lea, Hall, Douglas T., Wang, Lan, & Briscoe, Jon P. (2015). 12. Protean career orientation: a review of existing and emerging research. Flourishing in life, work and careers: Individual wellbeing and career experiences, 235.

Index

Pages in *italics* refer to figures and pages in **bold** refer to tables.

For Product Safety Concerns and Information please contact our EU representative GPSR@taylorandfrancis.com Taylor & Francis Verlag GmbH, Kaufingerstraße 24, 80331 München, Germany